Novel Nanomaterials and Nanotechnology: From Fabrication Methods and Improvement Strategies to Applications in Biosensing and Biomedicine

Novel Nanomaterials and Nanotechnology: From Fabrication Methods and Improvement Strategies to Applications in Biosensing and Biomedicine

Guest Editors

Juan Yan
Bin Zhao
Tianxi Yang

Basel • Beijing • Wuhan • Barcelona • Belgrade • Novi Sad • Cluj • Manchester

Guest Editors

Juan Yan
College of Food Science &
Technology
Shanghai Ocean University
Shanghai
China

Bin Zhao
Department of Biochemistry
and Molecular Biology
The University of British
Columbia
Vancouver
Canada

Tianxi Yang
Food, Nutrition and Health,
Faculty of Land and Food
Systems
The University of British
Columbia
Vancouver
Canada

Editorial Office
MDPI AG
Grosspeteranlage 5
4052 Basel, Switzerland

This is a reprint of the Special Issue, published open access by the journal *Biosensors* (ISSN 2079-6374), freely accessible at: https://www.mdpi.com/journal/biosensors/special_issues/M4BE5O47HG.

For citation purposes, cite each article independently as indicated on the article page online and as indicated below:

Lastname, A.A.; Lastname, B.B. Article Title. *Journal Name* **Year**, *Volume Number*, Page Range.

ISBN 978-3-7258-3197-5 (Hbk)
ISBN 978-3-7258-3198-2 (PDF)
https://doi.org/10.3390/books978-3-7258-3198-2

© 2025 by the authors. Articles in this book are Open Access and distributed under the Creative Commons Attribution (CC BY) license. The book as a whole is distributed by MDPI under the terms and conditions of the Creative Commons Attribution-NonCommercial-NoDerivs (CC BY-NC-ND) license (https://creativecommons.org/licenses/by-nc-nd/4.0/).

Contents

Hiram Martin Valenzuela-Amaro, Alberto Aguayo-Acosta,
Edgar Ricardo Meléndez-Sánchez, Orlando de la Rosa, Perla Guadalupe Vázquez-Ortega,
Mariel Araceli Oyervides-Muñoz, et al.
Emerging Applications of Nanobiosensors in Pathogen Detection in Water and Food
Reprinted from: *Biosensors* 2023, *13*, 922, https://doi.org/10.3390/bios13100922 1

Lele Wang, Yanli Wen, Lanying Li, Xue Yang, Wen Li, Meixia Cao, et al.
Development of Optical Differential Sensing Based on Nanomaterials for Biological Analysis
Reprinted from: *Biosensors* 2024, *14*, 170, https://doi.org/10.3390/bios14040170 25

Xinran Li, Haoqian Wang, Xin Qi, Yi Ji, Fukai Li, Xiaoyun Chen, et al.
PCR Independent Strategy-Based Biosensors for RNA Detection
Reprinted from: *Biosensors* 2024, *14*, 200, https://doi.org/10.3390/bios14040200 47

Minja Mladenović, Stefan Jarić, Mirjana Mundžić, Aleksandra Pavlović, Ivan Bobrinetskiy
and Nikola Ž. Knežević
Biosensors for Cancer Biomarkers Based on Mesoporous Silica Nanoparticles
Reprinted from: *Biosensors* 2024, *14*, 326, https://doi.org/10.3390/bios14070326 72

Xinyu Wang, Huiyuan Wang, Hongmin Zhang, Tianxi Yang, Bin Zhao and Juan Yan
Investigation of the Impact of Hydrogen Bonding Degree in Long Single-Stranded DNA
(ssDNA) Generated with Dual Rolling Circle Amplification (RCA) on the Preparation and
Performance of DNA Hydrogels
Reprinted from: *Biosensors* 2023, *13*, 755, https://doi.org/10.3390/bios13070755 96

Xiaoxiao Zhou, Shouhui Chen, Yi Pan, Yuanfeng Wang, Naifeng Xu, Yanwen Xue, et al.
High-Performance Au@Ag Nanorods Substrate for SERS Detection of Malachite Green in
Aquatic Products
Reprinted from: *Biosensors* 2023, *13*, 766, https://doi.org/10.3390/bios13080766 111

Chao Li, Zichao Guo, Sisi Pu, Chaohui Zhou, Xi Cheng, Ren Zhao and Nengqin Jia
Molybdenum Disulfide-Integrated Iron Organic Framework Hybrid Nanozyme-Based
Aptasensor for Colorimetric Detection of Exosomes
Reprinted from: *Biosensors* 2023, *13*, 800, https://doi.org/10.3390/bios13080800 122

Xiaoli Xu, Xiaohui Lin, Lingling Wang, Yixin Ma, Tao Sun and Xiaojun Bian
A Novel Dual Bacteria-Imprinted Polymer Sensor for Highly Selective and Rapid Detection of
Pathogenic Bacteria
Reprinted from: *Biosensors* 2023, *13*, 868, https://doi.org/10.3390/bios13090868 136

Leontýna Varvařovská, Petr Kudrna, Bruno Sopko and Taťána Jarošíková
The Development of a Specific Nanofiber Bioreceptor for Detection of *Escherichia coli* and
Staphylococcus aureus from Air
Reprinted from: *Biosensors* 2024, *14*, 234, https://doi.org/10.3390/bios14050234 150

Xinru Yin, Cheng Zhao, Yong Zhao and Yongheng Zhu
Parallel Monitoring of Glucose, Free Amino Acids, and Vitamin C in Fruits Using a
High-Throughput Paper-Based Sensor Modified with Poly(carboxybetaine acrylamide)
Reprinted from: *Biosensors* 2023, *13*, 1001, https://doi.org/10.3390/bios13121001 162

Ivan Lopez Carrasco, Gianaurelio Cuniberti, Jörg Opitz and Natalia Beshchasna
Evaluation of Transducer Elements Based on Different Material Configurations for Aptamer-Based Electrochemical Biosensors
Reprinted from: *Biosensors* **2024**, *14*, 341, https://doi.org/10.3390/bios14070341 . . . . **178**

Baining Sun, Chenxiang Zheng, Dun Pan, Leer Shen, Wan Zhang, Xiaohua Chen, et al.
Using AuNPs-DNA Walker with Fluorophores Detects the Hepatitis Virus Rapidly
Reprinted from: *Biosensors* **2024**, *14*, 370, https://doi.org/10.3390/bios14080370 . . . . **196**

Review

Emerging Applications of Nanobiosensors in Pathogen Detection in Water and Food

Hiram Martin Valenzuela-Amaro [1,2,†], Alberto Aguayo-Acosta [1,2,†], Edgar Ricardo Meléndez-Sánchez [1,2], Orlando de la Rosa [1,2], Perla Guadalupe Vázquez-Ortega [3], Mariel Araceli Oyervides-Muñoz [1,2], Juan Eduardo Sosa-Hernández [1,2,*] and Roberto Parra-Saldívar [1,2,*]

1. Tecnologico de Monterrey, Institute of Advanced Materials for Sustainable Manufacturing, Monterrey 64849, Mexico; amaro.hiram@tec.mx (H.M.V.-A.); aguayo.alberto@tec.mx (A.A.-A.); edgar.rmelendez@tec.mx (E.R.M.-S.); orlando.delarosa@tec.mx (O.d.l.R.); mariel.oyervides@tec.mx (M.A.O.-M.)
2. Tecnologico de Monterrey, School of Engineering and Sciences, Monterrey 64849, Mexico
3. Tecnologico Nacional de México, Instituto Tecnológico de Durango, Durango 34080, Mexico; pvazquez@itdurango.edu.mx
* Correspondence: eduardo.sosa@tec.mx (J.E.S.-H.); r.parra@tec.mx (R.P.-S.)
† These authors contributed equally to this work.

Abstract: Food and waterborne illnesses are still a major concern in health and food safety areas. Every year, almost 0.42 million and 2.2 million deaths related to food and waterborne illness are reported worldwide, respectively. In foodborne pathogens, bacteria such as *Salmonella*, Shiga-toxin producer *Escherichia coli*, *Campylobacter*, and *Listeria monocytogenes* are considered to be high-concern pathogens. High-concern waterborne pathogens are *Vibrio cholerae*, leptospirosis, *Schistosoma mansoni*, and *Schistosima japonicum*, among others. Despite the major efforts of food and water quality control to monitor the presence of these pathogens of concern in these kinds of sources, foodborne and waterborne illness occurrence is still high globally. For these reasons, the development of novel and faster pathogen-detection methods applicable to real-time surveillance strategies are required. Methods based on biosensor devices have emerged as novel tools for faster detection of food and water pathogens, in contrast to traditional methods that are usually time-consuming and are unsuitable for large-scale monitoring. Biosensor devices can be summarized as devices that use biochemical reactions with a biorecognition section (isolated enzymes, antibodies, tissues, genetic materials, or aptamers) to detect pathogens. In most cases, biosensors are based on the correlation of electrical, thermal, or optical signals in the presence of pathogen biomarkers. The application of nano and molecular technologies allows the identification of pathogens in a faster and high-sensibility manner, at extremely low-pathogen concentrations. In fact, the integration of gold, silver, iron, and magnetic nanoparticles (NP) in biosensors has demonstrated an improvement in their detection functionality. The present review summarizes the principal application of nanomaterials and biosensor-based devices for the detection of pathogens in food and water samples. Additionally, it highlights the improvement of biosensor devices through nanomaterials. Nanomaterials offer unique advantages for pathogen detection. The nanoscale and high specific surface area allows for more effective interaction with pathogenic agents, enhancing the sensitivity and selectivity of the biosensors. Finally, biosensors' capability to functionalize with specific molecules such as antibodies or nucleic acids facilitates the specific detection of the target pathogens.

Keywords: nanobiosensors; nanomaterials; foodborne diseases; waterborne diseases; food safety

1. Introduction

Every year, contaminated food is responsible for 420,000 deaths and 600 million cases of foodborne illnesses caused by spoiled food [1]. This is not just a problem in low–middle-income countries, high-income countries also have several troubles related to foodborne pathogens. In the U.S. alone, there are more than 9.4 million deaths per year due to the ingestion of pathogenic bacteria in food [2]. During 2010, 420,000 people (one-third of them being children under the age of five) died from illnesses related to salmonellosis and *Escherichia coli* infections [3]. Foodborne illnesses arise from the presence of pathogens, toxins, or contaminants in food products, and are typically associated with gastrointestinal symptoms (diarrhea, vomiting, abdominal pain, and fever), and other adverse effects on human health such as neurological, hepatic, and renal complications, even becoming a life-threatening issue if not appropriately addressed [4,5]. In recent years, the majority of reported foodborne illness outbreaks were caused by pathogens such as Norovirus [5], *Campylobacter* [6], *Salmonella* [5,6], *Listeria monocytogenes* [7], and Shiga toxin-producing *E. coli* [8]. Less frequently reported but still of concern are the pathogens *Staphylococcus aureus* [9], *Clostridium* species [10], *Bacillus cereus* [11], and *Yersinia enterocolitica* [12].

Similarly to food safety, the presence of pathogens in water is a major issue for public health [13]. It is estimated that 663 million people consume unsafe water from surface or groundwater sources [14]. More than 2.2 million deaths per year and more cases of illness (diarrhea, gastrointestinal, and systematic diseases) are linked to contaminated water ingestion [15]; the pathogens of greatest concern are *Salmonella, Shigella, Campylobacter, S. aureus*, and *E. coli* [16,17]. However, viruses and parasites are becoming a problem for water security [18]. Parasites and viruses linked to waterborne outbreaks include *Vibrio cholerae, Leptospira, Schistosoma mansoni*, and *Schistosoma japonicum* [16,19,20].

Monitoring the presence of pathogens in water is particularly important as a disease-preventive measure from waterborne illnesses and to monitor water quality. This can be achieved through applying wastewater-based surveillance protocols, which allow the detection of pathogens using molecular biology tools [21,22], which can be applied to verify the discharged water quality and indicate the treatment required to prevent adverse effects on the environment; ensuring water sustainability for future generations.

Pathogen-detection methods play a crucial role in ensuring food and water safety; however, actual monitoring methods are time-consuming processes that usually take days to obtain a precise result [23], making them ineffective for real-time monitoring [24]. In fact, the identification of pathogens such as bacteria and viruses is carried out by gold-standard methodologies, which are traditional techniques such as viable plate counts, flow cytometry, and staining methods, among others [25–27]. Nevertheless, the detection time is one of the major limitations of this technique because these techniques require the growth of the microorganism in laboratory conditions (this has not been a limitation per se), which can take several days to produce a result, hindering the response time for the control of pathogens [26]. Techniques based on molecular biology that are used for pathogen detection involve [28] polymerase chain reaction techniques (PCR) [21,29–31], multiplex polymerase chain reaction (mPCR) [32], quantitative polymerase chain reaction (qPCR) [33], digital droplet PCR (ddPCR) [34], fluorescence in situ hybridization (FISH) [31], enzyme-linked immunosorbent assay (ELISA) [35], surface-enhanced Raman spectroscopy (SERS) [36], immunological methods [37], next-generation sequencing [38], whole-genome sequencing [39], flow cytometry [40], and surface plasmon resonance imaging (SPR) [41]; these techniques have already been applied as detection methodologies of pathogens in food and water matrices [5,26].

Despite the application of molecular-biology techniques in food and water security, if we consider the technological development of the health sector related to pathogen detection, this sector has already developed advanced technologies such as biosensors with nanomaterials and the incorporation of informatic technologies [42]. Efforts are being conducted in the hope of bringing about more specific and faster methodologies to produce a rapid-response diagnosis and prevent outbreaks, focusing on nanomaterials such as glyconanomaterials [43], nanoparticles [44], ZnO nanorods, nanoconjugate (Au–Fe$_3$O$_4$), silicon nanonet FET, nanosphere (RNs@Au) in a biosensor device, combined with molecular detection methods (ELISA, qPCR) and also incorporated with informatic technologies, which are used to create more-sensible and appropriate in situ detection systems for pathogens of major concern. This technology has been applied in order to improve the health care system's response to pathogen-presence emergencies, (as reviewed by Jian et al., 2021) [45] for HIV and Influenza A virus. These technologies have also been applied to Ebola [46], Malaria [47], Dengue virus [43], and in recent years in SARS-CoV-2 monitoring protocols [42]. Considering the advances made in health security and the demands for improved food and water safety, these existing technologies in the health sector should be transferred to other sectors such as food and water security.

For the above mentioned, and the increase in pathogens related to food and waterborne illnesses, the development of pathogen-detection methods is becoming an urgent step to ensuring health and safety [48]. Unfortunately, and despite recent advances in new pathogen-detection approaches, the application of nanomaterials and biosensors is still limited, this is why technologies capable of obtaining better results, in a fast and affordable way, have been studied, resulting in novel technologies, such as biosensor devices, with "rapid, sensitive and specific" protocol for pathogen detection, resolving the priority assignment of ensuring health security, preventing food- and water-ingestion-related outbreaks [49], with even more affordable technology with the inclusion of the use of biosensors and NPs in recent years [44,50,51].

The previously mentioned methods help to perform faster monitoring (real-time surveillance systems) [52], reducing response times of pathogen detection in water [53]. Additionally, the use of biosensors improved with NPs enhanced the detection performance of the device making it a faster, more specific, and portable device [54]. In fact, due to the diversity of the detection capabilities of nanoparticles, they are the subject of many studies that attempt to understand their role when incorporated into pathogen-detection systems [55].

The basic components of a biosensor device are a biorecognition element, a transducer, an amplifier, and a processor component. The biorecognition element recognizes the analyte of interest, the transducer generates a signal from the recognition of the biomarker into a measurable signal, then the signal is processed using the processor and amplifier component, to obtain a signal output [56,57]. In summary, it is a bioanalytical device that detects specific biomarkers using biochemical reactions [58], mediated by isolated enzymes, antibodies, tissues, organelles, or whole cells for pathogen detection, using electrical, thermal, or optical signals [59], which are able to correlate the presence of specific pathogen and signal emission measures [50].

As is already mentioned, the biosensor application has garnered attention in the field of pathogen detection due to their attractive characteristics, such as precision, selectivity and fast analysis [27]. Nevertheless, it is necessary to mention that these methodologies have certain disadvantages, such as the use of expensive enzymes and equipment, including the extensive workflow required for the device's development. However, these technologies have a promising future due to their potential application in pathogen-rapid-detection methods [53]. Currently, biosensor-based technology has proved its worth due to its unique sensitivity, low detection limit, and simple operation [60].

In the last decade, the biosensors' structure has been focused on the miniaturization of the devices without affecting the detection efficacy. To achieve this, NPs have been included in the biosensor architecture, resulting in the development of a nanoscale platform. Indeed, in the different sections of the biosensor, NPs are used as signal transducers to convert a biomolecular interaction into an electrical, optical, or magnetic signal [61]. This functionality inside the biosensor is because of unique properties at the nanometric scales (surface area, small size, affinity for some biomolecules, catalytic activity, and autofluorescence) [62,63].

Like traditional biosensor devices, the nanobiosensors are composed of three main sections: a biorecognition probe, transducer, and amplifier [64] (Figure 1). The NPs are often in the transducer's component, helping to enhance the biochemical, electrical, magnetic, or optical signal transduction [61]. Also, these signals can be read simply and effectively as a result of the incorporation of functionalized NPs into the biorecognition component [65].

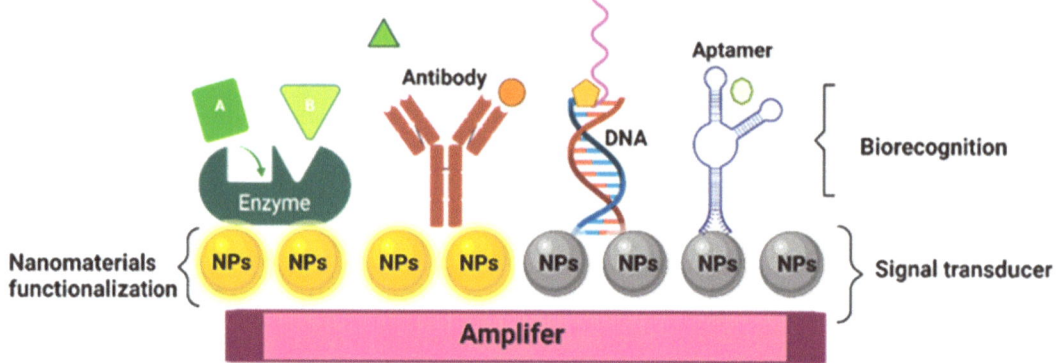

Figure 1. General structure of nanobiosensor with different agents of biorecognition.

In fact, nanomaterials have been identified as candidates to enhance biosensors' sensitivity, improving the detection limits and increasing detection specificity [54,55]. The foregoing is based on the fact that the specificity of signal recognition results in the adequate selection of functionalized ligands with NPs, improving the biomarker attraction; also, NPs convert signals from one form to another or act as detectors of the generated signals [66,67]. Biosensors have several methodologies to acquire relevant signals; for example, the electrochemical biosensors work under the method of capitalizing on reactions between immobilized biomolecules and the biomarker, resulting in electron/ion generation/consumption, modifying the electrical properties of the solution, and resulting in a measurable electrical current [68]. On the other hand, optical biosensors work under the method of discerning variations in light properties (absorption, transmission, and reflection), triggered by physical or chemical interactions with biorecognition elements. These biosensors are categorized into two major groups: label-free, where signals arise directly from analyte interactions, and label-based, employing techniques such as calorimetry, fluorescence, or luminescence to produce detectable optical signals. Both methodologies are available to be applied in diverse areas for pathogen detection [69–71].

Other possible classifications of biosensors are based on the type of biorecognition immobilized on the nanomaterial [72], which is divided into the following: enzymes [73], antibodies [74], antigens [75], DNA-RNA [76], organelle [77], cell membrane [78], and phage particle [79]. The conversion of this signal can be achieved using different methods, and this can be classified according to the type of conversion used [80]. Finally, the signal conversion section can include the following optical systems: [69] electrochemical nanobiosensors [81], thermoelectric [82], and piezoelectric [83].

2. Biofunctionalization of Nanostructured Surfaces for Interaction with Biorecognition Agents

In order to allow protein adsorption without altering the natural structure of the bioactive molecule, it is essential for the biomaterial surface to be biocompatible [84]. To ensure this, a bioconjugation protocol is applied in the biosensor. Bioconjugation involves the interaction of chemical or functional groups between NPs and biomolecules [85,86]. Additionally, it is important to mention that the properties can affect the efficiency of the connection with the biomarkers, including the optimal distance between biorecognition molecules and the nanostructure, the pH of the storage buffer used, as well as potential modifications to the biological and antigenic properties of biomolecules after conjugation. Hence, it is relevant to develop different approaches to nanostructure conjugation [87,88], using custom functional groups (such as primary amines, carboxylates, cis-diols, and sulfhydryls) on nanosurfaces [62,89].

Therefore, the adsorption methods of bioactive molecules into biosensor and nanobiosensor devices can be classified into the following categories: (i) Non-covalent immobilization strategies, based on electrostatic interaction, hydrophobic adsorption, and coordination bond formation between biomarker and the surface of nanomaterials (Figure 2) [86,90]. (ii) Covalent immobilization strategies, involving chemical bonds, chemically activate and modify biorecognition molecules, achieving a stable binding. They enhance sensitivity and selectivity in pathogen detection, ensuring robust interaction [91]. Finally, (iii) a combination of the above-mentioned techniques [87,92].

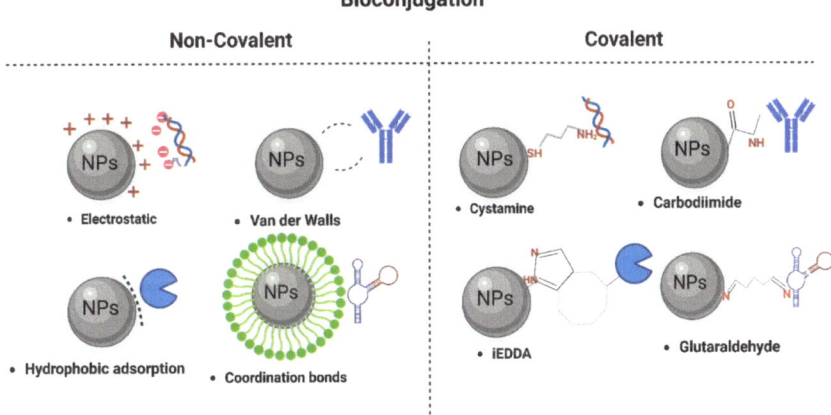

Figure 2. Different techniques of bioconjugation in nanomaterials.

2.1. Biorecognition Section of Bionsensor Device: Enzymes Applications

Enzymatic reactions in the biodetection processes involve the following steps: Firstly, enzymes recognize and bind to the target molecule in the environment/solution, through specific binding sites or active sites on the enzyme. Enzymes are typically immobilized on the surface, electrode, or substrate of the sensor to provide ideal conditions to react with a target molecule and produce a detectable signal [93]. Subsequently, enzymatic activity can be used as a signal through variations in the concentration of protons, entrance or exit of gases, and heat emission [94]. Finally, the signal generated is detected and quantified using a biosensor-detection component (electrochemical and fluorescence techniques) [93,94]. In fact, a biosensor base in enzymes immobilized on Au-NP was used to detect *Campylobacter jejuni* in chicken breast samples; in this biosensor, the nuclease, enzymes, and deoxyribozymes were immobilized to detect the pathogen for the reaction of the scission–enzyme, generating a heteroduplex of DNA–RNA, which finally induced a detectable signal based in a fluorescence-detection model with a limit of detection (LOD) of 10 pM. The viability of this method of DNA detection is assessed as an ultra-sensitive analysis; also, the authors remark that the Au-NP-based detection method can reach the lowest LOD (1 pM) of DNA in samples, one fold less than that required by the already mentioned ultra-sensitive fluorescence-detection method [95]. Also, a comparative analysis of the Influenza A virus detection using a biosensor-based technique showed an LOD of 1 pM mL^{-1} [96], which is extremely low compared to the concentration required in qPCR detection methods [97].

To enhance the biosensor-recognition capacities, the immobilization and stabilization of the enzymes are normal processes. The enzyme immobilization techniques commonly applied to nanostructures are covalent binding, crosslinking, and self-assembled monolayers [98]. However, using enzymes as biorecognition agents has certain disadvantages for in situ applications. Enzymatic activity can be influenced by environmental conditions such as temperature and pH, which affect the stability of the biosensor [99].

2.2. Antibody Applications

Antibodies (immunoglobulins) are proteins produced by cells of the immune system called B lymphocytes [100]. They consist of a basic structure composed of four polypeptide chains: two identical heavy chains and two identical light chains with their typical "Y" shape [100,101]. Antibodies are biological molecules, derived from animals, that have gained importance in pathogen-biomarker detection methods, due to their high specificity and in vivo uniqueness [102], designing monoclonal antibodies to precisely target antigens or receptors [103]. An antibody-based biosensor was presented by Majid et al. (2019) [104]. In this type of biosensor, the immobilization of antibodies in gold-NP or nanomaterial are related to weak electrostatic, hydrophobic, or van de Waals force interactions [105]. On the other hand, the preferred method for immobilizing antibodies on NPs and other surfaces is through covalent bonding, specifically using carbodiimide chemistry and maleimide conjugation. This approach allows for longer-lasting and reusable devices, as well as better control over antibody orientation, resulting in enhanced detection capabilities [105,106].

This type of antibody-functionalized biosensor has been used for different pathogen detection, as reported by Guo et al. (2020) [107], who developed a method using NP etching. These techniques allow for the specific detection of *Salmonella Typhimurium*, using catalase-modified antibodies that bind to the bacteria and catalyze the conversion of H_2O_2 to H_2O. In the absence of *S. Typhimurium*, the catalase-modified antibodies do not bind to the bacteria, resulting in a significant accumulation of residual H_2O_2. Horseradish peroxidase (HRP) triggers the production of •OH, causing a color change in the Au nanowires from dark blue to pink. The linear detection range is between 18 CFU mL^{-1} and 1.8×10^5 CFU mL^{-1}, with a detection limit of 35 CFU mL^{-1} [107]. However, antibody-functionalized biosensors have limitations including lack of specificity, long-term stability, high production cost, challenges in antibody immobilization, potential cross-reactivity, limited antibody availability, and batch-to-batch variability [50].

2.3. DNA Applications

Nucleic acid-based biosensors, such as DNA, stand out as biorecognition elements due to their simplicity, speed, and high specificity. For this reason, they are widely used for the detection of pathogens and other substances of interest in various biodetection applications [88]. These characteristics make DNA a powerful and versatile tool in the field of biodetection [108,109]. These molecular probes can be used in different ways in methods such as DNAzyme [110], DNA hairpin [111], DNA hybridization [112], and DNA origami [113]. It is widely recognized that DNA and its assembly structure can be applied to detect specific targets, including nucleic acids, proteins, metal ions, and small biological molecules. Common bioreceptors in this category include deoxyribonucleic acid (DNA), ribonucleic acid (RNA), and peptide nucleic acids (PNA) [114].

These biomarkers have been functionalized with nanomaterials to enhance their selectivity and durability in pathogen biodetection. An application of pathogenic bacteria detection in milk was developed using a combination of photo-induced electron transfer (PET) between a G-quadruplex DNAzyme and silver nanocluster-labeled DNA, along with exponential circular amplification based on the hairpin probe, achieving an ultra-low detection limit of 8 CFU mL^{-1} for *S. Typhimurium*. This strategy represents a promising platform for highly sensitive and specific detection of pathogenic bacteria in food analysis [115]. In another study, a fluorescent DNA hairpin template was developed by designing two hairpin probes with Au-NPs for the detection of *S. aureus* 16S rRNA. HP1 was biofunctionalized with thiol groups and a fluorescent chromophore, and a thiol group was attached to the NP surface. The addition of HP2 causes the target sequence to walk along the surface of the Au-NPs, thus opening the hairpin structure of HP2 and enabling the recycling of the target sequence. They achieved a LOD of 7.73 CFU mL^{-1} with an FM of 4.36×10^{-5}, demonstrating a novel and efficient method for the detection of *S. aureus* [116].

2.4. Aptamer Applications

Aptamers are short sequences of RNA or DNA (oligonucleotides), capable of folding into unique three-dimensional structures and binding to targets such as proteins, lipids, ions, small-molecular-weight metabolites, even whole cells with high specificity and affinity [117]. To produce aptamers, the SELEX (Systematic Evolution of Ligands by Exponential Enrichment) process is utilized. In this process, aptamers are generated through in vitro synthesis of combinatorial libraries with diverse sequences [53]. Through an iterative selection process, aptamers with higher affinity for the desired target are enriched and amplified, while those with lower affinity are discarded. This enables the generation of highly specific and high-affinity aptamers for various biomolecular targets [118]. The analytes of aptamer-based biosensors can vary in size and complexity as it can detect specific molecules such as proteins or more complex analytes such as whole cells. Aptamers modify their structure once reacting with a specific analyte and the conformational change can be transduced using different types of signals such as optical or electrochemical [49,119].

Some of the applications of aptamer-based sensors were developed for the detection of *S. aureus*, *E. coli* and *C. jejuni* pathogen biomarkers [53]. Another example was the results in *S. aureus* detection protocols, where they designed an ultra-sensitive magnetic fluorescence aptasensor based on fluorescence-resonance energy transfer, and the aptamers were placed on the surface of Fe_3O_4 and modified carbon dots (CDs). CDs were used as the fluorescence donor and Fe_3O_4 as the "off-on" sensor receptor. Due to the strong affinity of the aptamers to bacteria, the presence of target bacteria led to the disassembly of the Fe_3O_4/CDs aptasensor, resulting in the recovery of CDs fluorescence with a range of detection exhibited between 50×10^7 CFU mL^{-1} and 8 CFU mL^{-1} [120].

Another example is the application of *E. coli* detection using graphene oxide (GO)-modified Au-NPs, enhanced with aptamers; an E8 aptamer was used for *E. coli* detection. The detection limit was found to be 10 cells/mL in water and coconut water-enriched samples. Furthermore, the aptamer-based nanosensor exhibited selectivity towards its target without any cross-reactivity with other bacteria. The color changes from red to blue, based on aggregation, can be easily seen by the naked eye [121]. Another protocol for *E. coli* detection in water was the nanobiosensor using QDs functionalized with aptamer II and coated with magnetic NPs. Fluorescence values were recorded for 100, 200, 300, 400, and 500 CFU, each with CdTe-MPA QDs at 100 μg mL^{-1}, resulting in digital signals of 29.3 mV, 34.18 mV, 39.06 mV, 43.94 mV, and 48.82 mV, respectively, demonstrating that CdTe-MPA QDs conjugated with aptamer II were capable of selectively capturing and detecting *E. coli* [122].

Aptamers exhibit significant advantages to their application in pathogen detection, including lower molecular weight, easier and more cost-effective production methods, and good chemical stability [53]. Moreover, their ability to be generated against a wide range of targets ranging from small molecules to large proteins, and even whole live cells [123], has led to their utilization in various pathogen-detection nanobiosensor-based technologies, combined with different technologies including surface plasmon resonance (SPR), electrochemistry, piezoelectric effect, and chemiluminescence [80].

For example, SPR sensors utilize the reflection of light on a modified metal surface to detect changes using the biomarker binding in the refractive index, resulting in the precise and sensitive detection of the biomarker target [80]. In the case of electrochemical sensors, the analyte interaction is translated into an electrical signal, providing a quantitative means of detection and enabling real-time measurements [124]. Piezoelectric sensors, on the other hand, leverage the piezoelectric effect to convert the mechanical energy generated by analyte interaction into electrical energy, allowing highly sensitive and accurate detection [125]. Furthermore, chemiluminescence is another technology used in biosensors, where the analyte interaction triggers a chemical reaction that generates light. This emitted chemiluminescent light can be measured to detect and quantify the analyte, providing highly sensitive and specific detection [71]. Functionalizing nanomaterials with aptamers has allowed the combination of various signal-transduction strategies for detecting foodborne pathogens.

This detection versality provides the capability to utilize different approaches according to specification of each application, enhancing the sensitivity and selectivity of detection systems. Thus, aptamers are a powerful and promising tool in the fight against food contamination and the protection of public health [126]. These technologies enable the detection and quantification of substances in biological samples, providing versatile and efficient options for applications in fields such as medical diagnosis, food safety, and environmental monitoring (Figure 3).

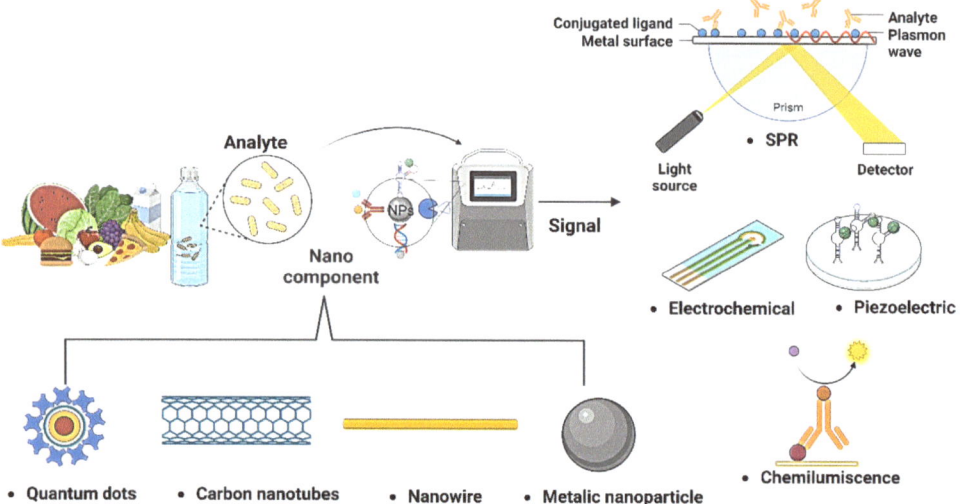

Figure 3. Combination of different nanomaterials and sensors for the detection of pathogens.

2.5. Molecularly Imprinted Polymers (MIPs)

MIPs are defined as a group of biomimic compounds that replicate the natural interactions between a biorecognition section (antibody, antigen, or enzyme) and a biomarker; these compounds have a "lock and key" bonding mechanism to interact with the molecule of interest [127]. MIP development methods can be divided into the following: (a) covalent, (b) semi-covalent, and (c) non covalent methods. These are in concordance with the site of action-binding modes; in general, the methods are as follows: bulk, suspension, emulsion, precipitation, multi-step swelling, and surface imprinting electrochemical polymerization [128].

In recent years the application of MIPs-based techniques has been applied to detect pathogens related to foodborne illnesses. In fact, in the case of bacteria detection, MIPs can be divided depending on the detection target (whole cells or cell membrane subunits), and subdivided in to microcontact/stamp imprinting (with a LOD of 70 CFU mL^{-1} of *E. coli*) [129], drop coating (with a LOD of 1.6×10^8 cells mL^{-1} of *E. coli* strain) [130], Pickering emulsion interfacial imprinting (with a LOD of 1×10^3 CFU mL^{-1} of *L. monocytogenes* strain) [131], and electropolymerization (with a LOD of 4 CFU mL^{-1} of *S. aureus* strain), among other methods that have been proved, through their LODs, to have a high detection sensitivity for foodborne pathogen [129].

3. Optical and Electrochemical Nanobiosensors

Nanobiosensors play a crucial role in the detection of biomolecules in food and water through two distinct phenomena. Optical nanobiosensors are based on the phenomenon of the interaction of optical nanostructures with light. When specific biomolecules bind to the analyte in the sample, they trigger changes in the optical properties of light, such as absorbance or fluorescence [132]. These changes are detected and quantified to determine the presence and concentration of the analyte. Optical nanobiosensors offer high sensitivity and selectivity by harnessing this phenomenon of light–matter interaction, ensuring precise detection in food and water quality-control applications [69,70]. The Surface Plasmon Resonance (SPR) phenomenon has fundamental applications in the detection of pathogens in food and water. This sensitive and specific optical technique is used to identify the presence of pathogens (bacteria and viruses) in food and water samples. Changes in the SPR resonance angle reveal the interaction between surface biomolecules and pathogens, allowing for rapid and accurate detection of potential microbiological contaminants in these critical products for public health [132,133] (Figure 4A). On the other hand, electrochemical nanobiosensors rely on the phenomenon of electrochemical reactions on nanostructured electrodes. These electrodes provide a large surface area, enabling profoundly sensitive and selective detection. When specific biomolecules in the sample interact with the analyte, changes in electrical current or electrical potential occur on the electrode's surface, modifying the electrical properties of the solution and generating a detectable signal (Figure 4B) [68].

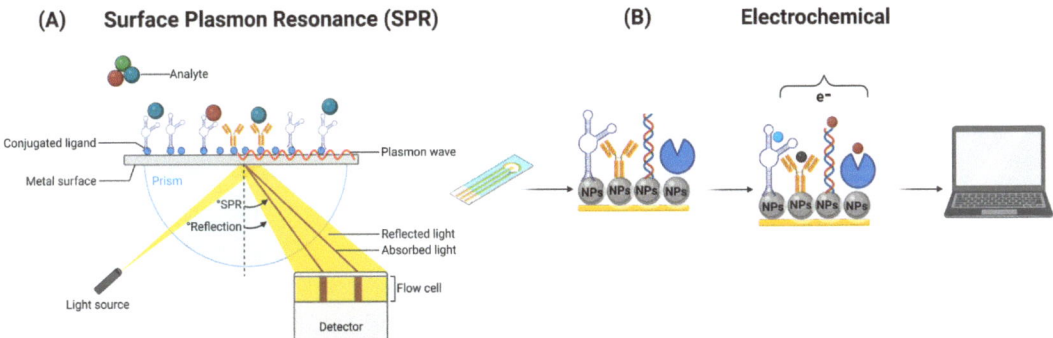

Figure 4. Operation of an optical SPR (**A**) and an electrochemical (**B**) biosensor, respectively.

Both types of nanobiosensors can be miniaturized for portable applications and are essential for ensuring the safety and quality of food and water, with the choice of nanobiosensor type depending on the specific properties of the analyte and the goals of the application at hand. Leticia Tessaro et al. (2022) [133] delve into the utilization of AuNPs in an SPR nanobiosensor designed for SARS-CoV-2 detection. While this method boasts sensitivity and precision on par with traditional RT-qPCR techniques, the cost associated with AuNPs may hinder widespread adoption. Nevertheless, it has achieved a detection time of 100 min and an LOD of 1 ng mL^{-1} (equivalent to 2.7×10^3 copies per µL), establishing itself as capable of detecting the virus on food surfaces, thus emphasizing its potential in safeguarding food during pandemics.

In a similar applications, Jiayun Hu et al. (2018) [134] demonstrated the exceptional plasmonic properties of AuNPs for LSPR-based detection, offering high sensitivity with an LOD of 10 CFU mL^{-1} in *Pseudomonas aeruginosa* detection. The cost aspect remains a concern for large-scale applications. Moreover, the versatile LSPR whole-cell detection scheme demonstrated they can be extended to other microorganisms, including various bacteria and viruses, through the use of different affinity agents. This robust LSPR detection platform holds promise for clinical applications, owing to its rapid detection capability of approximately 3 h, making it suitable for point-of-care and field-based applications. Ajinkya Hariram Dabhade et al. (2023) [135] introduce AgNPs in an electrochemical biosensor for *E. coli* detection, showcasing cost-effectiveness and simplicity. This sensor demonstrates good selectivity and stability, with an LOD of 150 CFU mL^{-1}. The ease of synthesis and their reproducibility make AgNPs a practical choice for on-site, real-time detection applications. Faezeh Shahdost-Fard et al., 2023 [136], introduce a unique nanocomposite-comprising sponge, copper tungsten oxide hydroxide, and AgNPs. While the synthesis process may be complex, this nanomaterial exhibits impressive performance in *S. aureus* detection. The nanocomposite-based electrochemical aptasensor offers a low LOD of 1 CFU mL^{-1} and high specificity. Its applicability in clinical samples underscores its potential for addressing nosocomial infections. The work carried out by Singh et al.,

2018 [137], employed gold nanoparticles (GNPs) in a rapid-pathogen-detection assay, capitalizing on colistin's interaction with lipopolysaccharides and the optical properties of the nanomaterial. This cost-effective approach eliminates tedious sample-preparation steps, offering a rapid, sensitive, visual detection method within 5 min. While GNPs are versatile, their sensitivity for pathogen detection at low concentrations may require further optimization for specific applications, as it exhibited a LOD of 10 cells mL^{-1} in tap water and 100 cells mL^{-1} in lake water samples. Zeynep Altintas et al., 2018 [138], present a fully automated microfluidic electrochemical biosensor designed for real-time bacteria detection. It employs immunoassays, including nanomaterial-amplified assays, to quantify E. coli concentrations. The sensor achieved a LOD of 1.99×10^4 CFU mL^{-1} using nanomaterial amplification.

Srijit Nair et al., 2018 [139], introduce a novel approach for detecting uropathogenic E. coli (UPEC) using crossed surface-relief gratings (CSRGs) as nanometallic sensors. This optical-sensing platform leverages SPR-based light energy exchange for real-time, selective, and label-free UPEC detection. The LOD is reported at 10^5 CFU mL^{-1}, which is clinically relevant to urinary tract infection (UTI) diagnosis.

Olja Simoska et al., 2019 [140], focus on real-time electrochemical detection of phenazine metabolites produced by *P. aeruginosa*. Transparent carbon ultramicroelectrode arrays (T-CUAs) are used to monitor the concentrations of pyocyanin (PYO) and other metabolites. Although this work primarily centers on metabolite detection, it offers valuable insights into real-time monitoring. The study provides detailed information about phenazine dynamics over time.

4. Nanomaterials for the Detection of Pathogens in Water and Food

As is mentioned above, one of the major concerns in food and water safety is the precise detection of pathogens, this has led, in combination with novel sensor technologies, to an increasing exploration of nanomaterials in combination with highly efficient aptamers to revolutionize the pathogen detection in water and food. This fusion of nanotechnology and aptamers opens new possibilities for more effective control and quicker responses to potential public health risks. The following Table 1 summarizes the last five years of nanobiosensor production for the detection of viruses, bacteria, and parasites using aptamers in complex matrices.

Table 1. NPs application for detection of pathogenic bacteria in food and water matrices.

Nanomaterial	Pathogen	Matrix	LOD	Signal	Bioconjugate Material	Reference
Iron core gold NPs	S. enteritidis	Beverage samples	32 Salmonella mL^{-1}	Fluorescence	Antibody	[63]
FeO-NPS and Quantum dots	E. coli	Water	1×10^2 CFU	Fluorescence	Aptamer	[122]
NAC (N-acetylcysteine) monomer	L. monocytogenes	Milk and pork meat	1×10^3 CFU mL^{-1}	Fluorescence	MPIs	[131]
Au-N triangles	P. aeruginosa	Water	1 cell	LSPR	Aptamer	[134]
Ag-NPs	E. coli	Water	150 CFU mL^{-1}	Electrochemical	Aptamer	[135]
AgNPs	S. aureus	Bacterial suspension and human serum	1.0 CFU mL^{-1}	Electrochemical	Aptamer	[136]
Au-NPs	S. aureus	Tap water	10^1 to 10^4 CFU mL^{-1}	Fluorescence	Aptamer	[141]
AuNPs	S. aureus	Luria-Bertani media	1.5×10^7 cells mL^{-1}	Colorimetric	Aptamer	[142]
AuNPs	Ochratoxin A	Peanut, soybean, and corn	28.18 pg/mL	Colorimetric	Aptamer	[143]

Table 1. *Cont.*

Nanomaterial	Pathogen	Matrix	LOD	Signal	Bioconjugate Material	Reference
AuNPs	E. coli	Flour	2.5 ng µL^{-1}	Colorimetric	Probe	[144]
Graphene oxide coated AuNPs	E. coli S. Typhimurium	Bacterial suspension	1×10^3 CFU	Colorimetric	Antibody	[145]
Ag-NPs	S. aureus	Water	1.0 CFU mL^{-1}	Electrochemical	Aptamer	[146]
Chitosan-AgNPs	Glipopolysaccharide	Bacterial suspension	248 CFU mL^{-1}	Electrochemical	-	[147]
AgNPs	E. coli	Pork, cabbage and milk	2.0 CFU mL^{-1}	Photoelectrochemical	Peptide Magainin	[148]
Au-NPs and oxide of graphene NPs	E. coli	Water	9.34 CFU mL^{-1}	Electrochemical	Aptamer	[149]
Multiwalled carbon nanotubes	E. coli	Water	0.8 CFU mL^{-1}	Electrochemical	Antibody	[150]
Graphene and carbon nanotubes	Salmonella enteritidis	Water	10^2–10^8 CFU mL^{-1}	Colorimetric	Antibody	[151]
Quantum dots	S. aureus, S. Typhimurium	Water	16–28 CFU mL^{-1}	Colorimetric	Aptamers	[152]
SiNPs	E. coli	Bacterial suspension	10^3 CFU mL^{-1}	Electrochemical	Polyclonalantibodies	[153]
SiNPs	E. coli	Bacterial suspension	8 CFU mL^{-1}	Fluorescence	Rhodamine B	[154]
SiNPs	AFB1 from filamentous fungi	Peanut, maize, and badam	0.214 pg mL^{-1}	Fluorescence	Aptamer	[155]
MNPs	S. aureus	Milk, Romaine lettuce, ham, and sausage	2.5 ng µL^{-1}	Colorimetric	Probes	[156]
Iron oxide MNPs assisted AuNPs	B. cereus and Shigella flexneri	Inoculated media	12 cells mL^{-1} and 3 cells mL^{-1}	Electrochemical	Vancomycin	[157]
Magnetic NPs	S. Typhimurium	Food	53 UFC/mL	Fluorescence	Oligonucleotides	[158]
Iron oxide encapsulated quantum dots	Hepatitis E virus Norovirus	Clinical samples	56 RNA copies mL^{-1} 69 RNA copies mL^{-1}	Fluorescence Electrochemical	Antibody	[159]
QDs	S. Typhimurium	Chicken meats	43 CFU mL^{-1}	Fluorescence	Antibody	[160]
QDs	S. Typhimurium and V. parahaemolyticus	Aquatic samples	10 CFU mL^{-1} 10^2 CFU mL^{-1}	Fluorescence	Aptamer	[161]
QDs nanobeads	S. Typhimurium	Potable water, orange juice, lettuce, and chicken	10^{-1} CFU mL^{-1}	Fluorescence	Antibody	[162]
TAA *, TBA **, TMA *** and TE ****	S. aureus	Lettuce/Shrimp	4 CFU mL^{-1}	Electrochemical/Fluorescence	MPIs	[163]

Abbreviations are referred to the following compounds: * 3-thiopheneacetic acid, ** 3-thiopheneboronic acid, *** 3-thiophenemethylamine, **** 3-thiopheneethanol.

Nanobiosensors, due to their small size and high sensitivity, enable the real-time detection of low concentrations of biomarkers, a crucial characteristic in applications of food and water monitoring. This versatility allows them to adapt to various molecules and technologies, such as artificial intelligence incorporation. Moreover, they are more cost-effective and environmentally friendly than conventional techniques. Their miniaturization capability makes them ideal for portable devices and on-site diagnostic systems, providing quick and efficient access to quality testing and analysis in food and water. This makes them promising tools in various scientific and technological applications (Figure 5).

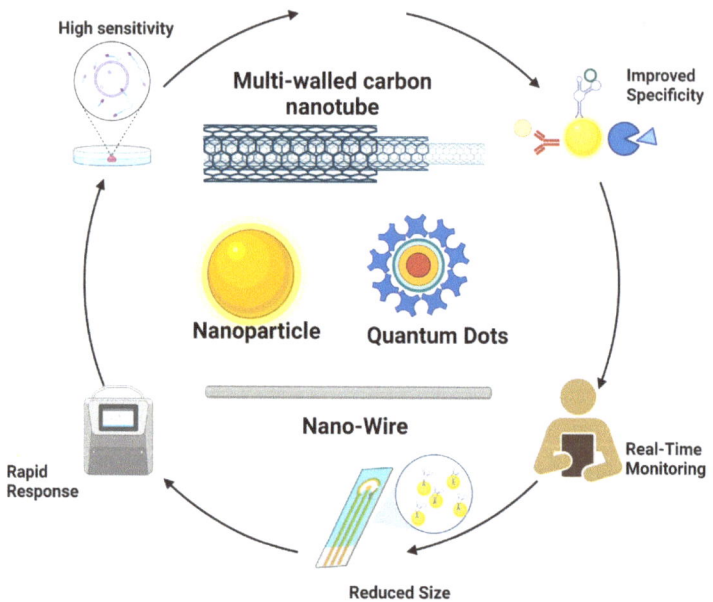

Figure 5. Strengths of nanobiosensors.

4.1. Gold Nanoparticles (Au-NPs)

Among the different types of NPs, metallic nanoparticles (MNPs) exhibit many useful characteristics such as high surface-to-volume ratio, conductivity, selectivity, and excellent optical and chemical properties, for their application in the biotechnology field [164,165]. The application can vary depending on the metal used, size, shape, surface properties, and functionalization of the MNPs [166]. On one hand, Au-NPs have been successfully used in pathogen detection because they can easily be conjugated with recognition and biorecognition elements such as aptamers, DNA, antibodies, carbohydrates, and proteins, which can enhance the reactivity and selectivity of the NPs towards specific pathogens [51,167].

In fact, Au-NPs are one of the most stable MNPs, not to mention their unique characteristics such as good chemical reactivity, conductivity, and high resistance, which have attracted attention for their use in biosensor development [168]. The surface of Au-NPs has been functionalized with various biocomponents [169]. These nanobiosensors have a very low LOD for different chemical and biological analytes, not to mention their high stability against oxidation [168]. Also, their characteristics, such as stability, conjugation, amplification properties, and their ability to serve as colorimetric biosensors [170–172] are especially relevant in the case of Au-NPs due to their localized surface plasmon resonance, which is a phenomenon that gives unique optical properties to MNPs, particularly Au-NPs. This is due to the interaction of electromagnetic waves with NPs of specific sizes and shapes, resulting in differential absorption of the light spectrum and different colors exhibited by the NPs [50,51]. These properties can be altered in the presence of different analytes, making Au-NPs highly suitable for biosensor development.

4.2. Silver Nanoparticles (Ag-NPs)

Ag-NPs stand out for their wide range of applications. These nanomaterials have been incorporated into textiles, healthcare products, consumer goods, medical devices, and biodetection applications, among others [173]. These materials are highly attractive in diagnostics field due to high conductivity, catalytic activity, and plasmonic properties presented, which may be leveraged to enhance the biosensor's performance [174]. Sensitivity is a crucial factor for biosensors to detect low concentrations of biomarkers. Ag-NPs have been used to increase the electroactive surface area of electrodes, enhancing the electron-transfer rate and improving biosensor sensitivity [173,174]. In the incorporation of Ag-NPs in biosensor structures, Ag-NPs can amplify signals or improve the detection of nucleic acids. Their plasmonic resonance absorption band, below 500 nm, confers selective absorption in the visible and near-infrared spectrum [168]. In connection with pathogen detection, the phenomenon of surface plasmon resonance (SPR) works using the electrons on the surface of a metal, which are excited by photons of specific wavelengths and incidence angles [175,176] and applied to target detection based on the refractive index [175]. This is achieved when the biomarker is bound to a biorecognition element of the biosensor, the recognition event between the biomarker and the biorecognition element results in a change in the SPR resonance angle [31]. Conjugated polymers, such as those that include silver nanoparticles are promising materials for addressing the current and emerging issues such as pandemic monitoring [177], and pathogen detection both in food [148] and water [146].

4.3. Carbon-Based Nanoparticles

Similar to Au-NPs, carbon-based NPs are useful for the implementation of detection techniques for pathogen monitoring in water [119,178]. Carbon-based NPs such as carbon nanotubes, graphene, and carbon nanodots have great potential in the biosensing of pathogens because of their ability to be coated with different biomolecules for the association of molecular patterns from pathogens and to generate a signal for specific pathogens as functionalized NPs can mimic the specific surface structure of pathogens [179]. Carbon NPs have been used in the fluorescence resonance energy transfer (FRET) technique with quantum dots as donors modified with aptamers for the detection of *Vibrio parahaemolyticus* and *S. Typhimurium* in the range of 25 to 35 CFU mL^{-1} and up to between 50 and 10^6 CFU mL^{-1}, respectively [180]. Also, these NPs can be used in combination with aptamers to amplify the sensitivity and specificity of the device.

4.4. Magnetic Nanomaterials (MNPs)

Magnetic NPs possess their own versatility when used for biosensing pathogens, because of their specific attributes, particularly fast separation and concentration, that makes them easy tools for pathogen detection [181]. MNPs have been used for detecting pathogens using nucleic acid detection and quantification in devices for point-of-care testing in the detection of the Hepatitis B virus (LOD of 50 IU mL^{-1}) and SARS-CoV-2 (500 copies mL^{-1}) [182]. Magnetic NPs (MNPs) can conform to a section of the transducer part of the biosensor, or be suspended in solution in direct contact with the analyte of interest [183]. When the MNPs are in contact with the sample, they bind to the target molecule through the interaction of the label in the NPs (a functional group) and a protein; once the complex of MNPs and target is formed, an external magnetic field attracts it to the active-detection surface, and after a wash of the unbinding molecules, targets are detected [184].

When talking about magnetic NPs in biosensing, it is important to mention the magnetic relaxation switching mechanism (MRS). This phenomenon describes the incidence when cross-linking occurs between the MNPs in the binding and recognition of targets. When these MNPs clusters are formed, a change in the transverse relaxation of the sample is reflected as motional averaging or static dephasing according to the MNPs cluster size and this change can be monitored using nuclear magnetic resonance [185].

4.5. Silica Nanoparticles (Si-NPs)

Si-NPs have applications in the biomedical field [186], and they present good optical properties and good biocompatibility [187]. NPs are mesoporous, so in combination with other metals, have attractive and profitable characteristics for biosensing purposes [188]. Their uniformity and easily changed pore size among the gating mechanism makes it very useful in biosensing for drug delivery, for example [189]. Another important characteristic of Si-NPs is that they are considered as a GRAS (generally recognized as safe) material by the FDA [190,191]. The mesoporous nature of the Si-NPs is characteristic of a large interest, this feature can be employed to separate bacteria from complex samples even preserving its viability, and colloidal stabilization of magnetic NPs for the same purpose. Also, the silanol functional groups from SiNPs make possible the use and design of various bio-recognition systems that help to increase their sensibility and selectivity while reducing the detection time of different pathogens [190].

4.6. Quantum Dots (QD)

Quantum dots (QD) are colloidal nanocrystalline semiconductors that possess properties such as a quantum confinement effect, allowing them to emit and absorb light at specific wavelengths [191]. Because of this, QDs exhibit excellent optical properties, including a broad absorption spectrum, a narrow emission spectrum, and tunable luminescence, which show great prospects in biodetection [192]. QD-based biosensors include but may not be limited to fluorescence, bioluminescent, chemiluminescent, and photoelectrochemical approaches [193]. Some of the characteristics that make the use of quantum dots attractive for biosensing applications are that they possess high-quantum yield, better photobleaching resistance, wide absorption spectra, a narrow emission spectrum and their specificity with biologic targets in comparison with common fluorophores and dyes [194]. Also, it is very remarkable that its surface is easily functionalized with biologic components in order to integrate QD probes [193]. In the field of nanomaterials, the use of combinations of magnetic compounds displays attractive characteristics for current applications; these nanocomposites, besides maintaining complementary magnetic behavior, add functional proprieties to the final product [159].

As presented above, numerous studies focus their determinations on *S. Typhimurium* mainly because it is the most common pathogen related to food poisoning in Western countries causing gastroenteritis [195]. If well-used as the model or the target of the experimentations, the modifications in for example primers' design or binding proteins may allow the replication of studies carried with this strain to any other food pathogens [161,162,196].

5. Prospects and Limitations to Detecting Pathogens with DNA Using Nanobiosensors

The importance of exceptionally responsive devices is essential to advancing biosensors applied in pathogen detection. Insufficient sensitivity and affinity towards biomarkers can significantly impact the performance of the device and prevent the pathogen detection. In fact, some of the biomarkers are at an ultra-low concentration in the samples (pM), and for this reason, it is necessary that the device is available to detect these ultra-low concentrations [197]. These concentrations of pathogen biomarkers in the samples present a limitation to the performance of the detection methods, this is related to the source and nature of the target biomarker itself [198]. However, through using a genetic and whole-cell-based biomarker target, and adequate sensor technologies, some of the biosensors have LOD of three genetic copies per sample [199], 1 CFU mL^{-1} [200] or have even reached an LOD of 3×10^6 gene copies per sample [9], or 5×10^4 CFU mL^{-1} biomarker concentration in the sample, this is considered an ultrasensitive detection range [201].

As is mentioned above, the choice of signal recognition technology can determine the sensitivity level required to identify biochemical, genetic, or whole cell biomarker concentration in the sample [132,202] and affect the performance of the device. In fact, the detection's ultra-low concentration of pathogen-disease biomarkers concentration in samples is a mandatory requirement for early detection in clinical diagnoses [203]. Some of the disadvantages of biosensor-based pathogen detection are as follows: The factor to be determined is the target molecule to be sensed where sensitivity and specificity are compromised by the biomarker choice [109]. The use of genetic markers leads to a more sensitive device. However, it implicates complex systems and laborious sample/re-agent handling procedures [204].

On the other hand, signal emission technology is another important factor in pathogen detection. Colorimetric-based biosensors have several factors that may alter their detection capability, such has colorimetric substrate, incubation time, and even the temperature at which the signal is measured [205]. Particularly in DNA-based biosensors, other factors are lack of ability to form a complex, complication in large-scale patterns, reaction induction by mistake, and high sensitivity to enzymatic degradation and oxidation [109]. However, despite the disadvantages, there are several applied technologies in the biorecognition element of the biosensor device. For example, gene-sequence biomarkers such as CRISPR/Cas9-based technology, where the lowest LOD was three genetic copies per sample, reaching up to 3×10^6 gene copies per sample [9]. In this study, the detection signal was the dose-response intensity. CRISPR/Cas12 based lateral flow, where the lowest LOD was four gene copies in the sample [199]. Other limitations to consider in the application of this device is the ability of the biosensor to discriminate between live/dead cells (LOD 1 CFU mL^{-1}); the use of functionalized NPs with bacteriophages as a biorecognition agent is a solution applied for successful discrimination between live/dead [200].

One of the most astonishing advancements in the field of biosensors is the implementation of artificial intelligence and other informatic technologies in pathogen detection. The combination of artificial intelligence and biosensors has created an interdisciplinary concept of AI biosensors. The basic architecture of AI biosensors consists of three main elements: information gathering, signal conversion, and AI data processing [206]. A nanobiosensor with AI offers advantages in terms of sensitivity, speed, and analytical capability compared to conventional biosensors. This makes it suitable for application where highly precise and rapid detection of biomarkers is required, such as advanced medical diagnostics, environmental monitoring of molecular-level contaminants, and nanoscale quality control in the food industry [206,207]. The study conducted by Taniguchi et al. (2021) [208] revealed that by utilizing nanopores in conjunction with artificial intelligence, the identification of similarly sized coronaviruses is achievable. This capability has the potential to differentiate between various types of coronaviruses, such as HCoV-229E, SARS-CoV, MERS-CoV, and SARS-CoV-2. Furthermore, this technique demonstrated its effectiveness by successfully detecting SARS-CoV-2 in saliva samples. In summary, solid-state virus-immunodetection techniques hold a promising outlook for the development of versatile, adaptable, and cost-effective diagnostic tools in the future [202].

6. Conclusions

Over the years, significant advancements have been made in the field of biosensors, particularly in areas related to food safety and the monitoring of pathogenic microorganisms associated with food and waterborne illnesses. Despite these achievements, progress in technologies for the development of pathogen-detecting biosensors remains a highly promising area of study. This is due to the presence of various nanomaterials (MNPs, QD, carbon nanotubes, among others) with specific properties that enable the identification of specific pathogens and enhance the performance of the devices.

The nanomaterials used in biosensors offer unique advantages for pathogen detection. Thanks to their small size and large specific surface area, they facilitate more effective interaction with pathogen biomarkers, enhancing the sensitivity and selectivity of biosensors. Furthermore, their capacity to be functionalized with specific molecules, such as antibodies, nucleic acids, or aptamers, provides intrinsic advantages in the selectivity and sensitivity of the devices. Particularly with the aptamers, due to their chain-like structure, they offer greater flexibility and ease of design, making them highly selective and sensitive agents for the precise detection of pathogens. In fact, aptamers, functionalized aptamers, and other genetic-based biomarker-detection technologies have promising applications for the enhanced specificity and selectivity that has been proved in pathogen monitoring. Also, this technology compared with antibody techniques has several advantages such as improved specificity and the ability to detect genetic material, rather than proteinic, of structural biomarkers. In addition, pathogen detection through biosensors has a substantial impact on public health. The presented revision shows how nanobiosensors' technology contributes to the precise and rapid identification of pathogenic agents in food and water, and if applied correctively, can prevent disease outbreaks and prompt appropriate measures to ensure consumer safety.

In summary, the nanomaterials employed in biosensors present diverse advantages for pathogen detection. Their reduced size, high-specific surface area, functionalization capacity, and signal amplification properties contribute to the sensitivity, selectivity, and precision of biosensors. Future possible applications of DNA-based technologies in combination with nanoparticles' formulation, particularity the application of aptamer technologies and nanoparticles with DNA probes, will have more sensitive and specific detection techniques. In addition, this type of biosensor has the lowest capacity for detection limits by using a genetic fingerprint to discriminate between pathogens, and in the future between non-pathogenic strains, and strains of concern. These qualities, combined with the potential to use detection techniques such as fluorescence, and the application of digital technologies such as IA models, has huge potential to improve the detection capacity of the monitoring methods, creating nanomaterial-based and aptamer-based biosensors, and promising tools for pathogen monitoring and detection, enhancing safety in the food industry and public health overall.

Author Contributions: Conceptualization: H.M.V.-A., A.A.-A., E.R.M.-S., J.E.S.-H. and R.P.-S.; Writing—original draft preparation: H.M.V.-A., A.A.-A., E.R.M.-S., P.G.V.-O., J.E.S.-H. and R.P.-S.; Data curation: H.M.V.-A., A.A.-A., E.R.M.-S., P.G.V.-O., O.d.l.R. and M.A.O.-M.; Visualization: H.M.V.-A., A.A.-A., E.R.M.-S. and P.G.V.-O.; Writing—review and editing: H.M.V.-A., E.R.M.-S.; A.A.-A., O.d.l.R., M.A.O.-M., J.E.S.-H. and R.P.-S.; Administration: A.A.-A., O.d.l.R., M.A.O.-M., J.E.S.-H. and R.P.-S.; Funding acquisition: J.E.S.-H. and R.P.-S.; Supervision. J.E.S.-H. and R.P.-S. All authors have read and agreed to the published version of the manuscript.

Funding: We are thankful for the funding from the Fundación FEMSA with the project entitled "Unidad de respuesta rápida al monitoreo de COVID-19 por agua residual" (grant number NA) and Tecnologico de Monterrey internal funding through the project Challenge-Based Research Funding Program 2022 (Muestreador Pasivo I026-IAMSM005-C4-11-1).

Institutional Review Board Statement: Not applicable.

Informed Consent Statement: Not applicable.

Data Availability Statement: Not applicable.

Acknowledgments: Figures were created with Biorender.com (accessed on 21 August 2023).

Conflicts of Interest: The authors declare no conflict of interest.

References

1. Estimating the Burden of Foodborne Diseases. Available online: https://www.who.int/activities/estimating-the-burden-of-foodborne-diseases (accessed on 14 March 2023).
2. Riu, J.; Giussani, B. Electrochemical Biosensors for the Detection of Pathogenic Bacteria in Food. *TrAC Trends Anal. Chem.* **2020**, *126*, 115863. [CrossRef]
3. Haughton, P. La OMS Intensifica sus Esfuerzos para Mejorar la Salubridad de los Alimentos y Proteger a la Población de las Enfermedades. Available online: https://www.who.int/es/news/item/07-06-2021-who-steps-up-action-to-improve-food-safety-and-protect-people-from-disease (accessed on 10 May 2023).
4. AL-Mamun, M.; Chowdhury, T.; Biswas, B.; Absar, N. Food Poisoning and Intoxication: A Global Leading Concern for Human Health. In *Food Safety and Preservation*; Elsevier: Amsterdam, The Netherlands, 2018; pp. 307–352, ISBN 978-0-12-814956-0.
5. Chin, N.A.; Salihah, N.T.; Shivanand, P.; Ahmed, M.U. Recent Trends and Developments of PCR-Based Methods for the Detection of Food-Borne Salmonella Bacteria and Norovirus. *J. Food Sci. Technol.* **2022**, *59*, 4570–4582. [CrossRef]
6. Thomas, K.M.; De Glanville, W.A.; Barker, G.C.; Benschop, J.; Buza, J.J.; Cleaveland, S.; Davis, M.A.; French, N.P.; Mmbaga, B.T.; Prinsen, G.; et al. Prevalence of Campylobacter and Salmonella in African Food Animals and Meat: A Systematic Review and Meta-Analysis. *Int. J. Food Microbiol.* **2020**, *315*, 108382. [CrossRef]

7. Dos Santos, J.S.; Biduski, B.; Dos Santos, L.R. Listeria Monocytogenes: Health Risk and a Challenge for Food Processing Establishments. *Arch. Microbiol.* **2021**, *203*, 5907–5919. [CrossRef]
8. EFSA BIOHAZ Panel; Koutsoumanis, K.; Allende, A.; Alvarez-Ordóñez, A.; Bover-Cid, S.; Chemaly, M.; Davies, R.; De Cesare, A.; Herman, L.; Hilbert, F.; et al. Pathogenicity Assessment of Shiga Toxin-producing *Escherichia coli* (STEC) and the Public Health Risk Posed by Contamination of Food with STEC. *EFS2* **2020**, *18*, e05967. [CrossRef]
9. Zhou, J.; Yin, L.; Dong, Y.; Peng, L.; Liu, G.; Man, S.; Ma, L. CRISPR-Cas13a Based Bacterial Detection Platform: Sensing Pathogen Staphylococcus Aureus in Food Samples. *Anal. Chim. Acta* **2020**, *1127*, 225–233. [CrossRef]
10. Mora, Z.V.L.; Macías-Rodríguez, M.E.; Arratia-Quijada, J.; Gonzalez-Torres, Y.S.; Nuño, K.; Villarruel-López, A. Clostridium Perfringens as Foodborne Pathogen in Broiler Production: Pathophysiology and Potential Strategies for Controlling Necrotic Enteritis. *Animals* **2020**, *10*, 1718. [CrossRef]
11. Enosi Tuipulotu, D.; Mathur, A.; Ngo, C.; Man, S.M. Bacillus Cereus: Epidemiology, Virulence Factors, and Host–Pathogen Interactions. *Trends Microbiol.* **2021**, *29*, 458–471. [CrossRef]
12. Gupta, V.; Gulati, P.; Bhagat, N.; Dhar, M.S.; Virdi, J.S. Detection of Yersinia Enterocolitica in Food: An Overview. *Eur. J. Clin. Microbiol. Infect. Dis.* **2015**, *34*, 641–650. [CrossRef]
13. Pichel, N.; Vivar, M.; Fuentes, M. The Problem of Drinking Water Access: A Review of Disinfection Technologies with an Emphasis on Solar Treatment Methods. *Chemosphere* **2019**, *218*, 1014–1030. [CrossRef]
14. Adelodun, B.; Ajibade, F.O.; Ighalo, J.O.; Odey, G.; Ibrahim, R.G.; Kareem, K.Y.; Bakare, H.O.; Tiamiyu, A.O.; Ajibade, T.F.; Abdulkadir, T.S.; et al. Assessment of Socioeconomic Inequality Based on Virus-Contaminated Water Usage in Developing Countries: A Review. *Environ. Res.* **2021**, *192*, 110309. [CrossRef]
15. Ramírez-Castillo, F.; Loera-Muro, A.; Jacques, M.; Garneau, P.; Avelar-González, F.; Harel, J.; Guerrero-Barrera, A. Waterborne Pathogens: Detection Methods and Challenges. *Pathogens* **2015**, *4*, 307–334. [CrossRef]
16. Cissé, G. Food-Borne and Water-Borne Diseases under Climate Change in Low- and Middle-Income Countries: Further Efforts Needed for Reducing Environmental Health Exposure Risks. *Acta Trop.* **2019**, *194*, 181–188. [CrossRef]
17. Mahagamage, M.G.Y.L.; Pathirage, M.V.S.C.; Manage, P.M. Contamination Status of *Salmonella* Spp., *Shigella* Spp. and *Campylobacter* Spp. in Surface and Groundwater of the Kelani River Basin, Sri Lanka. *Water* **2020**, *12*, 2187. [CrossRef]
18. Schoenen, D. Role of Disinfection in Suppressing the Spread of Pathogens with Drinking Water: Possibilities and Limitations. *Water Res.* **2002**, *36*, 3874–3888. [CrossRef]
19. Wen, X.; Chen, F.; Lin, Y.; Zhu, H.; Yuan, F.; Kuang, D.; Jia, Z.; Yuan, Z. Microbial Indicators and Their Use for Monitoring Drinking Water Quality—A Review. *Sustainability* **2020**, *12*, 2249. [CrossRef]
20. Semenza, J.C.; Rocklöv, J.; Ebi, K.L. Climate Change and Cascading Risks from Infectious Disease. *Infect. Dis. Ther.* **2022**, *11*, 1371–1390. [CrossRef]
21. Jahne, M.A.; Schoen, M.E.; Kaufmann, A.; Pecson, B.M.; Olivieri, A.; Sharvelle, S.; Anderson, A.; Ashbolt, N.J.; Garland, J.L. Enteric Pathogen Reduction Targets for Onsite Non-Potable Water Systems: A Critical Evaluation. *Water Res.* **2023**, *233*, 119742. [CrossRef]
22. Parra-Arroyo, L.; Martínez-Ruiz, M.; Lucero, S.; Oyervides-Muñoz, M.A.; Wilkinson, M.; Melchor-Martínez, E.M.; Araújo, R.G.; Coronado-Apodaca, K.G.; Bedran, H.V.; Buitrón, G.; et al. Degradation of Viral RNA in Wastewater Complex Matrix Models and Other Standards for Wastewater-Based Epidemiology: A Review. *TrAC Trends Anal. Chem.* **2023**, *158*, 116890. [CrossRef]
23. Kaya, H.O.; Cetin, A.E.; Azimzadeh, M.; Topkaya, S.N. Pathogen Detection with Electrochemical Biosensors: Advantages, Challenges and Future Perspectives. *J. Electroanal. Chem.* **2021**, *882*, 114989. [CrossRef]
24. Kumar, H.; Kuča, K.; Bhatia, S.K.; Saini, K.; Kaushal, A.; Verma, R.; Bhalla, T.C.; Kumar, D. Applications of Nanotechnology in Sensor-Based Detection of Foodborne Pathogens. *Sensors* **2020**, *20*, 1966. [CrossRef]
25. Kabiraz, M.P.; Majumdar, P.R.; Mahmud, M.M.C.; Bhowmik, S.; Ali, A. Conventional and Advanced Detection Techniques of Foodborne Pathogens: A Comprehensive Review. *Heliyon* **2023**, *9*, e15482. [CrossRef]
26. Clais, S.; Boulet, G.; Van Kerckhoven, M.; Lanckacker, E.; Delputte, P.; Maes, L.; Cos, P. Comparison of Viable Plate Count, Turbidity Measurement and Real-time PCR for Quantification of *Porphyromonas Gingivalis*. *Lett. Appl. Microbiol.* **2015**, *60*, 79–84. [CrossRef]
27. Rajapaksha, P.; Elbourne, A.; Gangadoo, S.; Brown, R.; Cozzolino, D.; Chapman, J. A Review of Methods for the Detection of Pathogenic Microorganisms. *Analyst* **2019**, *144*, 396–411. [CrossRef]
28. López, M.M.; Ilop, P. Noales Are Molecular Tools Solving the Challenges Posed by Detection of Plant Pathogenic Bacteria and Viruses? *Curr. Issues Mol. Biol.* **2009**, *11*, 13–46. [CrossRef]
29. Aw, T.G.; Rose, J.B. Detection of Pathogens in Water: From Phylochips to qPCR to Pyrosequencing. *Curr. Opin. Biotechnol.* **2012**, *23*, 422–430. [CrossRef]
30. Fu, Y.; Peng, H.; Liu, J.; Nguyen, T.H.; Hashmi, M.Z.; Shen, C. Occurrence and Quantification of Culturable and Viable but Non-Culturable (VBNC) Pathogens in Biofilm on Different Pipes from a Metropolitan Drinking Water Distribution System. *Sci. Total Environ.* **2021**, *764*, 142851. [CrossRef]
31. Srivastava, K.R.; Awasthi, S.; Mishra, P.K.; Srivastava, P.K. Biosensors/Molecular Tools for Detection of Waterborne Pathogens. In *Waterborne Pathogens*; Elsevier: Amsterdam, The Netherlands, 2020; pp. 237–277, ISBN 978-0-12-818783-8.

32. Sun, Y.-J.; Chen, G.-F.; Zhang, C.-Y.; Guo, C.-L.; Wang, Y.-Y.; Sun, R. Development of a Multiplex Polymerase Chain Reaction Assay for the Parallel Detection of Harmful Algal Bloom-Forming Species Distributed along the Chinese Coast. *Harmful Algae* **2019**, *84*, 36–45. [CrossRef]
33. Kim, J.-H.; Oh, S.-W. Rapid and Sensitive Detection of *E. coli* O157:H7 and *S. typhimurium* in Iceberg Lettuce and Cabbage Using Filtration, DNA Concentration, and qPCR without Enrichment. *Food Chem.* **2020**, *327*, 127036. [CrossRef]
34. Chen, B.; Jiang, Y.; Cao, X.; Liu, C.; Zhang, N.; Shi, D. Droplet Digital PCR as an Emerging Tool in Detecting Pathogens Nucleic Acids in Infectious Diseases. *Clin. Chim. Acta* **2021**, *517*, 156–161. [CrossRef] [PubMed]
35. Zhao, Y.; Zeng, D.; Yan, C.; Chen, W.; Ren, J.; Jiang, Y.; Jiang, L.; Xue, F.; Ji, D.; Tang, F.; et al. Rapid and Accurate Detection of *Escherichia coli* O157:H7 in Beef Using Microfluidic Wax-Printed Paper-Based ELISA. *Analyst* **2020**, *145*, 3106–3115. [CrossRef]
36. Liu, S.; Hu, Q.; Li, C.; Zhang, F.; Gu, H.; Wang, X.; Li, S.; Xue, L.; Madl, T.; Zhang, Y.; et al. Wide-Range, Rapid, and Specific Identification of Pathogenic Bacteria by Surface-Enhanced Raman Spectroscopy. *ACS Sens.* **2021**, *6*, 2911–2919. [CrossRef]
37. Bu, T.; Jia, P.; Liu, J.; Liu, Y.; Sun, X.; Zhang, M.; Tian, Y.; Zhang, D.; Wang, J.; Wang, L. Diversely Positive-Charged Gold Nanoparticles Based Biosensor: A Label-Free and Sensitive Tool for Foodborne Pathogen Detection. *Food Chem. X* **2019**, *3*, 100052. [CrossRef]
38. DeMone, C.; Hwang, M.-H.; Feng, Z.; McClure, J.T.; Greenwood, S.J.; Fung, R.; Kim, M.; Weese, J.S.; Shapiro, K. Application of next Generation Sequencing for Detection of Protozoan Pathogens in Shellfish. *Food Waterborne Parasitol.* **2020**, *21*, e00096. [CrossRef]
39. Eyre, D.W. Infection Prevention and Control Insights from a Decade of Pathogen Whole-Genome Sequencing. *J. Hosp. Infect.* **2022**, *122*, 180–186. [CrossRef]
40. Dąbrowiecki, Z.; Dąbrowiecka, M.; Olszański, R.; Siermontowski, P. Developing a Methodology for Testing and Preliminary Determination of the Presence of and in Environmental Water Samples by Immunomagnetic Separation Combined with Flow Cytometry. *Pol. Hyperb. Res.* **2019**, *68*, 71–92. [CrossRef]
41. Bulard, E.; Bouchet-Spinelli, A.; Chaud, P.; Roget, A.; Calemczuk, R.; Fort, S.; Livache, T. Carbohydrates as New Probes for the Identification of Closely Related *Escherichia coli* Strains Using Surface Plasmon Resonance Imaging. Available online: https://pubs.acs.org/doi/pdf/10.1021/ac5037704 (accessed on 26 April 2023).
42. Ahmed, S.; Ansari, A.; Siddiqui, M.A.; Imran, M.; Kumari, B.; Khan, A.; Ranjan, P. Electrochemical and Optical-Based Systems for SARS-CoV-2 and Various Pathogens Assessment. *Adv. Nat. Sci. Nanosci. Nanotechnol.* **2023**, *14*, 033001. [CrossRef]
43. Losada-Garcia, N.; Garcia-Sanz, C.; Andreu, A.; Velasco-Torrijos, T.; Palomo, J.M. Glyconanomaterials for Human Virus Detection and Inhibition. *Nanomaterials* **2021**, *11*, 1684. [CrossRef]
44. Wen, C.-Y.; Liang, X.; Liu, J.; Zhao, T.-Y.; Li, X.; Zhang, Y.; Guo, G.; Zhang, Z.; Zeng, J. An Achromatic Colorimetric Nanosensor for Sensitive Multiple Pathogen Detection by Coupling Plasmonic Nanoparticles with Magnetic Separation. *Talanta* **2023**, *256*, 124271. [CrossRef]
45. Jain, S.; Nehra, M.; Kumar, R.; Dilbaghi, N.; Hu, T.; Kumar, S.; Kaushik, A.; Li, C. Internet of Medical Things (IoMT)-Integrated Biosensors for Point-of-Care Testing of Infectious Diseases. *Biosens. Bioelectron.* **2021**, *179*, 113074. [CrossRef]
46. Salama, A.M.; Yasin, G.; Zourob, M.; Lu, J. Fluorescent Biosensors for the Detection of Viruses Using Graphene and Two-Dimensional Carbon Nanomaterials. *Biosensors* **2022**, *12*, 460. [CrossRef] [PubMed]
47. Nate, Z.; Gill, A.A.S.; Chauhan, R.; Karpoormath, R. Recent Progress in Electrochemical Sensors for Detection and Quantification of Malaria. *Anal. Biochem.* **2022**, *643*, 114592. [CrossRef] [PubMed]
48. Nnachi, R.C.; Sui, N.; Ke, B.; Luo, Z.; Bhalla, N.; He, D.; Yang, Z. Biosensors for Rapid Detection of Bacterial Pathogens in Water, Food and Environment. *Environ. Int.* **2022**, *166*, 107357. [CrossRef]
49. Sharifi, S.; Vahed, S.Z.; Ahmadian, E.; Dizaj, S.M.; Eftekhari, A.; Khalilov, R.; Ahmadi, M.; Hamidi-Asl, E.; Labib, M. Detection of Pathogenic Bacteria via Nanomaterials-Modified Aptasensors. *Biosens. Bioelectron.* **2020**, *150*, 111933. [CrossRef]
50. Hegde, M.; Pai, P.; Shetty, M.G.; Babitha, K.S. Gold Nanoparticle Based Biosensors for Rapid Pathogen Detection: A Review. *Environ. Nanotechnol. Monit. Manag.* **2022**, *18*, 100756. [CrossRef]
51. Sadanandan, S.; Meenakshi, V.S.; Ramkumar, K.; Pillai, N.P.; Anuvinda, P.; Sreelekshmi, P.J.; Devika, V.; Ramanunni, K.; Jeevan Sankar, R.; Sreejaya, M.M. Biorecognition Elements Appended Gold Nanoparticle Biosensors for the Detection of Food-Borne Pathogens—A Review. *Food Control* **2023**, *148*, 109510. [CrossRef]
52. Cho, I.-H.; Ku, S. Current Technical Approaches for the Early Detection of Foodborne Pathogens: Challenges and Opportunities. *IJMS* **2017**, *18*, 2078. [CrossRef] [PubMed]
53. Li, Y.; Liu, L.; Huang, Q.; Tong, T.; Zhou, Y.; Li, Z.; Bai, Q.; Liang, H.; Chen, L. Recent Advances on Aptamer-Based Biosensors for Detection of Pathogenic Bacteria. *World J. Microbiol. Biotechnol.* **2021**, *37*, 45. [CrossRef]
54. Chamundeeswari, M.; Jeslin, J.; Verma, M.L. Nanocarriers for Drug Delivery Applications. *Environ. Chem. Lett.* **2019**, *17*, 849–865. [CrossRef]
55. Ghorbani, F.; Abbaszadeh, H.; Mehdizadeh, A.; Ebrahimi-Warkiani, M.; Rashidi, M.-R.; Yousefi, M. Biosensors and Nanobiosensors for Rapid Detection of Autoimmune Diseases: A Review. *Microchim. Acta* **2019**, *186*, 838. [CrossRef]
56. Ali, A.A.; Altemimi, A.B.; Alhelfi, N.; Ibrahim, S.A. Application of Biosensors for Detection of Pathogenic Food Bacteria: A Review. *Biosensors* **2020**, *10*, 58. [CrossRef] [PubMed]
57. Kuswandi, B. Nanobiosensor Approaches for Pollutant Monitoring. *Environ. Chem. Lett.* **2019**, *17*, 975–990. [CrossRef]

58. Chandra, P.; Prakash, R. (Eds.) *Nanobiomaterial Engineering: Concepts and Their Applications in Biomedicine and Diagnostics*; Springer: Singapore, 2020; ISBN 978-981-329-839-2.
59. Fracchiolla, N.; Artuso, S.; Cortelezzi, A. Biosensors in Clinical Practice: Focus on Oncohematology. *Sensors* **2013**, *13*, 6423–6447. [CrossRef] [PubMed]
60. Zhang, Z.; Zhou, J.; Du, X. Electrochemical Biosensors for Detection of Foodborne Pathogens. *Micromachines* **2019**, *10*, 222. [CrossRef]
61. Saha, K.; Agasti, S.S.; Kim, C.; Li, X.; Rotello, V.M. Gold Nanoparticles in Chemical and Biological Sensing. *Chem. Rev.* **2012**, *112*, 2739–2779. [CrossRef]
62. Tuteja, S.K.; Mutreja, R.; Neethirajan, S.; Ingebrandt, S. Bioconjugation of Different Nanosurfaces With Biorecognition Molecules for the Development of Selective Nanosensor Platforms. In *Advances in Nanosensors for Biological and Environmental Analysis*; Elsevier: Amsterdam, The Netherlands, 2019; pp. 79–94, ISBN 978-0-12-817456-2.
63. Zhao, X.; Smith, G.; Javed, B.; Dee, G.; Gun'ko, Y.K.; Curtin, J.; Byrne, H.J.; O'Connor, C.; Tian, F. Design and Development of Magnetic Iron Core Gold Nanoparticle-Based Fluorescent Multiplex Assay to Detect Salmonella. *Nanomaterials* **2022**, *12*, 3917. [CrossRef]
64. Verma, M.L.; Rani, V. Biosensors for Toxic Metals, Polychlorinated Biphenyls, Biological Oxygen Demand, Endocrine Disruptors, Hormones, Dioxin, Phenolic and Organophosphorus Compounds: A Review. *Environ. Chem. Lett.* **2021**, *19*, 1657–1666. [CrossRef]
65. de Morais Mirres, A.C.; da Silva, B.E.P.d.M.; Tessaro, L.; Galvan, D.; de Andrade, J.C.; Aquino, A.; Joshi, N.; Conte-Junior, C.A. Recent Advances in Nanomaterial-Based Biosensors for Pesticide Detection in Foods. *Biosensors* **2022**, *12*, 572. [CrossRef]
66. Munawar, A.; Ong, Y.; Schirhagl, R.; Tahir, M.A.; Khan, W.S.; Bajwa, S.Z. Nanosensors for Diagnosis with Optical, Electric and Mechanical Transducers. *RSC Adv.* **2019**, *9*, 6793–6803. [CrossRef]
67. Javaid, M.; Haleem, A.; Singh, R.P.; Rab, S.; Suman, R. Exploring the Potential of Nanosensors: A Brief Overview. *Sens. Int.* **2021**, *2*, 100130. [CrossRef]
68. Irkham, I.; Ibrahim, A.U.; Pwavodi, P.C.; Al-Turjman, F.; Hartati, Y.W. Smart Graphene-Based Electrochemical Nanobiosensor for Clinical Diagnosis: Review. *Sensors* **2023**, *23*, 2240. [CrossRef] [PubMed]
69. Song, M.; Yang, M.; Hao, J. Pathogenic Virus Detection by Optical Nanobiosensors. *Cell Rep. Phys. Sci.* **2021**, *2*, 100288. [CrossRef] [PubMed]
70. Naresh, V.; Lee, N. A Review on Biosensors and Recent Development of Nanostructured Materials-Enabled Biosensors. *Sensors* **2021**, *21*, 1109. [CrossRef]
71. Saleh Ibrahim, Y.; Alexis Ramírez-Coronel, A.; Kumar Sain, D.; Haleem Al-qaim, Z.; Hassan Jawhar, Z.; Yaseen Mahmood Alabdali, A.; Hayif Jasim Ali, S.; Althomali, R.H.; Fakri Mustafa, Y.; Mireya Romero-Parra, R. Advances in Nanomaterials-Based Chemiluminescence (Bio)Sensor for Specific and Sensitive Determination of Pathogenic Bacteria. *Microchem. J.* **2023**, *191*, 108860. [CrossRef]
72. Selvolini, G.; Marrazza, G. MIP-Based Sensors: Promising New Tools for Cancer Biomarker Determination. *Sensors* **2017**, *17*, 718. [CrossRef]
73. Hroncekova, S.; Lorencova, L.; Bertok, T.; Hires, M.; Jane, E.; Bučko, M.; Kasak, P.; Tkac, J. Amperometric Miniaturised Portable Enzymatic Nanobiosensor for the Ultrasensitive Analysis of a Prostate Cancer Biomarker. *JFB* **2023**, *14*, 161. [CrossRef]
74. Farrokhnia, M.; Amoabediny, G.; Ebrahimi, M.; Ganjali, M.; Arjmand, M. Ultrasensitive Early Detection of Insulin Antibody Employing Novel Electrochemical Nano-Biosensor Based on Controllable Electro-Fabrication Process. *Talanta* **2022**, *238*, 122947. [CrossRef] [PubMed]
75. Zhang, L.; Mazouzi, Y.; Salmain, M.; Liedberg, B.; Boujday, S. Antibody-Gold Nanoparticle Bioconjugates for Biosensors: Synthesis, Characterization and Selected Applications. *Biosens. Bioelectron.* **2020**, *165*, 112370. [CrossRef] [PubMed]
76. Park, D.H.; Choi, M.Y.; Choi, J.-H. Recent Development in Plasmonic Nanobiosensors for Viral DNA/RNA Biomarkers. *Biosensors* **2022**, *12*, 1121. [CrossRef]
77. Rawat, N.K.; Ghosh, R. Chapter 8—Conducting Polymer–Based Nanobiosensors. In *Nanosensors for Smart Cities*; Han, B., Tomer, V.K., Nguyen, T.A., Farmani, A., Kumar Singh, P., Eds.; Micro and Nano Technologies; Elsevier: Amsterdam, The Netherlands, 2020; pp. 129–142; ISBN 978-0-12-819870-4.
78. Fang, R.H.; Jiang, Y.; Fang, J.C.; Zhang, L. Cell Membrane-Derived Nanomaterials for Biomedical Applications. *Biomaterials* **2017**, *128*, 69–83. [CrossRef]
79. Aliakbar Ahovan, Z.; Hashemi, A.; De Plano, L.M.; Gholipourmalekabadi, M.; Seifalian, A. Bacteriophage Based Biosensors: Trends, Outcomes and Challenges. *Nanomaterials* **2020**, *10*, 501. [CrossRef] [PubMed]
80. Solaimuthu, A.; Vijayan, A.N.; Murali, P.; Korrapati, P.S. Nano-Biosensors and Their Relevance in Tissue Engineering. *Curr. Opin. Biomed. Eng.* **2020**, *13*, 84–93. [CrossRef]
81. Negahdary, M.; Angnes, L. Electrochemical Nanobiosensors Equipped with Peptides: A Review. *Microchim. Acta* **2022**, *189*, 94. [CrossRef]
82. Tsao, Y.-H.; Husain, R.A.; Lin, Y.-J.; Khan, I.; Chen, S.-W.; Lin, Z.-H. A Self-Powered Mercury Ion Nanosensor Based on the Thermoelectric Effect and Chemical Transformation Mechanism. *Nano Energy* **2019**, *62*, 268–274. [CrossRef]
83. Yu, R.; Niu, S.; Pan, C.; Wang, Z.L. Piezotronic Effect Enhanced Performance of Schottky-Contacted Optical, Gas, Chemical and Biological Nanosensors. *Nano Energy* **2015**, *14*, 312–339. [CrossRef]

84. Tavakolian, M.; Jafari, S.M.; Van De Ven, T.G.M. A Review on Surface-Functionalized Cellulosic Nanostructures as Biocompatible Antibacterial Materials. *Nano-Micro Lett.* **2020**, *12*, 73. [CrossRef]
85. Borse, V.B.; Konwar, A.N.; Jayant, R.D.; Patil, P.O. Perspectives of Characterization and Bioconjugation of Gold Nanoparticles and Their Application in Lateral Flow Immunosensing. *Drug Deliv. Transl. Res.* **2020**, *10*, 878–902. [CrossRef]
86. Zhou, Y.; Fang, Y.; Ramasamy, R. Non-Covalent Functionalization of Carbon Nanotubes for Electrochemical Biosensor Development. *Sensors* **2019**, *19*, 392. [CrossRef]
87. Fratila, R.M.; Mitchell, S.G.; Del Pino, P.; Grazu, V.; De La Fuente, J.M. Strategies for the Biofunctionalization of Gold and Iron Oxide Nanoparticles. *Langmuir* **2014**, *30*, 15057–15071. [CrossRef]
88. Wang, D.-X.; Wang, J.; Wang, Y.-X.; Du, Y.-C.; Huang, Y.; Tang, A.-N.; Cui, Y.-X.; Kong, D.-M. DNA Nanostructure-Based Nucleic Acid Probes: Construction and Biological Applications. *Chem. Sci.* **2021**, *12*, 7602–7622. [CrossRef]
89. Yaraki, M.T.; Tan, Y.N. Bioconjugation of Different Nanosurfaces With Biorecognition Molecules for the Development of Selective Nanosensor Platforms. *Chem. Asian J.* **2020**, *15*, 3180–3208. [CrossRef]
90. Valenzuela-Amaro, H.M.; Vázquez Ortega, P.G.; Zazueta-Alvarez, D.E.; López-Miranda, J.; Rojas-Contreras, J.A. Síntesis verde de nanopartículas de magnetita (NPs-Fe_3O_4): Factores y limitaciones. *Mundo Nano Rev. Interdiscip. Nanociencias Nanotechnol.* **2022**, *16*, 1e–18e. [CrossRef]
91. Carnerero, J.M.; Jimenez-Ruiz, A.; Castillo, P.M.; Prado-Gotor, R. Covalent and Non-Covalent DNA–Gold-Nanoparticle Interactions: New Avenues of Research. *ChemPhysChem* **2017**, *18*, 17–33. [CrossRef] [PubMed]
92. Greca, L.G.; Lehtonen, J.; Tardy, B.L.; Guo, J.; Rojas, O.J. Biofabrication of Multifunctional Nanocellulosic 3D Structures: A Facile and Customizable Route. *Mater. Horiz.* **2018**, *5*, 408–415. [CrossRef]
93. Rocchitta, G.; Spanu, A.; Babudieri, S.; Latte, G.; Madeddu, G.; Galleri, G.; Nuvoli, S.; Bagella, P.; Demartis, M.; Fiore, V.; et al. Enzyme Biosensors for Biomedical Applications: Strategies for Safeguarding Analytical Performances in Biological Fluids. *Sensors* **2016**, *16*, 780. [CrossRef]
94. Rahmawati, I.; Einaga, Y.; Ivandini, T.A.; Fiorani, A. Enzymatic Biosensors with Electrochemiluminescence Transduction. *ChemElectroChem* **2022**, *9*, e202200175. [CrossRef]
95. McVey, C.; Huang, F.; Elliott, C.; Cao, C. Endonuclease Controlled Aggregation of Gold Nanoparticles for the Ultrasensitive Detection of Pathogenic Bacterial DNA. *Biosens. Bioelectron.* **2017**, *92*, 502–508. [CrossRef]
96. Zamora-Gálvez, A.; Morales-Narváez, E.; Mayorga-Martinez, C.C.; Merkoçi, A. Nanomaterials Connected to Antibodies and Molecularly Imprinted Polymers as Bio/Receptors for Bio/Sensor Applications. *Appl. Mater. Today* **2017**, *9*, 387–401. [CrossRef]
97. Jannetto, P.J.; Buchan, B.W.; Vaughan, K.A.; Ledford, J.S.; Anderson, D.K.; Henley, D.C.; Quigley, N.B.; Ledeboer, N.A. Real-Time Detection of Influenza A, Influenza B, and Respiratory Syncytial Virus A and B in Respiratory Specimens by Use of Nanoparticle Probes. *J. Clin. Microbiol.* **2010**, *48*, 3997–4002. [CrossRef]
98. Karthik, V.; Senthil Kumar, P.; Vo, D.-V.N.; Selvakumar, P.; Gokulakrishnan, M.; Keerthana, P.; Audilakshmi, V.; Jeyanthi, J. Enzyme-Loaded Nanoparticles for the Degradation of Wastewater Contaminants: A Review. *Environ. Chem. Lett.* **2021**, *19*, 2331–2350. [CrossRef]
99. Banakar, M.; Hamidi, M.; Khurshid, Z.; Zafar, M.S.; Sapkota, J.; Azizian, R.; Rokaya, D. Electrochemical Biosensors for Pathogen Detection: An Updated Review. *Biosensors* **2022**, *12*, 927. [CrossRef] [PubMed]
100. Chiu, M.L.; Goulet, D.R.; Teplyakov, A.; Gilliland, G.L. Antibody Structure and Function: The Basis for Engineering Therapeutics. *Antibodies* **2019**, *8*, 55. [CrossRef] [PubMed]
101. Zahavi, D.; Weiner, L. Monoclonal Antibodies in Cancer Therapy. *Antibodies* **2020**, *9*, 34. [CrossRef] [PubMed]
102. McKeague, M.; DeRosa, M.C. Challenges and Opportunities for Small Molecule Aptamer Development. *J. Nucleic Acids* **2012**, *2012*, 1–20. [CrossRef]
103. Chavda, V.P.; Balar, P.C.; Teli, D.; Davidson, M.; Bojarska, J.; Apostolopoulos, V. Antibody–Biopolymer Conjugates in Oncology: A Review. *Molecules* **2023**, *28*, 2605. [CrossRef]
104. Majdi, H.; Salehi, R.; Pourhassan-Moghaddam, M.; Mahmoodi, S.; Poursalehi, Z.; Vasilescu, S. Antibody Conjugated Green Synthesized Chitosan-Gold Nanoparticles for Optical Biosensing. *Colloid Interface Sci. Commun.* **2019**, *33*, 100207. [CrossRef]
105. Marques, A.C.; Costa, P.J.; Velho, S.; Amaral, M.H. Functionalizing Nanoparticles with Cancer-Targeting Antibodies: A Comparison of Strategies. *J. Control. Release* **2020**, *320*, 180–200. [CrossRef]
106. Lara, S.; Perez-Potti, A. Applications of Nanomaterials for Immunosensing. *Biosensors* **2018**, *8*, 104. [CrossRef]
107. Guo, R.; Huang, F.; Cai, G.; Zheng, L.; Xue, L.; Li, Y.; Liao, M.; Wang, M.; Lin, J. A Colorimetric Immunosensor for Determination of Foodborne Bacteria Using Rotating Immunomagnetic Separation, Gold Nanorod Indication, and Click Chemistry Amplification. *Microchim. Acta* **2020**, *187*, 197. [CrossRef]
108. Dolatabadi, J.E.N.; Mashinchian, O.; Ayoubi, B.; Jamali, A.A.; Mobed, A.; Losic, D.; Omidi, Y.; De La Guardia, M. Optical and Electrochemical DNA Nanobiosensors. *TrAC Trends Anal. Chem.* **2011**, *30*, 459–472. [CrossRef]
109. Hua, Y.; Ma, J.; Li, D.; Wang, R. DNA-Based Biosensors for the Biochemical Analysis: A Review. *Biosensors* **2022**, *12*, 183. [CrossRef]
110. Ma, X.; Ding, W.; Wang, C.; Wu, H.; Tian, X.; Lyu, M.; Wang, S. DNAzyme Biosensors for the Detection of Pathogenic Bacteria. *Sens. Actuators B Chem.* **2021**, *331*, 129422. [CrossRef]

111. Ahmed, M.; Patel, R. Electrochemical/Voltammetric/Amperometric Nanosensors for the Detection of Pathogenic Bacteria. In *Nanosensors for Point-of-Care Diagnostics of Pathogenic Bacteria*; Acharya, A., Singhal, N.K., Eds.; Springer Nature: Singapore, 2023; pp. 113–141, ISBN 978-981-9912-18-6.
112. Wu, Q.; Zhang, Y.; Yang, Q.; Yuan, N.; Zhang, W. Review of Electrochemical DNA Biosensors for Detecting Food Borne Pathogens. *Sensors* 2019, *19*, 4916. [CrossRef] [PubMed]
113. D'Agata, R.; Bellassai, N.; Jungbluth, V.; Spoto, G. Recent Advances in Antifouling Materials for Surface Plasmon Resonance Biosensing in Clinical Diagnostics and Food Safety. *Polymers* 2021, *13*, 1929. [CrossRef] [PubMed]
114. Bhardwaj, N.; Bhardwaj, S.K.; Nayak, M.K.; Mehta, J.; Kim, K.-H.; Deep, A. Fluorescent Nanobiosensors for the Targeted Detection of Foodborne Bacteria. *TrAC Trends Anal. Chem.* 2017, *97*, 120–135. [CrossRef]
115. Leng, X.; Wang, Y.; Li, R.; Liu, S.; Yao, J.; Pei, Q.; Cui, X.; Tu, Y.; Tang, D.; Huang, J. Circular Exponential Amplification of Photoinduced Electron Transfer Using Hairpin Probes, G-Quadruplex DNAzyme and Silver Nanocluster-Labeled DNA for Ultrasensitive Fluorometric Determination of Pathogenic Bacteria. *Microchim. Acta* 2018, *185*, 168. [CrossRef]
116. Zhou, Y.; Wang, Z.; Zhang, S.; Deng, L. An Ultrasensitive Fluorescence Detection Template of Pathogenic Bacteria Based on Dual Catalytic Hairpin DNA Walker@Gold Nanoparticles Enzyme-Free Amplification. *Spectrochim. Acta Part A Mol. Biomol. Spectrosc.* 2022, *277*, 121259. [CrossRef] [PubMed]
117. Chinnappan, R.; AlZabn, R.; Abu-Salah, K.M.; Zourob, M. An Aptamer Based Fluorometric Microcystin-LR Assay Using DNA Strand-Based Competitive Displacement. *Microchim. Acta* 2019, *186*, 435. [CrossRef]
118. Tessaro, L.; Aquino, A.; De Almeida Rodrigues, P.; Joshi, N.; Ferrari, R.G.; Conte-Junior, C.A. Nucleic Acid-Based Nanobiosensor (NAB) Used for Salmonella Detection in Foods: A Systematic Review. *Nanomaterials* 2022, *12*, 821. [CrossRef]
119. Bakhshandeh, B.; Sorboni, S.G.; Haghighi, D.M.; Ahmadi, F.; Dehghani, Z.; Badiei, A. New Analytical Methods Using Carbon-Based Nanomaterials for Detection of Salmonella Species as a Major Food Poisoning Organism in Water and Soil Resources. *Chemosphere* 2022, *287*, 132243. [CrossRef]
120. Cui, F.; Sun, J.; De Dieu Habimana, J.; Yang, X.; Ji, J.; Zhang, Y.; Lei, H.; Li, Z.; Zheng, J.; Fan, M.; et al. Ultrasensitive Fluorometric Angling Determination of *Staphylococcus Aureus* in Vitro and Fluorescence Imaging in Vivo Using Carbon Dots with Full-Color Emission. *Anal. Chem.* 2019, *91*, 14681–14690. [CrossRef] [PubMed]
121. Gupta, R.; Kumar, A.; Kumar, S.; Pinnaka, A.K.; Singhal, N.K. Naked Eye Colorimetric Detection of *Escherichia coli* Using Aptamer Conjugated Graphene Oxide Enclosed Gold Nanoparticles. *Sens. Actuators B Chem.* 2021, *329*, 129100. [CrossRef]
122. Pandit, C.; Alajangi, H.K.; Singh, J.; Khajuria, A.; Sharma, A.; Hassan, M.S.; Parida, M.; Semwal, A.D.; Gopalan, N.; Sharma, R.K.; et al. Development of Magnetic Nanoparticle Assisted Aptamer-Quantum Dot Based Biosensor for the Detection of *Escherichia coli* in Water Samples. *Sci. Total Environ.* 2022, *831*, 154857. [CrossRef] [PubMed]
123. Sande, M.G.; Rodrigues, J.L.; Ferreira, D.; Silva, C.J.; Rodrigues, L.R. Novel Biorecognition Elements against Pathogens in the Design of State-of-the-Art Diagnostics. *Biosensors* 2021, *11*, 418. [CrossRef] [PubMed]
124. Singh, A.; Sharma, A.; Ahmed, A.; Sundramoorthy, A.K.; Furukawa, H.; Arya, S.; Khosla, A. Recent Advances in Electrochemical Biosensors: Applications, Challenges, and Future Scope. *Biosensors* 2021, *11*, 336. [CrossRef]
125. Feyziazar, M.; Amini, M.; Jahanban-Esfahlan, A.; Baradaran, B.; Oroojalian, F.; Kamrani, A.; Mokhtarzadeh, A.; Soleymani, J.; De La Guardia, M. Recent Advances on the Piezoelectric, Electrochemical, and Optical Biosensors for the Detection of Protozoan Pathogens. *TrAC Trends Anal. Chem.* 2022, *157*, 116803. [CrossRef]
126. Sun, F.; Zhang, J.; Yang, Q.; Wu, W. Quantum Dot Biosensor Combined with Antibody and Aptamer for Tracing Food-Borne Pathogens. *Food Qual. Saf.* 2021, *5*, fyab019. [CrossRef]
127. Ramajayam, K.; Ganesan, S.; Ramesh, P.; Beena, M.; Kokulnathan, T.; Palaniappan, A. Molecularly Imprinted Polymer-Based Biomimetic Systems for Sensing Environmental Contaminants, Biomarkers, and Bioimaging Applications. *Biomimetics* 2023, *8*, 245. [CrossRef]
128. Yang, Y.; Shen, X. Preparation and Application of Molecularly Imprinted Polymers for Flavonoids: Review and Perspective. *Molecules* 2022, *27*, 7355. [CrossRef]
129. Zhang, J.; Wang, Y.; Lu, X. Molecular Imprinting Technology for Sensing Foodborne Pathogenic Bacteria. *Anal. Bioanal. Chem.* 2021, *413*, 4581–4598. [CrossRef]
130. Samardzic, R.; Sussitz, H.F.; Jongkon, N.; Lieberzeit, P.A. Quartz Crystal Microbalance In-Line Sensing of *Escherichia coli* in a Bioreactor Using Molecularly Imprinted Polymers. Available online: https://www.ingentaconnect.com/contentone/asp/senlet/2014/00000012/f0020006/art00040 (accessed on 12 September 2023).
131. Zhao, X.; Cui, Y.; Wang, J.; Wang, J. Preparation of Fluorescent Molecularly Imprinted Polymers via Pickering Emulsion Interfaces and the Application for Visual Sensing Analysis of Listeria Monocytogenes. *Polymers* 2019, *11*, 984. [CrossRef]
132. Sivakumar, R.; Lee, N.Y. Recent Advances in Airborne Pathogen Detection Using Optical and Electrochemical Biosensors. *Anal. Chim. Acta* 2022, *1234*, 340297. [CrossRef]
133. Tessaro, L.; Aquino, A.; Panzenhagen, P.; Ochioni, A.C.; Mutz, Y.S.; Raymundo-Pereira, P.A.; Vieira, I.R.S.; Belem, N.K.R.; Conte-Junior, C.A. Development and Application of an SPR Nanobiosensor Based on AuNPs for the Detection of SARS-CoV-2 on Food Surfaces. *Biosensors* 2022, *12*, 1101. [CrossRef]
134. Hu, J.; Fu, K.; Bohn, P.W. Whole-Cell Pseudomonas Aeruginosa Localized Surface Plasmon Resonance Aptasensor. *Anal. Chem.* 2018, *90*, 2326–2332. [CrossRef]

135. Dabhade, A.H.; Verma, R.P.; Paramasivan, B.; Kumawat, A.; Saha, B. Development of Silver Nanoparticles and Aptamer Conjugated Biosensor for Rapid Detection of *E. coli* in a Water Sample. *3 Biotech* **2023**, *13*, 244. [CrossRef]
136. Shahdost-Fard, F.; Faridfar, S.; Keihan, A.H.; Aghaei, M.; Petrenko, I.; Ahmadi, F.; Ehrlich, H.; Rahimi-Nasrabadi, M. Applicability of a Green Nanocomposite Consisted of Spongin Decorated $Cu_2WO_4(OH)_2$ and AgNPs as a High-Performance Aptasensing Platform in Staphylococcus Aureus Detection. *Biosensors* **2023**, *13*, 271. [CrossRef]
137. Singh, P.; Gupta, R.; Choudhary, M.; Pinnaka, A.K.; Kumar, R.; Bhalla, V. Drug and Nanoparticle Mediated Rapid Naked Eye Water Test for Pathogens Detection. *Sens. Actuators B Chem.* **2018**, *262*, 603–610. [CrossRef]
138. Altintas, Z.; Akgun, M.; Kokturk, G.; Uludag, Y. A Fully Automated Microfluidic-Based Electrochemical Sensor for Real-Time Bacteria Detection. *Biosens. Bioelectron.* **2018**, *100*, 541–548. [CrossRef]
139. Nair, S.; Gomez-Cruz, J.; Manjarrez-Hernandez, Á.; Ascanio, G.; Sabat, R.G.; Escobedo, C. Selective Uropathogenic *E. coli* Detection Using Crossed Surface-Relief Gratings. *Sensors* **2018**, *18*, 3634. [CrossRef]
140. Simoska, O.; Sans, M.; Fitzpatrick, M.D.; Crittenden, C.M.; Eberlin, L.S.; Shear, J.B.; Stevenson, K.J. Real-Time Electrochemical Detection of *Pseudomonas Aeruginosa* Phenazine Metabolites Using Transparent Carbon Ultramicroelectrode Arrays. *ACS Sens.* **2019**, *4*, 170–179. [CrossRef]
141. Sun, R.; Zou, H.; Zhang, Y.; Zhang, X.; Chen, L.; Lv, R.; Sheng, R.; Du, T.; Li, Y.; Wang, H.; et al. Vancomycin Recognition and Induced-Aggregation of the Au Nanoparticles through Freeze-Thaw for Foodborne Pathogen *Staphylococcus aureus* Detection. *Anal. Chim. Acta* **2022**, *1190*, 339253. [CrossRef]
142. Lim, S.H.; Ryu, Y.C.; Hwang, B.H. Aptamer-Immobilized Gold Nanoparticles Enable Facile and On-Site Detection of *Staphylococcus aureus*. *Biotechnol. Bioprocess Eng.* **2021**, *26*, 107–113. [CrossRef]
143. Yang, X.; Huang, R.; Xiong, L.; Chen, F.; Sun, W.; Yu, L. A Colorimetric Aptasensor for Ochratoxin A Detection Based on Tetramethylrhodamine Charge Effect-Assisted Silver Enhancement. *Biosensors* **2023**, *13*, 468. [CrossRef]
144. Dester, E.; Kao, K.; Alocilja, E.C. Detection of Unamplified *E. coli* O157 DNA Extracted from Large Food Samples Using a Gold Nanoparticle Colorimetric Biosensor. *Biosensors* **2022**, *12*, 274. [CrossRef]
145. Kaushal, S.; Pinnaka, A.K.; Soni, S.; Singhal, N.K. Antibody Assisted Graphene Oxide Coated Gold Nanoparticles for Rapid Bacterial Detection and near Infrared Light Enhanced Antibacterial Activity. *Sens. Actuators B Chem.* **2021**, *329*, 129141. [CrossRef]
146. Abbaspour, A.; Norouz-Sarvestani, F.; Noori, A.; Soltani, N. Aptamer-Conjugated Silver Nanoparticles for Electrochemical Dual-Aptamer-Based Sandwich Detection of *Staphylococcus aureus*. *Biosens. Bioelectron.* **2015**, *68*, 149–155. [CrossRef]
147. Imran, M.; Ehrhardt, C.J.; Bertino, M.F.; Shah, M.R.; Yadavalli, V.K. Chitosan Stabilized Silver Nanoparticles for the Electrochemical Detection of Lipopolysaccharide: A Facile Biosensing Approach for Gram-Negative Bacteria. *Micromachines* **2020**, *11*, 413. [CrossRef]
148. Yin, M.; Liu, C.; Ge, R.; Fang, Y.; Wei, J.; Chen, X.; Chen, Q.; Chen, X. Paper-Supported near-Infrared-Light-Triggered Photoelectrochemical Platform for Monitoring *Escherichia coli* O157:H7 Based on Silver Nanoparticles-Sensitized-Upconversion Nanophosphors. *Biosens. Bioelectron.* **2022**, *203*, 114022. [CrossRef]
149. Qaanei, M.; Taheri, R.A.; Eskandari, K. Electrochemical Aptasensor for *Escherichia coli* O157:H7 Bacteria Detection Using a Nanocomposite of Reduced Graphene Oxide, Gold Nanoparticles and Polyvinyl Alcohol. *Anal. Methods* **2021**, *13*, 3101–3109. [CrossRef]
150. Ertaş, T.; Dinç, B.; Üstünsoy, R.; Eraslan, H.; Ergenç, A.F.; Bektaş, M. Novel Electrochemical Biosensor for *Escherichia coli* Using Gold-Coated Tungsten Wires and Antibody Functionalized Short Multiwalled Carbon Nanotubes. *Instrum. Sci. Technol.* **2023**, *2*, 1–16. [CrossRef]
151. Wang, Z.; Yao, X.; Wang, R.; Ji, Y.; Yue, T.; Sun, J.; Li, T.; Wang, J.; Zhang, D. Label-Free Strip Sensor Based on Surface Positively Charged Nitrogen-Rich Carbon Nanoparticles for Rapid Detection of Salmonella Enteritidis. *Biosens. Bioelectron.* **2019**, *132*, 360–367. [CrossRef]
152. Kurt, H.; Yüce, M.; Hussain, B.; Budak, H. Dual-Excitation Upconverting Nanoparticle and Quantum Dot Aptasensor for Multiplexed Food Pathogen Detection. *Biosens. Bioelectron.* **2016**, *81*, 280–286. [CrossRef]
153. Mathelié-Guinlet, M.; Cohen-Bouhacina, T.; Gammoudi, I.; Martin, A.; Béven, L.; Delville, M.-H.; Grauby-Heywang, C. Silica Nanoparticles-Assisted Electrochemical Biosensor for the Rapid, Sensitive and Specific Detection of *Escherichia coli*. *Sens. Actuators B Chem.* **2019**, *292*, 314–320. [CrossRef]
154. Jenie, S.N.A.; Kusumastuti, Y.; Krismastuti, F.S.H.; Untoro, Y.M.; Dewi, R.T.; Udin, L.Z.; Artanti, N. Rapid Fluorescence Quenching Detection of *Escherichia coli* Using Natural Silica-Based Nanoparticles. *Sensors* **2021**, *21*, 881. [CrossRef]
155. Wu, Z.; Sun, D.-W.; Pu, H.; Wei, Q. A Dual Signal-on Biosensor Based on Dual-Gated Locked Mesoporous Silica Nanoparticles for the Detection of Aflatoxin B1. *Talanta* **2023**, *253*, 124027. [CrossRef]
156. Boodoo, C.; Dester, E.; David, J.; Patel, V.; Kc, R.; Alocilja, E.C. Multi-Probe Nano-Genomic Biosensor to Detect *S. aureus* from Magnetically-Extracted Food Samples. *Biosensors* **2023**, *13*, 608. [CrossRef]
157. Diouani, M.F.; Sayhi, M.; Djafar, Z.R.; Ben Jomaa, S.; Belgacem, K.; Gharbi, H.; Ghita, M.; Popescu, L.-M.; Piticescu, R.; Laouini, D. Magnetic Separation and Centri-Chronoamperometric Detection of Foodborne Bacteria Using Antibiotic-Coated Metallic Nanoparticles. *Biosensors* **2021**, *11*, 205. [CrossRef]
158. Wen, J.; Ren, L.; He, Q.; Bao, J.; Zhang, X.; Pi, Z.; Chen, Y. Contamination-Free V-Shaped Ultrafast Reaction Cascade Transferase Signal Amplification Driven CRISPR/Cas12a Magnetic Relaxation Switching Biosensor for Bacteria Detection. *Biosens. Bioelectron.* **2023**, *219*, 114790. [CrossRef]

159. Ganganboina, A.B.; Chowdhury, A.D.; Khoris, I.M.; Doong, R.; Li, T.-C.; Hara, T.; Abe, F.; Suzuki, T.; Park, E.Y. Hollow Magnetic-Fluorescent Nanoparticles for Dual-Modality Virus Detection. *Biosens. Bioelectron.* **2020**, *170*, 112680. [CrossRef]
160. Kulkarni, M.B.; Ayachit, N.H.; Aminabhavi, T.M. Biosensors and Microfluidic Biosensors: From Fabrication to Application. *Biosensors* **2022**, *12*, 543. [CrossRef]
161. Liu, L.; Hong, J.; Wang, W.; Xiao, S.; Xie, H.; Wang, Q.; Gan, N. Fluorescent Aptasensor for Detection of Live Foodborne Pathogens Based on Multicolor Perovskite-Quantum-Dot-Encoded DNA Probes and Dual-Stirring-Bar-Assisted Signal Amplification. *J. Pharm. Anal.* **2022**, *12*, 913–922. [CrossRef]
162. Shang, Y.; Cai, S.; Ye, Q.; Wu, Q.; Shao, Y.; Qu, X.; Xiang, X.; Zhou, B.; Ding, Y.; Chen, M.; et al. Quantum Dot Nanobeads-Labelled Lateral Flow Immunoassay Strip for Rapid and Sensitive Detection of Salmonella Typhimurium Based on Strand Displacement Loop-Mediated Isothermal Amplification. *Engineering* **2022**, *19*, 62–70. [CrossRef]
163. Wang, L.; Lin, X.; Liu, T.; Zhang, Z.; Kong, J.; Yu, H.; Yan, J.; Luan, D.; Zhao, Y.; Bian, X. Reusable and Universal Impedimetric Sensing Platform for the Rapid and Sensitive Detection of Pathogenic Bacteria Based on Bacteria-Imprinted Polythiophene Film. *Analyst* **2022**, *147*, 4433–4441. [CrossRef]
164. Alafeef, M.; Moitra, P.; Pan, D. Nano-Enabled Sensing Approaches for Pathogenic Bacterial Detection. *Biosens. Bioelectron.* **2020**, *165*, 112276. [CrossRef]
165. Ayodhya, D. Recent Progress on Detection of Bivalent, Trivalent, and Hexavalent Toxic Heavy Metal Ions in Water Using Metallic Nanoparticles: A Review. *Results Chem.* **2023**, *5*, 100874. [CrossRef]
166. Patel, R.; Mitra, B.; Vinchurkar, M.; Adami, A.; Patkar, R.; Giacomozzi, F.; Lorenzelli, L.; Baghini, M.S. A Review of Recent Advances in Plant-Pathogen Detection Systems. *Heliyon* **2022**, *8*, e11855. [CrossRef]
167. Loiseau, A.; Asila, V.; Boitel-Aullen, G.; Lam, M.; Salmain, M.; Boujday, S. Silver-Based Plasmonic Nanoparticles for and Their Use in Biosensing. *Biosensors* **2019**, *9*, 78. [CrossRef]
168. Ibrahim, N.; Jamaluddin, N.D.; Tan, L.L.; Mohd Yusof, N.Y. A Review on the Development of Gold and Silver Nanoparticles-Based Biosensor as a Detection Strategy of Emerging and Pathogenic RNA Virus. *Sensors* **2021**, *21*, 5114. [CrossRef]
169. Yu, X.; Jiao, Y.; Chai, Q. Applications of Gold Nanoparticles in Biosensors. *Nano LIFE* **2016**, *6*, 1642001. [CrossRef]
170. Baetsen-Young, A.M.; Vasher, M.; Matta, L.L.; Colgan, P.; Alocilja, E.C.; Day, B. Direct Colorimetric Detection of Unamplified Pathogen DNA by Dextrin-Capped Gold Nanoparticles. *Biosens. Bioelectron.* **2018**, *101*, 29–36. [CrossRef]
171. Hui, C.; Hu, S.; Yang, X.; Guo, Y. A Panel of Visual Bacterial Biosensors for the Rapid Detection of Genotoxic and Oxidative Damage: A Proof of Concept Study. *Mutat. Res./Genet. Toxicol. Environ. Mutagen.* **2023**, *888*, 503639. [CrossRef] [PubMed]
172. Liu, J.; Xu, J.-Z.; Rao, Z.-M.; Zhang, W.-G. An Enzymatic Colorimetric Whole-Cell Biosensor for High-Throughput Identification of Lysine Overproducers. *Biosens. Bioelectron.* **2022**, *216*, 114681. [CrossRef] [PubMed]
173. Miranda, R.R.; Sampaio, I.; Zucolotto, V. Exploring Silver Nanoparticles for Cancer Therapy and Diagnosis. *Colloids Surf. B Biointerfaces* **2022**, *210*, 112254. [CrossRef] [PubMed]
174. Douaki, A.; Demelash Abera, B.; Cantarella, G.; Shkodra, B.; Mushtaq, A.; Ibba, P.; Inam, A.S.; Petti, L.; Lugli, P. Flexible Screen Printed Aptasensor for Rapid Detection of Furaneol: A Comparison of CNTs and AgNPs Effect on Aptasensor Performance. *Nanomaterials* **2020**, *10*, 1167. [CrossRef]
175. Li, G. *Nano-Inspired Biosensors for Protein Assay with Clinical Applications*; Elsevier: Amsterdam, The Netherlands, 2018; ISBN 978-0-12-815054-2.
176. Varghese Alex, K.; Tamil Pavai, P.; Rugmini, R.; Shiva Prasad, M.; Kamakshi, K.; Sekhar, K.C. Green Synthesized Ag Nanoparticles for Bio-Sensing and Photocatalytic Applications. *ACS Omega* **2020**, *5*, 13123–13129. [CrossRef] [PubMed]
177. Nguyen, T.N.; Phung, V.-D.; Tran, V.V. Recent Advances in Conjugated Polymer-Based Biosensors for Virus Detection. *Biosensors* **2023**, *13*, 586. [CrossRef]
178. Sharma, A.; Sharma, N.; Kumari, A.; Lee, H.-J.; Kim, T.; Tripathi, K.M. Nano-Carbon Based Sensors for Bacterial Detection and Discrimination in Clinical Diagnosis: A Junction between Material Science and Biology. *Appl. Mater. Today* **2020**, *18*, 100467. [CrossRef]
179. Bhattacharya, K.; Mukherjee, S.P.; Gallud, A.; Burkert, S.C.; Bistarelli, S.; Bellucci, S.; Bottini, M.; Star, A.; Fadeel, B. Biological Interactions of Carbon-Based Nanomaterials: From Coronation to Degradation. *Nanomed. Nanotechnol. Biol. Med.* **2016**, *12*, 333–351. [CrossRef]
180. Duan, N.; Wu, S.; Dai, S.; Miao, T.; Chen, J.; Wang, Z. Simultaneous Detection of Pathogenic Bacteria Using an Aptamer Based Biosensor and Dual Fluorescence Resonance Energy Transfer from Quantum Dots to Carbon Nanoparticles. *Microchim. Acta* **2015**, *182*, 917–923. [CrossRef]
181. Shen, Y.; Zhang, Y.; Gao, Z.F.; Ye, Y.; Wu, Q.; Chen, H.-Y.; Xu, J.-J. Recent Advances in Nanotechnology for Simultaneous Detection of Multiple Pathogenic Bacteria. *Nano Today* **2021**, *38*, 101121. [CrossRef]
182. Fang, Y.; Wang, Y.; Zhu, L.; Liu, H.; Su, X.; Liu, Y.; Chen, Z.; Chen, H.; He, N. A Novel Cartridge for Nucleic Acid Extraction, Amplification and Detection of Infectious Disease Pathogens with the Help of Magnetic Nanoparticles. *Chin. Chem. Lett.* **2023**, *34*, 108092. [CrossRef]
183. Hussain, C.M. *Analytical Applications of Functionalized Magnetic Nanoparticles*; Royal Society of Chemistry: London, UK, 2021; ISBN 978-1-83916-276-3.
184. Huang, H.T.; Garu, P.; Li, C.H.; Chang, W.C.; Chen, B.W.; Sung, S.Y.; Lee, C.M.; Chen, J.Y.; Hsieh, T.F.; Sheu, W.J.; et al. Magnetoresistive Biosensors for Direct Detection of Magnetic Nanoparticle Conjugated Biomarkers on a Chip. *SPIN* **2019**, *9*, 1940002. [CrossRef]

185. Min, C.; Shao, H.; Liong, M.; Yoon, T.-J.; Weissleder, R.; Lee, H. Mechanism of Magnetic Relaxation Switching Sensing. *ACS Nano* **2012**, *6*, 6821–6828. [CrossRef]
186. Arya, S.; Singh, A.; Sharma, A.; Gupta, V. 11—Silicon-Based Biosensor. In *Silicon-Based Hybrid Nanoparticles*; Thomas, S., Nguyen, T.A., Ahmadi, M., Yasin, G., Joshi, N., Eds.; Micro and Nano Technologies; Elsevier: Amsterdam, The Netherlands, 2022; pp. 247–267, ISBN 978-0-12-824007-6.
187. Bagheri, E.; Ansari, L.; Sameiyan, E.; Abnous, K.; Taghdisi, S.M.; Ramezani, M.; Alibolandi, M. Sensors Design Based on Hybrid Gold-Silica Nanostructures. *Biosens. Bioelectron.* **2020**, *153*, 112054. [CrossRef] [PubMed]
188. Costanzo, H.; Gooch, J.; Frascione, N. Nanomaterials for Optical Biosensors in Forensic Analysis. *Talanta* **2023**, *253*, 123945. [CrossRef] [PubMed]
189. Aggett, P.; Aguilar, F.; Crebelli, R.; Dusemund, B.; Filipič, M.; Frutos, M.J.; Galtier, P.; Gott, D.; Gundert-Remy, U.; Kuhnle, G.G.; et al. Re-evaluation of Silicon Dioxide (E 551) as a Food Additive. *EFS2* **2018**, *16*, e05088. [CrossRef]
190. Şen Karaman, D.; Pamukçu, A.; Karakaplan, M.B.; Kocaoglu, O.; Rosenholm, J.M. Recent Advances in the Use of Mesoporous Silica Nanoparticles for the Diagnosis of Bacterial Infections. *IJN* **2021**, *16*, 6575–6591. [CrossRef]
191. Wen, L.; Qiu, L.; Wu, Y.; Hu, X.; Zhang, X. Aptamer-Modified Semiconductor Quantum Dots for Biosensing Applications. *Sensors* **2017**, *17*, 1736. [CrossRef]
192. Sun, H.; Zhou, P.; Su, B. Electrochemiluminescence of Semiconductor Quantum Dots and Its Biosensing Applications: A Comprehensive Review. *Biosensors* **2023**, *13*, 708. [CrossRef]
193. Ma, F.; Li, C.; Zhang, C. Development of Quantum Dot-Based Biosensors: Principles and Applications. *J. Mater. Chem. B* **2018**, *6*, 6173–6190. [CrossRef]
194. Pourmadadi, M.; Rahmani, E.; Rajabzadeh-Khosroshahi, M.; Samadi, A.; Behzadmehr, R.; Rahdar, A.; Ferreira, L.F.R. Properties and Application of Carbon Quantum Dots (CQDs) in Biosensors for Disease Detection: A Comprehensive Review. *J. Drug Deliv. Sci. Technol.* **2023**, *80*, 104156. [CrossRef]
195. Smith, R.P.; Barraza, I.; Quinn, R.J.; Fortoul, M.C. Chapter One—The Mechanisms and Cell Signaling Pathways of Programmed Cell Death in the Bacterial World. In *Cell Death Regulation in Health and Disease—Part B*; Spetz, J.K.E., Galluzzi, L., Eds.; International Review of Cell and Molecular Biology; Academic Press: Cambridge, MA, USA, 2020; Volume 352, pp. 1–53.
196. Hao, L.; Xue, L.; Huang, F.; Cai, G.; Qi, W.; Zhang, M.; Han, Q.; Wang, Z.; Lin, J. A Microfluidic Biosensor Based on Magnetic Nanoparticle Separation, Quantum Dots Labeling and MnO$_2$ Nanoflower Amplification for Rapid and Sensitive Detection of Salmonella Typhimurium. *Micromachines* **2020**, *11*, 281. [CrossRef]
197. Schmitz, F.R.W.; Valério, A.; De Oliveira, D.; Hotza, D. An Overview and Future Prospects on Aptamers for Food Safety. *Appl. Microbiol. Biotechnol.* **2020**, *104*, 6929–6939. [CrossRef] [PubMed]
198. Pirzada, M.; Altintas, Z. Nanomaterials for Healthcare Biosensing Applications. *Sensors* **2019**, *19*, 5311. [CrossRef]
199. Mukama, O.; Wu, J.; Li, Z.; Liang, Q.; Yi, Z.; Lu, X.; Liu, Y.; Liu, Y.; Hussain, M.; Makafe, G.G.; et al. An Ultrasensitive and Specific Point-of-Care CRISPR/Cas12 Based Lateral Flow Biosensor for the Rapid Detection of Nucleic Acids. *Biosens. Bioelectron.* **2020**, *159*, 112143. [CrossRef] [PubMed]
200. Li, Y.; Xie, G.; Qiu, J.; Zhou, D.; Gou, D.; Tao, Y.; Li, Y.; Chen, H. A New Biosensor Based on the Recognition of Phages and the Signal Amplification of Organic-Inorganic Hybrid Nanoflowers for Discriminating and Quantitating Live Pathogenic Bacteria in Urine. *Sens. Actuators B Chem.* **2018**, *258*, 803–812. [CrossRef]
201. Zhang, X.; Li, G.; Wu, D.; Li, X.; Hu, N.; Chen, J.; Chen, G.; Wu, Y. Recent Progress in the Design Fabrication of Metal-Organic Frameworks-Based Nanozymes and Their Applications to Sensing and Cancer Therapy. *Biosens. Bioelectron.* **2019**, *137*, 178–198. [CrossRef]
202. Ahmed, A.; Rushworth, J.V.; Hirst, N.A.; Millner, P.A. Biosensors for Whole-Cell Bacterial Detection. *Clin. Microbiol. Rev.* **2014**, *27*, 631–646. [CrossRef]
203. Echeverri, D.; Orozco, J. Glycan-Based Electrochemical Biosensors: Promising Tools for the Detection of Infectious Diseases and Cancer Biomarkers. *Molecules* **2022**, *27*, 8533. [CrossRef]
204. Puchkova, A.; Vietz, C.; Pibiri, E.; Wünsch, B.; Sanz Paz, M.; Acuna, G.P.; Tinnefeld, P. DNA Origami Nanoantennas with over 5000-Fold Fluorescence Enhancement and Single-Molecule Detection at 25 µM. *Nano Lett.* **2015**, *15*, 8354–8359. [CrossRef]
205. Chen, J.; Alcaine, S.D.; Jiang, Z.; Rotello, V.M.; Nugen, S.R. Detection of *Escherichia coli* in Drinking Water Using T7 Bacteriophage-Conjugated Magnetic Probe. *Anal. Chem.* **2015**, *87*, 8977–8984. [CrossRef] [PubMed]
206. Jin, X.; Liu, C.; Xu, T.; Su, L.; Zhang, X. Artificial Intelligence Biosensors: Challenges and Prospects. *Biosens. Bioelectron.* **2020**, *165*, 112412. [CrossRef] [PubMed]
207. Vashistha, R.; Dangi, A.K.; Kumar, A.; Chhabra, D.; Shukla, P. Futuristic Biosensors for Cardiac Health Care: An Artificial Intelligence Approach. *3 Biotech* **2018**, *8*, 358. [CrossRef] [PubMed]
208. Taniguchi, M.; Minami, S.; Ono, C.; Hamajima, R.; Morimura, A.; Hamaguchi, S.; Akeda, Y.; Kanai, Y.; Kobayashi, T.; Kamitani, W.; et al. Combining Machine Learning and Nanopore Construction Creates an Artificial Intelligence Nanopore for Coronavirus Detection. *Nat. Commun.* **2021**, *12*, 3726. [CrossRef]

Disclaimer/Publisher's Note: The statements, opinions and data contained in all publications are solely those of the individual author(s) and contributor(s) and not of MDPI and/or the editor(s). MDPI and/or the editor(s) disclaim responsibility for any injury to people or property resulting from any ideas, methods, instructions or products referred to in the content.

Review

Development of Optical Differential Sensing Based on Nanomaterials for Biological Analysis

Lele Wang, Yanli Wen *, Lanying Li, Xue Yang, Wen Li, Meixia Cao, Qing Tao, Xiaoguang Sun and Gang Liu *

Key Laboratory of Bioanalysis and Metrology for State Market Regulation, Shanghai Institute of Measurement and Testing Technology, 1500 Zhang Heng Road, Shanghai 201203, China; wangll@simt.com.cn (L.W.); lily@simt.com.cn (L.L.); yangxue@simt.com.cn (X.Y.); liw@simt.com.cn (W.L.); caomx@simt.com.cn (M.C.); taoq@simt.com.cn (Q.T.); sunxg@simt.com.cn (X.S.)
* Correspondence: wenyl@simt.com.cn (Y.W.); liug@simt.com.cn (G.L.)

Abstract: The discrimination and recognition of biological targets, such as proteins, cells, and bacteria, are of utmost importance in various fields of biological research and production. These include areas like biological medicine, clinical diagnosis, and microbiology analysis. In order to efficiently and cost-effectively identify a specific target from a wide range of possibilities, researchers have developed a technique called differential sensing. Unlike traditional "lock-and-key" sensors that rely on specific interactions between receptors and analytes, differential sensing makes use of cross-reactive receptors. These sensors offer less specificity but can cross-react with a wide range of analytes to produce a large amount of data. Many pattern recognition strategies have been developed and have shown promising results in identifying complex analytes. To create advanced sensor arrays for higher analysis efficiency and larger recognizing range, various nanomaterials have been utilized as sensing probes. These nanomaterials possess distinct molecular affinities, optical/electrical properties, and biological compatibility, and are conveniently functionalized. In this review, our focus is on recently reported optical sensor arrays that utilize nanomaterials to discriminate bioanalytes, including proteins, cells, and bacteria.

Keywords: pattern recognition; nanomaterials; gold nanoparticle; graphene oxide; quantum dot

1. Introduction

In recent decades, natural/artificial specific receptors have been studied for the analysis of particular analytes based on the lock-and-key principle in many critical fields, including food safety [1–11], environmental monitoring [12–17], and medical diagnosis [18–31]. However, the production of highly specific receptors remains a challenge for a large range of analysis targets, especially when facing complex biological samples containing proteins, microorganisms, and cells.

Recently, pattern recognition has been intensively studied, also known as differential sensing or "artificial noses/tongues" [32–37]. Different from traditional molecular recognition based on one specific receptor, differential sensing was constructed on a receptor library of low-specific recognizing elements, each of which would respond to a certain target to different degrees [38–40]. By collecting the response signals, we can establish a fingerprint toward characteristic patterns for the individual analytes or complex mixtures. To perform differential sensing, a sensor array was constructed as the central component. Through array analysis, data from various sensing units could be gathered concurrently and subsequently scrutinized to facilitate target detection and recognition (Scheme 1). The number of channels within the array is a crucial factor influencing the discrimination capacity of the differential sensor. An illustrious example highlighting this principle is the olfactory system of a dog, which possesses approximately 4 billion olfactory receptor cells, an astonishing 45 times more than that of a human. The signals detected by these receptors have the potential to generate even larger quantities of interconnected data groups through their intricate associations with one another.

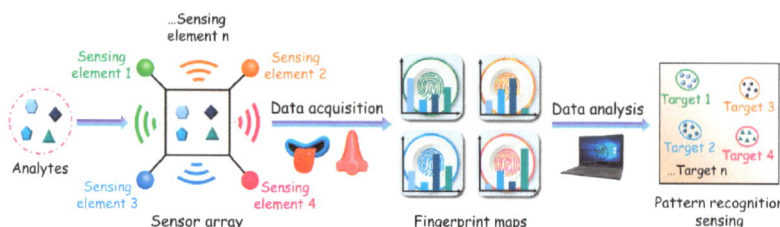

Scheme 1. Mechanisms of Sensor Arrays for Discrimination and Identification of Analytes. Reprinted with permission from [33]. Copyright 2023 American Chemical Society.

There are two main obstacles to the development of artificial sensors: Firstly, it is difficult to construct a large-scale array to collect adequate signals compared with natural systems. Secondly, the sensitivity is usually hindered by the relatively high blank noise signal or low signal read-out, especially in biological samples. Thus, there have been increasing research demands to develop novel biosensing strategies for higher sensitivity and larger scale of sensor arrays [37]. In recent decades, nanomaterials have become a shining star in the research of a growing number of biosensor strategies [41–50]. The emergence of fast-growing nanomaterials [51], such as metal nanoparticles [48,52–56], carbon nanomaterials [44,46,57,58], and quantum dots [59–61], has opened up exciting possibilities for novel sensor platforms [62–64]. These nanomaterials possess unique electronic, magnetic, and light properties, making them highly desirable for the field of differential sensing. Table 1 displays the main characteristics of the common nanomaterials studied for optical differential sensing.

Table 1. The main characteristics of the common nanomaterials studied for optical differential sensing.

Nanomaterials	Biological Interaction	Optical Signal
AuNPs	Competitive adsorption, Au-S modification	fluorescence quenching, Colorimetric signal due to aggregation
Graphene oxide (GO)	Competitive adsorption, Modification through -COOH	fluorescence quenching
QDs	Bind nonspecifically via electrostatic interactions	Fluorescence emission with different lengths and high quantum yield

In this review, we present an overview of the applications of functional nanomaterials in optical sensor arrays, including colorimetric and fluorescence methods. These arrays can be categorized into gold nanoparticle-based sensor arrays, graphene oxide (GO)-based sensor arrays, quantum dot (QD)-based sensor arrays and other metal nanoparticle-based sensor arrays. Table 2 presents the timeline for the historical development of optical differential sensing based on nanomaterials for biological analysis. Compared to the former literature, this review aims to provide a comprehensive understanding of the advancements, challenges, and future prospects in this rapidly evolving field. We here mainly focus on three main significant advantages and contributions of nanomaterials for the development of sensor arrays: Firstly, by manipulating their physical and chemical properties as well as surface modifications, functional nanomaterials enhance signal output, sensitivity, and selectivity. Secondly, the unique properties and interaction mechanisms of functional nanomaterials enable sensor arrays to detect multiple target molecules and achieve multiparameter analysis. Additionally, functional nanomaterials allow for efficient analysis of complex samples by integrating multiple sensing mechanisms such as fluorescence resonance energy transfer and surface plasmon resonance. Thus, the integration of functional nanomaterials into sensor arrays holds great promise in advancing the field of optical sensing, offering new avenues for exploring various detection technologies and expanding the range of potential applications.

Table 2. Development of optical differential sensing based on nanomaterials for biological analysis in different timelines.

Year	Development of Optical Differential Sensing Based on Nanomaterials for Biological Analysis
2007	Rotello's group developed a sensor array consisting of six non-covalent gold nanoparticle-fluorescent polymer conjugates for identification and quantitative differentiation of proteins [65]
2010	Rotello and co-workers developed enzyme-amplified array sensing (EAAS) with NPs to dramatically increase the sensitivity for protein identification [38]
2012	Rotello and co-workers also achieved colorimetric differentiation of proteins with catalytically active NPs used for both recognition and signal transduction/amplification [66]
2012	Rotello and co-workers also developed gold-nanoparticle green-fluorescent protein (NP–GFP)-based sensor arrays for the identification of mammalian cell types and cancer states [67]
2012	Dravid, Chou, and De developed nanoscale graphene oxide (nGO) as artificial receptors for array-based protein identification [68]
2012	Fan, Hu, and co-workers employed the combination of fluorescently labeled adaptive "ensemble aptamers" (ENSaptamers) and nGOs for high-precision identification of a wide range of bioanalytes, including proteins, cells, and bacteria [69]
2014	Ouyang and co-workers have synthesized novel blue-emitting ColAu NCs and Mac-Au NCs for discriminating proteins [70]
2014	Qu and Ren utilized a sensing array composed of seven luminescent nanodots, combined with graphene oxide, for protein recognition [71]
2015	He and Chang constructed an array-based protein discrimination system by using eight Au NDs as efficient protein receptors and competent signal transducers [72]
2016	Zhang and Tang develop a multicolor quantum dot (QD)-based multichannel sensing platform for rapid identification of multiple proteins [73]
2017	Shi and Wu employed a colorimetric sensor array consisting of four gold nanoparticles (AuNPs) with diverse surface properties for the rapid identification of microorganisms [74]
2018	Pu, Ren and Qu developed a sensitive and effective method for pattern recognition of proteins using nanozyme (g-C_3N_4) as a receptor [75]
2022	Li and Han utilized five fluorescent positively charged polymers (P1–P5) and negatively charged graphene oxide (GO) for differentiating between different proteins [76]
2022	Huang, Han and Li utilized three modified polyethyleneimine and negatively charged graphene oxide for differentiating different bacteria [77]
2023	Tian and Wu utilized silver nanoparticles for differentiating proteins in various osmolyte solutions [78]
2024	Yang employed DBCO-UCNPs for the differentiation of different pathogens in terms of phenotyping classification and antibiotic resistance identification [34]

2. Pattern Recognition Methods for Differential Sensing

Optical signals produced by the differential sensing array were analyzed by using pattern recognition methods such as linear discriminant analysis (LDA) [79], principal component analysis (PCA) [80], and hierarchical clustering analysis (HCA) [81]. A schematic representation of the above methods is shown in Scheme 2.

Linear discriminant analysis is a supervised pattern recognition method that can be used for both dimensionality reduction and classification [82]. The means and covariance matrices of the training data set are used to establish the discriminant functions. Once the discriminant functions are built, a prediction data set is tested by the discriminant functions to validate the classification accuracy. In order to ensure classification accuracy, the prediction data set should be different from the training data set; otherwise, LDA may produce optimistic results.

Principal component analysis (PCA) is an unsupervised method for dimensionality reduction of multivariate data. It can compress a multi-dimensional data set into a lower dimensional space and rank the new dimensions according to their importance. Often, a successful PCA may produce two or three principal components, which are convenient for producing score plots for the data set [83]. It is important to note that PCA is more suitable for the analysis of linear data; however, it is possible to fail the classification of nonlinear data.

Similar to PCA, hierarchical clustering analysis (HCA) is an unsupervised pattern recognition method. There are three basic steps for HCA: Firstly, the multivariate distances between all samples are calculated. Afterward, clustering is performed by establishing a hierarchy of points, in which similar distant points are joined. Finally, a two-dimensional dendrogram is shown that allows the visual examination of clustering relationships of all samples [84]. Because HCA employs all the sensor array data to represent the patterns, a poor result may be produced when the data set is noisy. HCA is most suitable for qualitative analysis of relationships in data.

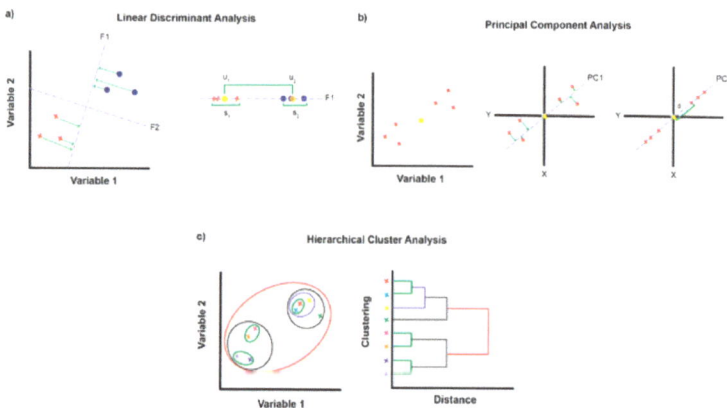

Scheme 2. Schematic representation of (**a**) the LDA method of projecting points onto a new vector F1 that fulfils the criteria of maximizing the ratio of between-class to within-class variance, (**b**) the PCA method of determining the center of the data, projecting points onto a new vector, and calculating the maximum variance and thus the best-fitting line, (**c**) the HCA bottom-up agglomerative approach and the resulting dendrogram illustrating the connectivity of data points. Reprinted with permission from [85]. Copyright 2021 American Chemical Society.

3. Gold Nanoparticle-Based Sensor Arrays

Gold nanoparticles (AuNPs) have been widely studied in the development of biosensors due to their unique optical and chemical properties, good biocompatibility, and easy surface functionalization [86–88]. Together with organic or biological molecules, AuNPs can produce differential response signals for target molecules [65,67,74,89–113].

3.1. Fluorescence Sensing Based on AuNPs

AuNPs are widely applied in biosensors as powerful fluorescence quenchers [114–120]. The competitive bindings between the analytes and the indicators to AuNPs lead to distinct fluorescence response fingerprints for many analytes, which could be identified by pattern recognition methods with a high degree of accuracy [121]. These AuNPs work as powerful fluorescence quenchers for fluorescence indicators, as well as the recognition elements for target analytes. The interactions between nanoparticle–indicators and nanoparticle-analytes could be tuned by modifying different groups on the surface of AuNPs.

A sensor array was developed for the differentiation of normal and cancerous cell lines, based on conjugates between three structurally related cationic AuNPs and the fluorescent polymer [90,122]. The nanoparticles quench the fluorescence of the polymer. In the presence of mammalian cells, there is competitive binding between nanoparticle-polymer complexes and cell types. The polymer was displaced with mammalian cells from the nanoparticle surface, generating a fluorescence response. Four different types of human cancer cells were discriminated by using LDA. The results showed a 100% accuracy of detection. The sensor array can also effectively differentiate isogenic cell types. Later, the same group designed a sensor array composed of AuNP-GFP complexes for discrimination between normal and metastatic cells and tissues [67]. Rather than using whole cells as the target analytes, the lysates isolated from tissues have the advantage of increased homogeneity of the test samples, which leads to reduced error in identification, increased reproducibility, and higher sensitivity. This sensing platform needed a small amount of sample (as little as 200 ng of cell- or tissue-lysed proteins).

The Rotello group synthesized two types of AuNPs, one with a cationic hydrophobic functional group and the other with a hydrophilic functional group [70] (Figure 1). Three fluorescent proteins with negative surface charge can bind to these particles through electrostatic interactions, resulting in fluorescence quenching. When exposed to bacteria biofilms, AuNP-fluorescent protein conjugates are disrupted to produce different colored fluorescence patterns. The multichannel sensor was able to completely differentiate six bacterial biofilms, including nonpathogenic and pathogenic bacteria. The performance of the sensor was further tested by the identification of biofilms in a mixed bacteria/mammalian cell in vitro wound model.

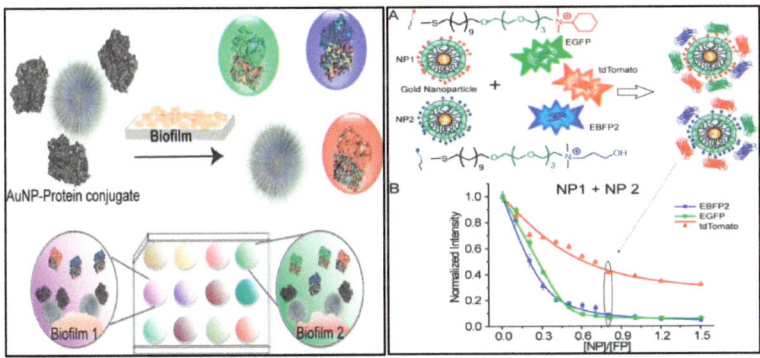

Figure 1. (**Left**) Schematic illustration of the multichannel sensor using AuNP-fluorescent protein conjugates that are disrupted in the presence of biofilms. (**Right**) The sensor composition. (**A**) Sensor elements and molecular structures of the functional ligands of NP1 and NP2. (**B**) Fluorescence titration with an equal molar mixture of NP1 and NP2. Reprinted from [70] with open access.

3.2. Colorimetric Sensing Based on AuNPs

The aggregation of AuNPs results in a visible color change from red to blue, which provides a versatile platform for colorimetric sensing of target analytes [96,123–125]. Zhang and co-workers created a colorimetric sensor array with aptamer-protected AuNPs as recognition elements [126]. The aptamer-protected AuNPs were able to resist aggregation in the presence of a high-concentration salt. Upon the addition of different target proteins, differential response patterns were obtained. This sensitive array sensing system can discriminate seven proteins with the naked eye at the 50 nM level. Similar approaches were also used for the analysis of many bioanalytes [127–129]. These sensor arrays exhibited an excellent ability to recognize proteins, bacteria, and mammalian cells.

Chen et al. constructed a DNA-AuNPs colorimetric sensor array for rapid and sensitive identification of proteins [128]. The sensor array composed of only two sensing elements could discriminate 12 proteins at the 50 nM level with the naked eye. Moreover, the proteins in human serum and protein mixtures were well-differentiated with 100% accuracy. Huang and co-workers also exploited DNA-AuNPs nanoconjugates to differentiate cell types [129]. The cross-reactive receptors (DNA-AuNPs) are employed to bind the different cells that produce differential color changes of AuNPs. The nanoplasmonic effect of AuNPs was enhanced via seeded growth, which resulted in the effective distinction of various cell lines with dark-field microscopy or even the naked eye. The results were analyzed by LDA, which showed 100% accuracy.

Wu and Shi [28] developed a colorimetric sensor array for rapid microorganism identification. The array utilized four distinct AuNPs as sensing elements, resulting in noticeable color shifts upon interaction with microorganisms. Through LDA, 15 microorganisms were successfully differentiated based on their unique response patterns. The sensor array also demonstrated the ability to discern mixtures of microorganisms. This straightforward and expedient method provides results within 5 s, making it suitable for applications in pathogen diagnosis and environmental monitoring.

A colorimetric sensor array was developed using D-amino acid (D-AA)-modified AuNPs as probes (Au/D-AA) for bacteria fingerprinting [130]. The aggregation of AuNPs is triggered by the metabolic activity of bacteria towards D-AA, allowing differentiation of eight types of bacteria and quantitative analysis of a single bacterium. The sensor array also enables rapid colorimetric antibiotic susceptibility testing (AST) by monitoring bacterial metabolic activity toward different antibiotic treatments, which has implications for clinical applications and antibiotic stewardship (Figure 2).

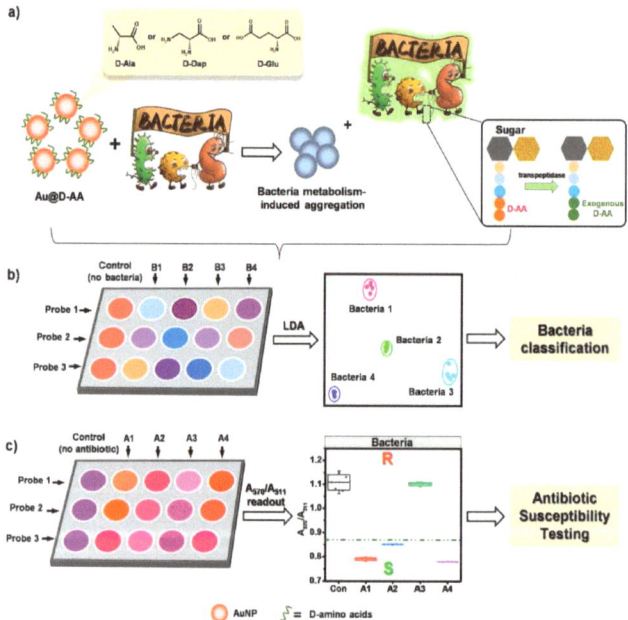

Figure 2. Principle of the Developed Assay Based on the Bacteria Metabolism-Triggered Consumption of dD-AA (**a**), Strategies for Multiple Bacteria Identifications through LDA (**b**) and AST through the Colorimetric Change of Probes after the Incubation with Bacteria and Antibiotic (**c**). Reprinted with permission from [130]. Copyright 2022 American Chemical Society.

Liu and co-workers presented an extensible multidimensional sensor using the conjugates of nonspecific dye-labeled DNA sequences and AuNPs as receptors [127]. The changes in the fluorescent and colorimetric signals were generated by the addition of the target proteins due to the competitive binding. The array has a strong ability to distinguish

11 protein analytes with a detection limit as low as 50 nM. Also, 10 proteins at 1.0 µM were well-identified when the proteins were spiked into the human urine sample.

3.3. Differential Sensing Based on Gold Nanoclusters (AuNCs)

More recently, AuNCs have attracted much interest in biosensing applications [131–134]. Compared with semiconductor quantum dots or other metal NDs, AuNCs possess several distinct features, such as photophysical/chemical properties, good stability, and excellent biocompatibility [135–142]. Several studies utilized AuNCs for the construction of differential sensing strategies [72,143–145]. Ouyang and co-workers designed a visual sensor array based on blue-emitting Col-AuNCs and Mac-AuNCs for the discrimination of proteins [70]. The colorimetric and fluorometric signal changes were recorded after the addition of the target proteins. Either or both proteins and protein mixtures after polyacrylamide electrophoresis were well-discriminated by LDA.

Luo's group also developed a protein sensing platform using six dual ligand functionalized AuNCs as sensing receptors [144], by functionalizing them with different amino acids. When they compared the relative fluorescence changes with the LDA method, ten proteins were successfully discriminated. Wu and co-workers [146] developed a fluorescence sensor array based on metal ion-AuNCs for the identification of proteins and bacteria. The sensor array successfully differentiated nine proteins with different concentrations and identified five different types of bacteria, demonstrating its potential for rapid and sensitive biomolecule sensing.

A pH-controlled histidine-templated AuNC (AuNCs@His) [147] was developed for a fluorescent sensor array that responds to reactive oxygen species (ROS) for distinguishing cancer cell types and their proliferation states. The sensor array exhibited excellent performance in accurately differentiating cancer cell types and their proliferation states, indicating great potential for precise cancer diagnosis (Figure 3). Li and Zhu [148] developed a multichannel sensor array for efficient identification of bacteria based on three antimicrobial agents (vancomycin, lysozyme, and bacitracin) functional AuNCs. This sensing platform successfully differentiated seven pathogenic bacteria, different concentrations of the same bacteria, and even bacterial mixtures, offering a rapid and reliable method for diagnosing urinary tract infections.

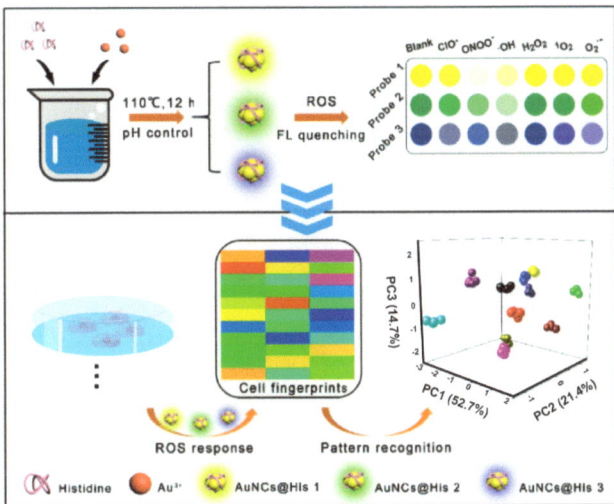

Figure 3. Schematic diagram of precise diagnosis of cancer via an ROS-responsive fluorescent sensor array based on pH-controlled multicolor histidine-templated AuNCs. Reprinted with permission from [147]. Copyright 2023 American Chemical Society.

In summary, gold nanoparticles are good candidates for the development of sensor arrays for biological analysis, and the main characteristics of the different sensor arrays are shown in Table 3.

Table 3. Summarization of Gold Nanoparticle-Based Sensor Array Construction Strategies with Different Artificial Receptors (ARs).

Nanomaterials	Strategies	Numbers of ARs	Signals	Data Analysis Methods	Analytes	LOD	Ref.
AuNPs	Competitive binding between nanoparticle-polymer complexes and cells	3	Fluorescence	LDA	Cells	n.a.	[90]
AuNPs	Competitive binding between NP-GFP complexes and cells	6	Fluorescence	LDA	Cells	5000 cells	[91]
AuNPs	Proteins displace β-Gal from the β-Gal/AuNP complex to restore its catalytic activity towards the fluorogenic substrate	6	Fluorescence	LDA	Proteins	1 nM	[38]
AuNPs	Competitive binding between GFP and analytes to the particle surface	8	Fluorescence	LDA	Cells, tissues	200 ng/ 1000 cells	[67]
AuNPs	Different aggregation behaviors and color changes when the aptamer-protected AuNPs mixed with proteins	3	Absorbance	LDA	Proteins	n.a.	[126]
AuNPs	Competitive interactions between bacterial species and the cationic AuNPs,	1	Fluorescence	LDA	Bacteria	n.a.	[149]
Col-Au NCs and Mac-Au NCs	Different interactions between proteins and the Au NCs surface	2	Fluorescence	LDA	Proteins	n.a.	[150]
AuNPs	Differential interactions between DNA-AuNPs and cells result in distinct Au growth reactions	6	Absorbance	LDA	Cells	n.a.	[129]
AuNPs	Competitive binding between DNA and proteins from the surface of AuNPs	3	Fluorescence, Absorbance	LDA, HCA	Proteins	50 nM	[127]
AuNDs	Differential interactions of proteins with AuNDs	8	Fluorescence	LDA, HCA	Proteins	n.a.	[72]
AuNPs	Competitive binding between the fluorescent proteins and the cell lysate analytes to BenzNPs	1	Fluorescence	LDA, HCA	Cells	1000 cells	[92]
AuNPs	Differential interactions of microorganisms and AuNPs caused aggregation of four sensing elements at different degrees	4	Absorbance	LDA	Microorganisms	n.a.	[28]
AuNCs	Differential interactions between free proteins and capping proteins on Au NCs	5	Phosphorescence	LDA, HCA	Proteins	n.a.	[143]
AuNPs	Different proteins triggered the DNA-protected AuNPs to exhibit different aggregation behaviors caused various solution color change	2	Absorbance	LDA, HCA	Proteins	50 nM	[128]
AuNCs	Differential binding between proteins and AuNCs resulting in the fluorescence change of AuNCs	6	Fluorescence	LDA	Proteins, serum	10 nM	[144]

Table 3. Cont.

Nanomaterials	Strategies	Numbers of ARs	Signals	Data Analysis Methods	Analytes	LOD	Ref.
AuNCs	Differential interactions between the protein and the metal ion-AuNCs	6	Fluorescence	LDA	Proteins, bacteria	n.a.	[146]
AuNPs	Aggregation of AuNPs induced by the differential metabolic capabilities of bacteria towards D-amino acids (D-AAs)	3	Absorbance	LDA, HCA	Bacteria	n.a.	[130]
AuNCs	Different oxidation of AuNCs@His by ROS	3	Fluorescence	PCA, HCA	Cells	n.a.	[147]
AuNCs	Fluorescence intensity of AuNCs was quenched to varying degrees by the bacteria	3	Fluorescence	LDA, HCA	Bacteria	10^5 CFU/mL	[148]

4. Graphene Oxide (GO)-Based Sensor Arrays

GO is a chemically exfoliated graphene derivative, which can be utilized as a fluorescence quencher for various fluorescent probes, such as fluorescent polymer [76,151,152], fluorescent protein [68], metal nanodots [153], and fluorescently labeled DNA [69,154–158]. More importantly, GO showed differential affinity toward different molecules or materials [159,160]. Thus, GO has been widely applied as an ideal artificial receptor for the construction of nose/tongue sensors [71,75,150,161–166], as shown in Table 4.

Table 4. Summarization of Graphene Oxide (GO)-Based Sensor Arrays Construction Strategies with Different Artificial Receptors (ARs).

Nanomaterials	Strategies	Numbers of ARs	Signals	Data Analysis Methods	Analytes	LOD	Ref.
nGO	Proteins displace fluorophores from the nGO surface through binding competition	5	Fluorescence	LDA	Proteins	10 nM	[68]
nGO	Competitive binding between ssDNA-nGO complexes and analytes	7	Fluorescence	LDA	Proteins, cells and bacteria	5 μM	[69]
GQDs, QDs, CDs-COOH, PEI-CDs, BSA-AuNCs, Lys-AuNCs, AgNCs and GO	Competitive binding between luminescent nanodots and analytes to GO surfaces	7	Fluorescence	LDA	Proteins, bacteria	n.a.	[71]
GQDs-COOH, GQDs-NH$_2$, PEI-CDs, QDs, BSA-AuNCs, Lys-AuNCs and GO	Competitive binding between luminescent nanodots and cells to GO surfaces	6	Fluorescence	LDA	Cells	200 cells	[153]
GO	Competitive interaction among GO, AIEgen and biomolecules	7	Fluorescence	PCA	Microbes	n.a.	[167]
GO	Competitive interaction among GO, fluorescent polymers and proteins	5	Fluorescence	LDA	Proteins	n.a.	[76]
GO	Competitive binding between bacteria and GO with fluorescent PEIs	1	Fluorescence	LDA	Bacteria	OD_{600} = 0.125	[77]

The differential sensor for protein detection was developed based on GO [68]. Initially, fluorescent reporters (eGFP, pyronin Y, rhodamine 6G, acridine orange, rhodamine B) were quenched when combined to GO, and then different proteins could displace the fluorophores and restored different levels of fluorescence signal according to the affinity between GO and the proteins. In their work, a novel kind of nanoscale GO (nGOs) with a near-uniform dimension of 20 nm was applied, showing much better recognition capability than conventional GO, because nGOs have a higher supramolecular response and replacement rate. Their results showed that the nGO arrays can discriminate eight different proteins at 100 nM and 10 nM, and the success rate was as high as 95% when analyzing 48 unknowns.

Fan and co-workers combined the adaptive "ensemble aptamers" (ENSaptamers) and nGOs to develop a sensor array for high-precision identification of proteins, bacteria, and cells [69]. Auguste and co-workers provided a sensing array for the identification of healthy, cancerous, and metastatic human breast cells using six luminescent nanodot-graphene oxide complexes as novel fluorescent nanoprobes [153]. The sensing system was disrupted in the presence of breast cells, producing a distinct fluorescence response pattern. The multichannel sensor was capable of effectively identifying healthy, cancerous, and metastatic human breast cells with as few as 200 cells. Tomita and co-workers constructed a cross-reactive DNA-based array for one-step identification of antibody degradation pathways. The signature-based sensing platform was able to identify a broad range of degraded antibodies, such as common features of native, denatured, and visibly aggregated antibodies, complicated degradation pathways of therapeutic omalizumab upon time-course heat-treatment, and the individual compositions of differently degraded omalizumab mixtures. Tang and Qin [167] developed a microbial lysate-responsive fluorescent sensor array using luminogens featuring aggregation-induced emission characteristics (AIEgens) and graphene oxide (GO). This combination effectively reduces background signals and enhances discrimination ability through competitive interactions among AIEgens, microbial lysates, and GO. The sensor array successfully identified six microbes, including fungi, Gram-positive bacteria, and Gram-negative bacteria.

Han and co-workers [77] developed a novel multichannel array using modified polyethyleneimine and graphene oxide. This complex system enabled the successful identification of 10 bacteria within minutes through electrostatic and hydrophobic interactions. The sensor array also demonstrated the ability to measure bacterial concentrations and identify mixed bacteria accurately. In biological samples such as urine, the array achieved high accuracy. Han and co-workers [76] also designed five positively charged poly(para-aryleneethynylene) (P1–P5) molecules to form electrostatic complexes (C1–C5) with negatively charged graphene oxide (GO), effectively distinguishing between 12 proteins while employing machine learning algorithms. Moreover, these sensor arrays accurately identified levels of Aβ40 and Aβ42 aggregates, including monomers, oligomers, and fibrils, offering an attractive strategy for early Alzheimer's disease diagnosis (Figure 4).

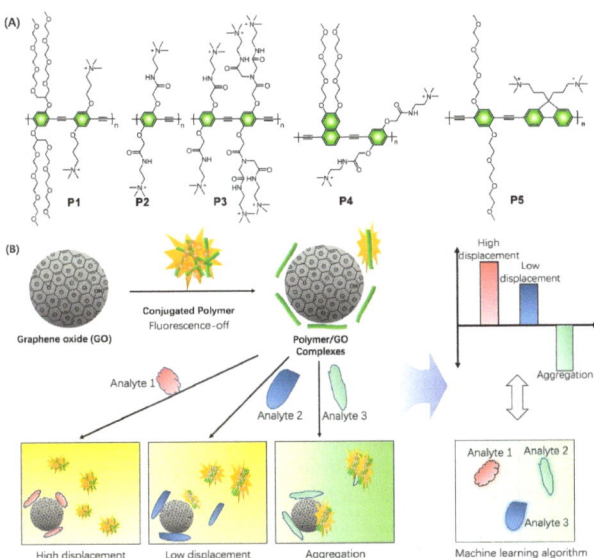

Figure 4. (**A**) Chemical structures of fluorescent polymers P1–P5. (**B**) Schematic illustration of the construction of electrostatic complexes from positively charged poly(para-aryleneethynylene)s and negatively charged GO and identification mechanism for multiple analytes. Reprinted with permission from [76]. Copyright 2022 American Chemical Society.

5. Quantum Dot (QD)-Based Sensor Arrays

Based on their distinguished characteristics of good photostability, high quantum yield, and long fluorescence lifetime, QDs have been extensively used in fluorescent bioanalysis [161,168–171]. Rotello and co-workers developed a QD-based sensor for sensing mammalian cell types and states [100]. The sensing system is composed of two quantum dots and one gold nanoparticle. The quantum dots serve as transducers, which can be quenched by the gold nanoparticle. Different cell types and states were successfully differentiated by the sensor array (Figure 5).

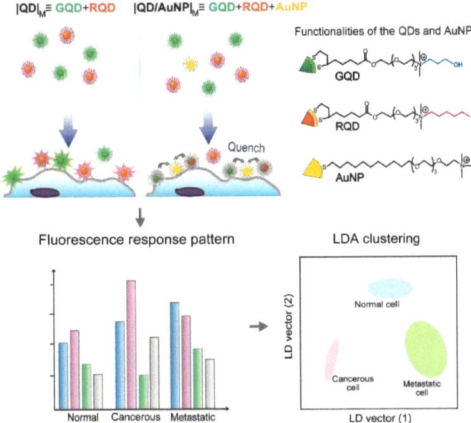

Figure 5. Illustration showing how nanoparticles interact with the cell surface in a sensing system, resulting in differential quenching and distinct patterns for distinguishing different cell types/states. Reprinted with permission from [100]. 2012 Elsevier Ireland Ltd. All rights reserved.

Wang and Chen developed a fluorescent sensor array using imidazolium ionic liquids (ILs) and ionic liquid-QD conjugates as semi-selective receptors for the discrimination of proteins [172]. The IL sensing system was able to differentiate eight proteins at a concentration of 500 nM with an accuracy of 91.7%. With the improvements of the sensitivity and discrimination accuracy, the IL@QDs/QDs sensing system could distinguish eight proteins with 100% accuracy at a very low concentration of 10 nM. Additionally, protein mixtures and proteins spiked in human urine were well-discriminated by the IL@QDs/QDs sensing system.

Yan and co-workers designed a multidimensional sensing device based on Mn-ZnS QDs for the discrimination of proteins [173]. The triple-channel optical properties (fluorescence, phosphorescence, light scattering) of Mn-ZnS QDs were utilized to achieve the output signals. After interaction with target proteins, the changes in the triple-channel optical properties of Mn–ZnS QDs were observed. The multidimensional sensing devices were able to generate distinct patterns for different proteins. Eight proteins added to human urine samples were successfully discriminated against with the aid of principal component analysis.

Combination of different nanomaterials, Wu and Zhang developed a nanoparticle quantum dot-based fluorescence sensor array for sensing proteins and cancer cells [174]. The sensor array consists of six types of nanoparticles (NPs, including CuO, ZnO, Eu_2O_3, AuNPs, AgNPs, Au-Ag core-shell) and CdSe quantum dots (Figure 6). These NPs can quench the fluorescence of CdSe quantum dots. The NP-QD interaction was disrupted by the addition of proteins, leading to fluorescence turn-on or further quenching. Eight proteins were readily differentiated by using LDA analysis. Moreover, protein quantification was achieved with the limits of detection below 2 µM in the range of 2–50 µM. Qu and Ren [71] designed seven fluorescent luminescent nanoprobes, including graphene quantum dots (GQDs), CdTe quantum dots (QDs), carboxyl-carbon dots (CDs-COOH), polyethyleneimine functionalized carbon dots (PEI-CDs), BSA-templated gold nanoclusters (BSA-AuNCs), lysozyme-templated gold nanoclusters (LysAuNCs), and DNA-templated silver nanoclusters (AgNCs), and they used graphene oxide (GO) as an excellent quencher with different affinity to proteins and the nanoprobes. The discrimination ability of this array was tested by analyzing eight proteins at low concentrations. Finally, 100% accuracy was achieved for the identification of 48 unknown protein samples. The summary of quantum dot (QD)-based sensor arrays is shown in Table 5.

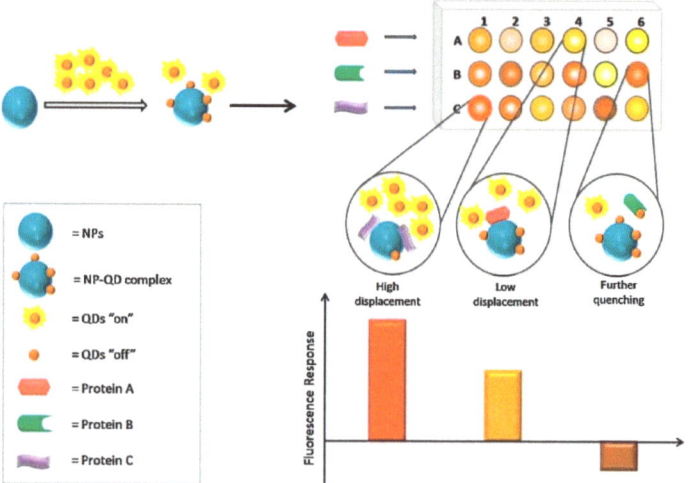

Figure 6. Schematic illustration of a fluorescence sensor array based on six types of NP-QD complexes. Reprinted with permission from [174]; 2017 Elsevier B.V. All rights reserved.

Table 5. Summarization of Quantum dot (QD)-based Sensor Arrays Construction Strategies with Different Artificial Receptors (ARs).

Nanomaterials	Strategies	Numbers of ARs	Signals	Data Analysis Methods	Analytes	LOD	Ref.
Mn–ZnS QDs	Different interactions of Mn–ZnS QDs with proteins	1	Fluorescence phosphorescence light scattering	PCA	Proteins	0.5 µM	[173]
QDs	Differential competitive and selective non-covalent interactions between nanoparticles and cell surface	2	Fluorescence	LDA	Cells	10,000 cells	[100]
CdTe QDs	Differential interactions between analytes and IL@CdTe QDs	5	Fluorescence	LDA	Proteins, bacteria	10 nM	[175]
CuO NPs, ZnO NPs, Eu$_2$O$_3$ NPs, AuNPs, AgNPs, Au-Ag core-shell and CdSe QDs	Protein presence disrupts nanoparticle-QD interactions, resulting in fluorescence turn on or further quenching	6	Fluorescence	LDA	Proteins, cells	5 µM	[174]

6. Other Metal Nanoparticle-Based Sensor Arrays

Other metal nanoparticles, such as Fe$_3$O$_4$ NPs, AgNPs, MoS$_2$, and CuS NPs, were also prepared to develop sensor arrays for the discrimination of proteins, bacteria, and cells [78,176–183]. Scientists fabricated dopamine and trimethylammonium functionalized Fe$_3$O$_4$ NPs, which were able to catalyze the oxidation of colorless ABTS to become a green product in the presence of H$_2$O$_2$ [66]. When analyte proteins were added into the mixture, the accessibility of reaction substrates to the NP surface was adjusted, leading to a change in the catalytic efficiency. The Fe$_3$O$_4$ NP-based sensor array can identify ten proteins at a concentration of 50 nM. Cui and co-workers developed a dynamically tunable, low-background, and highly reproducible CL system based on luminol-functionalized silver nanoparticles (luminol-AgNPs) for protein sensing [184]. Qu and Ren also utilized AgNPs to construct sensor arrays for the recognition of proteins [185]. Although AgNPs have some unique properties, their instability and toxicity limit their application in bioanalysis. Ren and Pu developed a sensor array for the identification of proteins and antibiotic-resistant bacteria utilizing CuS NPs and fluorescent dyes [186]. The sensing platform showed excellent discrimination ability between antibiotic-resistant and antibiotic-susceptible bacteria extracts.

Zhang and coworkers constructed quaternized magnetic nanoparticle (q-MNP)-fluorescent polymer systems for the detection and identification of bacteria [187]. The complexes of the q-MNP-fluorescent polymer were disrupted by the bacterial cell membranes, leading to a unique fluorescence response. Eight bacteria were quantitatively discriminated with LDA with an accuracy of 87.5% for 10^7 cfu/mL within 20 min. The sensor array was also used to identify 32 unknown bacteria samples with an accuracy of 96.8%. The summarization of the other metal nanoparticle-based sensor arrays is shown in Table 6.

Table 6. Summarization of the Sensor Arrays Construction Strategies with Different Artificial Receptors (ARs).

Nanomaterials	Strategies	Numbers of ARs	Signals	Data Analysis Methods	Analytes	LOD	Ref.
Fe$_3$O$_4$ NPs	Differential interactions of proteins with Fe$_3$O$_4$ NPs affected the accessibility of ABTS to the NP surface	2	Fluorescence	LDA	Proteins	50 nM	[66]

Table 6. Cont.

Nanomaterials	Strategies	Numbers of ARs	Signals	Data Analysis Methods	Analytes	LOD	Ref.
AgNPs	Different adsorption capacity of proteins onto luminol-AgNPs affected the accessibility of H_2O_2 to the NPs surface	1	Chemiluminescence	PCA	Proteins	n.a.	[184]
Quaternized magnetic nanoparticles (q-MNP)	Competitive binding between fluorescent polymer and bacteria to GO surfaces q-MNP	3	Fluorescence	LDA	Bacteria	n.a.	[187]
CuS NPs	Competitive binding between analytes and fluorescent dyes towards CuS NPs	4	Fluorescence	PCA	Proteins, bacteria	n.a.	[186]
AgNPs	The diversity in structure and properties of various proteins results in different effects on the synthesis of AgNPs under light irradiation, leading to AgNPs with distinct LSPR absorption spectra	3	Absorbance	PCA	Proteins	n.a.	[185]
DBCO-UCNPs	Different bacteria exhibit differences in metabolic capability, sensitivity to antibiotics, and surface properties and thus lead to discriminative responses	6	Fluorescence	PCA, HCA, LDA	Bacteria	10^5 CFU/mL	[34]

7. Conclusions

The integration of nanomaterials in optical differential sensors has provided a powerful platform for biosystems analysis [188]. In contrast to traditional lock-and-key biosensing, these sensors function as chemical noses with the ability to recognize a wide range of targets, including proteins, mammalian cells, and microorganisms [189].

The use of nanomaterials has expanded the design possibilities of analysis arrays in several significant ways. Firstly, more different molecular assembly modes and larger assembly quantities are now achievable using covalent bonding modifications or surface adsorption, etc. Secondly, nanomaterials themselves possess more diverse signal outputs, such as the abundant fluorescence signals of quantum dots at various wavelengths or the color changes of nanogold particles. Thirdly, nanomaterials provide a wider range of interaction mechanisms between nanointerfaces and biomolecules, reflecting surface charge and molecular structure, etc. Lastly, the application of hierarchical nanomaterials further enhances the capabilities of analysis arrays. By combining hierarchical nanomaterials, additional advantages for biosensing applications can be achieved. These materials improve signal intensity and enhance various energy transfer processes. The integration of hierarchical nanomaterials alongside other nanomaterials expands the design possibilities of analysis arrays, enabling even more diverse and efficient biosensing platforms [190]. Overall, the application of nanomaterials has dramatically improved the sensitivity and recognition range of pattern recognition detection, leading to more diverse array designs [191]. However, challenges remain in this field.

Future research directions and urgent issues include: (1) Further theoretical studies are needed to understand the signal mechanisms of most sensing arrays. (2) To radically improve the accuracy of pattern recognition, the stability and controllability of the nanomaterials are critical. (3) Further enhance the discrimination ability and sensitivity of pattern recognition sensors. (4) Efforts should be made to reduce the production cost of nanoprobes to decrease expenses associated with their use. (5) The application of interfacial self-assembly on micro/nanochip technology should be helpful for the high-throughput data collection for next-generation chemical noses. (6) The introduction of novel and superior nanomaterials would greatly improve the performance of the sensor array. For example, single-chirality carbon nanotubes are recently drawing a large amount of research interest for their near-infrared fluorescence signals and specific recognition and binding abilities for biomolecules. Addressing these challenges and capitalizing on emerging advancements will undoubtedly contribute to the continuous progress of this field.

Author Contributions: Conceptualization, L.W., L.L., X.Y. and Q.T.; resources and data curation, L.L., X.Y., M.C. and Q.T.; writing—original draft preparation, L.W., Q.T., X.S. and M.C.; writing—review and editing, L.W., Y.W., W.L. and G.L.; visualization, L.W. and X.S.; supervision, Y.W. and G.L.; project administration, W.L. and X.S.; funding acquisition, Y.W. and G.L. All authors have read and agreed to the published version of the manuscript.

Funding: This work was financially supported by the National Quality Infrastructure Program of China (2021YFF0600901 NQI) and the National Natural Science Foundation of China (No. 22074093).

Institutional Review Board Statement: Not applicable.

Informed Consent Statement: Not applicable.

Data Availability Statement: Not applicable.

Conflicts of Interest: The authors declare no conflict of interest.

References

1. Rodriguez, R.S.; O'Keefe, T.L.; Froehlich, C.; Lewis, R.E.; Sheldon, T.R.; Haynes, C.L. Sensing Food Contaminants: Advances in Analytical Methods and Techniques. *Anal. Chem.* **2021**, *93*, 23–40. [CrossRef]
2. Shruti, A.; Bage, N.; Kar, P. Nanomaterials based sensors for analysis of food safety. *Food Chem.* **2024**, *433*, 137284. [CrossRef]
3. Pan, M.F.; Yin, Z.J.; Liu, K.X.; Du, X.L.; Liu, H.L.; Wang, S. Carbon-Based Nanomaterials in Sensors for Food Safety. *Nanomaterials* **2019**, *9*, 1330. [CrossRef]
4. Lv, M.; Liu, Y.; Geng, J.H.; Kou, X.H.; Xin, Z.H.; Yang, D.Y. Engineering nanomaterials-based biosensors for food safety detection. *Biosens. Bioelectron.* **2018**, *106*, 122–128. [CrossRef]
5. Huang, Y.K.; Mei, L.; Chen, X.G.; Wang, Q. Recent Developments in Food Packaging Based on Nanomaterials. *Nanomaterials* **2018**, *8*, 830. [CrossRef]
6. Chu, H.Q.; Lu, Y.F. Application of Functional Nanomaterials in Food Safety. *Chin. J. Anal. Chem.* **2010**, *38*, 442–448. [CrossRef]
7. Wen, Y.; Wang, L.; Xu, L.; Li, L.; Ren, S.; Cao, C.; Jia, N.; Aldalbahi, A.; Song, S.; Shi, J.; et al. Electrochemical detection of PCR amplicons of Escherichia coli genome based on DNA nanostructural probes and polyHRP enzyme. *Analyst* **2016**, *141*, 5304–5310. [CrossRef]
8. Yang, X.; Wen, Y.; Wang, L.; Zhou, C.; Li, Q.; Xu, L.; Li, L.; Shi, J.; Lal, R.; Ren, S.; et al. PCR-Free Colorimetric DNA Hybridization Detection Using a 3D DNA Nanostructured Reporter Probe. *ACS Appl. Mater. Interfaces* **2017**, *9*, 38281–38287. [CrossRef]
9. Li, L.; Wang, L.; Xu, Q.; Xu, L.; Liang, W.; Li, Y.; Ding, M.; Aldalbahi, A.; Ge, Z.; Wang, L.; et al. Bacterial Analysis Using an Electrochemical DNA Biosensor with Poly-Adenine-Mediated DNA Self-Assembly. *ACS Appl. Mater. Interfaces* **2018**, *10*, 6895–6903. [CrossRef]
10. Bayramoglu, G.; Ozalp, V.C.; Dincbal, U.; Arica, M.Y. Fast and Sensitive Detection of Salmonella in Milk Samples Using Aptamer-Functionalized Magnetic Silica Solid Phase and MCM-41-Aptamer Gate System. *ACS Biomater. Sci. Eng.* **2018**, *4*, 1437–1444. [CrossRef]
11. Wang, L.; Wen, Y.; Yang, X.; Xu, L.; Liang, W.; Zhu, Y.; Wang, L.; Li, Y.; Li, Y.; Ding, M.; et al. Ultrasensitive Electrochemical DNA Biosensor Based on a Label-Free Assembling Strategy Using a Triblock polyA DNA Probe. *Anal. Chem.* **2019**, *91*, 16002–16009. [CrossRef] [PubMed]
12. da Costa, B.M.C.; Duarte, A.C.; Rocha-Santos, T.A.P. Environmental monitoring approaches for the detection of organic contaminants in marine environments: A critical review. *Trends Environ. Anal. Chem.* **2022**, *33*, e00154. [CrossRef]
13. Liang, M.M.; Guo, L.H. Application of Nanomaterials in Environmental Analysis and Monitoring. *J. Nanosci. Nanotechnol.* **2009**, *9*, 2283–2289. [CrossRef] [PubMed]
14. Farzin, L.; Shamsipur, M.; Sheibani, S. A review: Aptamer-based analytical strategies using the nanomaterials for environmental and human monitoring of toxic heavy metals. *Talanta* **2017**, *174*, 619–627. [CrossRef] [PubMed]
15. Abu, H.; Hossain, M.A.M.; Marlinda, A.; Al Mamun, M.; Simarani, K.; Johan, M.R. Nanomaterials based electrochemical nucleic acid biosensors for environmental monitoring: A review. *Appl. Surf. Sci. Adv.* **2021**, *4*, 100064. [CrossRef]
16. Wang, L.; Wen, Y.; Li, L.; Yang, X.; Jia, N.; Li, W.; Meng, J.; Duan, M.; Sun, X.; Liu, G. Sensitive and label-free electrochemical lead ion biosensor based on a DNAzyme triggered G-quadruplex/hemin conformation. *Biosens. Bioelectron.* **2018**, *115*, 91–96. [CrossRef] [PubMed]

17. Lotfi Zadeh Zhad, H.R.; Lai, R.Y. Application of Calcium-Binding Motif of E-Cadherin for Electrochemical Detection of Pb(II). *Anal. Chem.* **2018**, *90*, 6519–6525. [CrossRef]
18. Wang, L.Q.; Wang, X.J.; Wu, Y.G.; Guo, M.Q.; Gu, C.J.; Dai, C.H.; Kong, D.R.; Wang, Y.; Zhang, C.; Qu, D.; et al. Rapid and ultrasensitive electromechanical detection of ions, biomolecules and SARS-CoV-2 RNA in unamplified samples. *Nat. Biomed. Eng.* **2022**, *6*, 276–285. [CrossRef]
19. Song, P.; Shen, J.W.; Ye, D.K.; Dong, B.J.; Wang, F.; Pei, H.; Wang, J.B.; Shi, J.Y.; Wang, L.H.; Xue, W.; et al. Programming bulk enzyme heterojunctions for biosensor development with tetrahedral DNA framework. *Nat. Commun.* **2022**, *13*, 1917. [CrossRef]
20. Zhai, T.T.; Wei, Y.H.; Wang, L.H.; Li, J.; Fan, C.H. Advancing pathogen detection for airborne diseases. *Fundam. Res.* **2023**, *3*, 520–524. [CrossRef]
21. Wang, L.; Dong, L.; Liu, G.; Shen, X.; Wang, J.; Zhu, C.; Ding, M.; Wen, Y. Fluorometric determination of HIV DNA using molybdenum disulfide nanosheets and exonuclease III-assisted amplification. *Mikrochim. Acta* **2019**, *186*, 286. [CrossRef]
22. Zhang, Y.; Shuai, Z.; Zhou, H.; Luo, Z.; Liu, B.; Zhang, Y.; Zhang, L.; Chen, S.; Chao, J.; Weng, L.; et al. Single-Molecule Analysis of MicroRNA and Logic Operations Using a Smart Plasmonic Nanobiosensor. *J. Am. Chem. Soc.* **2018**, *140*, 3988–3993. [CrossRef]
23. Yang, F.; Li, Q.; Wang, L.; Zhang, G.J.; Fan, C. Framework-Nucleic-Acid-Enabled Biosensor Development. *ACS Sens.* **2018**, *3*, 903–919. [CrossRef]
24. Zhao, R.; Lv, M.; Li, Y.; Sun, M.; Kong, W.; Wang, L.; Song, S.; Fan, C.; Jia, L.; Qiu, S.; et al. Stable Nanocomposite Based on PEGylated and Silver Nanoparticles Loaded Graphene Oxide for Long-Term Antibacterial Activity. *ACS Appl. Mater. Interfaces* **2017**, *9*, 15328–15341. [CrossRef]
25. Yin, F.; Zhao, H.; Lu, S.; Shen, J.; Li, M.; Mao, X.; Li, F.; Shi, J.; Li, J.; Dong, B.; et al. DNA-framework-based multidimensional molecular classifiers for cancer diagnosis. *Nat. Nanotechnol.* **2023**, *18*, 677–686. [CrossRef]
26. Mao, D.; Dong, Z.; Liu, X.; Li, W.; Li, H.; Gu, C.; Chen, G.; Zhu, X.; Yang, Y. Intelligent DNA nanoreactor for in vivo easy-to-read tumor imaging and precise therapy. *Angew. Chem. Int. Ed. Engl.* **2023**, *63*, e202311309. [CrossRef]
27. Li, Y.; Wen, Y.; Wang, L.; Liang, W.; Xu, L.; Ren, S.; Zou, Z.; Zuo, X.; Fan, C.; Huang, Q.; et al. Analysis of telomerase activity based on a spired DNA tetrahedron TS primer. *Biosens. Bioelectron.* **2015**, *67*, 364–369. [CrossRef]
28. Qian, Y.; Fan, T.; Wang, P.; Zhang, X.; Luo, J.; Zhou, F.; Yao, Y.; Liao, X.; Li, Y.; Gao, F. A novel label-free homogeneous electrochemical immunosensor based on proximity hybridization-triggered isothermal exponential amplification induced G-quadruplex formation. *Sens. Actuators B Chem.* **2017**, *248*, 187–194. [CrossRef]
29. Zhang, L.; Wang, J.; Zhang, J.; Liu, Y.; Wu, L.; Shen, J.; Zhang, Y.; Hu, Y.; Fan, Q.; Huang, W.; et al. Individual Au-Nanocube Based Plasmonic Nanoprobe for Cancer Relevant MicroRNA Biomarker Detection. *ACS Sens.* **2017**, *2*, 1435–1440. [CrossRef] [PubMed]
30. Zhang, L.; Zhang, H.; Hu, Y.; Fan, Q.; Yang, W.; Li, A.; Li, S.; Huang, W.; Wang, L. Refractive index dependent real-time plasmonic nanoprobes on a single silver nanocube for ultrasensitive detection of the lung cancer-associated miRNAs. *Chem. Commun.* **2015**, *51*, 294–297. [CrossRef] [PubMed]
31. Zhou, X.; Zhao, M.; Duan, X.; Guo, B.; Cheng, W.; Ding, S.; Ju, H. Collapse of DNA Tetrahedron Nanostructure for "Off-On" Fluorescence Detection of DNA Methyltransferase Activity. *ACS Appl. Mater. Interfaces* **2017**, *9*, 40087–40093. [CrossRef] [PubMed]
32. Beard, D.J.; Perrine, S.A.; Phillips, E.; Hoque, S.; Conerly, S.; Tichenor, C.; Simmons, M.A.; Young, J.K. Conformational comparisons of a series of tachykinin peptide analogs. *J. Med. Chem.* **2007**, *50*, 6501–6506. [CrossRef] [PubMed]
33. Li, T.; Zhu, X.; Hai, X.; Bi, S.; Zhang, X. Recent Progress in Sensor Arrays: From Construction Principles of Sensing Elements to Applications. *ACS Sens.* **2023**, *8*, 994–1016. [CrossRef] [PubMed]
34. Wang, X.; Li, H.; Yang, W.; Chen, M.; Wang, J.; Yang, T. Chemical Nose Strategy with Metabolic Labeling and "Antibiotic-Responsive Spectrum" Enables Accurate and Rapid Pathogen Identification. *Anal. Chem.* **2024**, *96*, 427–436. [CrossRef] [PubMed]
35. Liu, J.B.; Li, G.; Yang, X.H.; Wang, K.M.; Li, L.; Liu, W.; Shi, X.; Guo, Y.L. Exciton Energy Transfer-Based Quantum Dot Fluorescence Sensing Array: "Chemical Noses" for Discrimination of Different Nucleobases. *Anal. Chem.* **2015**, *87*, 876–883. [CrossRef] [PubMed]
36. Li, L.; Li, G.; He, X.X.; Yang, X.H.; Liu, S.Y.; Tang, J.L.; Chen, Q.S.; Liu, J.B.; Wang, K.M. Protein- driven disassembly of surfactant-polyelectrolyte nanomicelles: Modulation of quantum dot/fluorochrome FRET for pattern sensing. *Sens. Actuators B-Chem.* **2018**, *272*, 393–399. [CrossRef]
37. Li, Z.; Askim, J.R.; Suslick, K.S. The Optoelectronic Nose: Colorimetric and Fluorometric Sensor Arrays. *Chem. Rev.* **2019**, *119*, 231–292. [CrossRef]
38. Miranda, O.R.; Chen, H.-T.; You, C.-C.; Mortenson, D.E.; Yang, X.-C.; Bunz, U.H.F.; Rotello, V.M. Enzyme-Amplified Array Sensing of Proteins in Solution and in Biofluids. *J. Am. Chem. Soc.* **2010**, *132*, 5285–5289. [CrossRef]
39. Yan, Q.; Ding, X.Y.; Chen, Z.H.; Xue, S.F.; Han, X.Y.; Lin, Z.Y.; Yang, M.; Shi, G.; Zhang, M. pH-Regulated Optical Performances in Organic/Inorganic Hybrid: A Dual-Mode Sensor Array for Pattern-Recognition-Based Biosensing. *Anal. Chem.* **2018**, *90*, 10536–10542. [CrossRef]
40. Mei, Y.; Zhang, Q.W.; Gu, Q.; Liu, Z.; He, X.; Tian, Y. Pillar[5]arene-Based Fluorescent Sensor Array for Biosensing of Intracellular Multi-neurotransmitters through Host-Guest Recognitions. *J. Am. Chem. Soc.* **2022**, *144*, 2351–2359. [CrossRef]
41. Yao, J.; Yang, M.; Duan, Y.X. Chemistry, Biology, and Medicine of Fluorescent Nanomaterials and Related Systems: New Insights into Biosensing, Bioimaging, Genomics, Diagnostics, and Therapy. *Chem. Rev.* **2014**, *114*, 6130–6178. [CrossRef]
42. Meng, H.M.; Liu, H.; Kuai, H.L.; Peng, R.Z.; Mo, L.T.; Zhang, X.B. Aptamer-integrated DNA nanostructures for biosensing, bioimaging and cancer therapy. *Chem. Soc. Rev.* **2016**, *45*, 2583–2602. [CrossRef]

43. Zhang, S.D.; Geryak, R.; Geldmeier, J.; Kim, S.; Tsukruk, V.V. Synthesis, Assembly, and Applications of Hybrid Nanostructures for Biosensing. *Chem. Rev.* **2017**, *117*, 12942–13038. [CrossRef]
44. Sun, H.J.; Zhou, Y.; Ren, J.S.; Qu, X.G. Carbon Nanozymes: Enzymatic Properties, Catalytic Mechanism, and Applications. *Angew. Chem.-Int. Ed.* **2018**, *57*, 9224–9237. [CrossRef]
45. Yadav, V.; Roy, S.; Singh, P.; Khan, Z.; Jaiswal, A. 2D MoS_2-Based Nanomaterials for Therapeutic, Bioimaging, and Biosensing Applications. *Small* **2019**, *15*, e1803706. [CrossRef]
46. Ji, C.Y.; Zhou, Y.Q.; Leblanc, R.M.; Peng, Z.L. Recent Developments of Carbon Dots in Biosensing: A Review. *ACS Sens.* **2020**, *5*, 2724–2741. [CrossRef]
47. Chung, S.; Revia, R.A.; Zhang, M.Q. Graphene Quantum Dots and Their Applications in Bioimaging, Biosensing, and Therapy. *Adv. Mater.* **2021**, *33*, e1904362. [CrossRef]
48. Xu, W.Q.; Jiao, L.; Wu, Y.; Hu, L.Y.; Gu, W.L.; Zhu, C.Z. Metal-Organic Frameworks Enhance Biomimetic Cascade Catalysis for Biosensing. *Adv. Mater.* **2021**, *33*, 2005172. [CrossRef]
49. Kumar, S.; Wang, Z.; Zhang, W.; Liu, X.C.; Li, M.Y.; Li, G.R.; Zhang, B.Y.; Singh, R. Optically Active Nanomaterials and Its Biosensing Applications-A Review. *Biosensors* **2023**, *13*, 85. [CrossRef]
50. Pini, F.; Francés-Soriano, L.; Andrigo, V.; Natile, M.M.; Hildebrandt, N. Optimizing Upconversion Nanoparticles for FRET Biosensing. *ACS Nano* **2023**, *17*, 4971–4984. [CrossRef]
51. Du, H.; Xie, Y.Q.; Wang, J. Nanomaterial-sensors for herbicides detection using electrochemical techniques and prospect applications. *Trac-Trends Anal. Chem.* **2021**, *135*, 116178. [CrossRef]
52. Zhai, T.T.; Zheng, H.R.; Fang, W.N.; Gao, Z.S.; Song, S.P.; Zuo, X.L.; Li, Q.; Wang, L.H.; Li, J.; Shi, J.Y.; et al. DNA-Encoded Gold-Gold Wettability for Programmable Plasmonic Engineering. *Angew. Chem.-Int. Ed.* **2022**, *61*, e202210377. [CrossRef]
53. Unser, S.; Bruzas, I.; He, J.; Sagle, L. Localized Surface Plasmon Resonance Biosensing: Current Challenges and Approaches. *Sensors* **2015**, *15*, 15684–15716. [CrossRef]
54. Wang, H.S. Metal-organic frameworks for biosensing and bioimaging applications. *Coord. Chem. Rev.* **2017**, *349*, 139–155. [CrossRef]
55. Loiseau, A.; Asila, V.; Boitel-Aullen, G.; Lam, M.; Salmain, M.; Boujday, S. Silver-Based Plasmonic Nanoparticles for and Their Use in Biosensing. *Biosensors* **2019**, *9*, 78. [CrossRef]
56. Xu, W.Q.; Jiao, L.; Yan, H.Y.; Wu, Y.; Chen, L.J.; Gu, W.L.; Du, D.; Lin, Y.H.; Zhu, C.Z. Glucose Oxidase-Integrated Metal-Organic Framework Hybrids as Biomimetic Cascade Nanozymes for Ultrasensitive Glucose Biosensing. *ACS Appl. Mater. Interfaces* **2019**, *11*, 22096–22101. [CrossRef]
57. De los Santos, Z.A.; Lin, Z.W.; Zheng, M. Optical Detection of Stereoselective Interactions with DNA-Wrapped Single-Wall Carbon Nanotubes. *J. Am. Chem. Soc.* **2021**, *143*, 20628–20632. [CrossRef]
58. Lim, S.Y.; Shen, W.; Gao, Z.Q. Carbon quantum dots and their applications. *Chem. Soc. Rev.* **2015**, *44*, 362–381. [CrossRef]
59. Ahn, N.; Livache, C.; Pinchetti, V.; Jung, H.; Jin, H.; Hahm, D.; Park, Y.S.; Klimov, V.I. Electrically driven amplified spontaneous emission from colloidal quantum dots. *Nature* **2023**, *617*, 79–85. [CrossRef]
60. Wu, P.; Yan, X.P. Doped quantum dots for chemo/biosensing and bioimaging. *Chem. Soc. Rev.* **2013**, *42*, 5489–5521. [CrossRef]
61. Wu, P.; Hou, X.D.; Xu, J.J.; Chen, H.Y. Ratiometric fluorescence, electrochemiluminescence, and photoelectrochemical chemo/biosensing based on semiconductor quantum dots. *Nanoscale* **2016**, *8*, 8427–8442. [CrossRef]
62. Sun, J.W.; Lu, Y.X.; He, L.Y.; Pang, J.W.; Yang, F.Y.; Liu, Y.Y. Colorimetric sensor array based on gold nanoparticles: Design principles and recent advances. *Trac-Trends Anal. Chem.* **2020**, *122*, 115754. [CrossRef]
63. Behera, P.; De, M. Nanomaterials in Optical Array-Based Sensing. In *Organic and Inorganic Materials Based Sensors*; Wiley: New York, NY, USA, 2024; pp. 495–533. [CrossRef]
64. Naresh, V.; Lee, N. A Review on Biosensors and Recent Development of Nanostructured Materials-Enabled Biosensors. *Sensors* **2021**, *21*, 1109. [CrossRef]
65. You, C.C.; Miranda, O.R.; Gider, B.; Ghosh, P.S.; Kim, I.B.; Erdogan, B.; Krovi, S.A.; Bunz, U.H.; Rotello, V.M. Detection and identification of proteins using nanoparticle-fluorescent polymer 'chemical nose' sensors. *Nat. Nanotechnol.* **2007**, *2*, 318–323. [CrossRef]
66. Li, X.N.; Wen, F.; Creran, B.; Jeong, Y.D.; Zhang, X.R.; Rotello, V.M. Colorimetric Protein Sensing Using Catalytically Amplified Sensor Arrays. *Small* **2012**, *8*, 3589–3592. [CrossRef]
67. Rana, S.; Singla, A.K.; Bajaj, A.; Elci, S.G.; Miranda, O.R.; Mout, R.; Yan, B.; Jirik, F.R.; Rotello, V.M. Array-Based Sensing of Metastatic Cells and Tissues Using Nanoparticle-Fluorescent Protein Conjugates. *ACS Nano* **2012**, *6*, 8233–8240. [CrossRef]
68. Chou, S.S.; De, M.; Luo, J.Y.; Rotello, V.M.; Huang, J.X.; Dravid, V.P. Nanoscale Graphene Oxide (nGO) as Artificial Receptors: Implications for Biomolecular Interactions and Sensing. *J. Am. Chem. Soc.* **2012**, *134*, 16725–16733. [CrossRef]
69. Pei, H.; Li, J.; Lv, M.; Wang, J.; Gao, J.; Lu, J.; Li, Y.; Huang, Q.; Hu, J.; Fan, C. A graphene-based sensor array for high-precision and adaptive target identification with ensemble aptamers. *J. Am. Chem. Soc.* **2012**, *134*, 13843–13849. [CrossRef]
70. Xu, S.; Lu, X.; Yao, C.; Huang, F.; Jiang, H.; Hua, W.; Na, N.; Liu, H.; Ouyang, J. A Visual Sensor Array for Pattern Recognition Analysis of Proteins Using Novel Blue-Emitting Fluorescent Gold Nanoclusters. *Anal. Chem.* **2014**, *86*, 11634–11639. [CrossRef]
71. Tao, Y.; Ran, X.; Ren, J.S.; Qu, X.G. Array-Based Sensing of Proteins and Bacteria By Using Multiple Luminescent Nanodots as Fluorescent Probes. *Small* **2014**, *10*, 3667–3671. [CrossRef]

72. Yuan, Z.; Du, Y.; Tseng, Y.T.; Peng, M.; Cai, N.; He, Y.; Chang, H.T.; Yeung, E.S. Fluorescent gold nanodots based sensor array for proteins discrimination. *Anal. Chem.* **2015**, *87*, 4253–4259. [CrossRef]
73. Xu, Q.; Zhang, Y.; Tang, B.; Zhang, C.-y. Multicolor Quantum Dot-Based Chemical Nose for Rapid and Array-Free Differentiation of Multiple Proteins. *Anal. Chem.* **2016**, *88*, 2051–2058. [CrossRef]
74. Li, B.Y.; Li, X.Z.; Dong, Y.H.; Wang, B.; Li, D.Y.; Shi, Y.M.; Wu, Y.Y. Colorimetric Sensor Array Based on Gold Nanoparticles with Diverse Surface Charges for Microorganisms Identification. *Anal. Chem.* **2017**, *89*, 10639–10643. [CrossRef]
75. Qiu, H.; Pu, F.; Ran, X.; Liu, C.; Ren, J.; Qu, X. Nanozyme as Artificial Receptor with Multiple Readouts for Pattern Recognition. *Anal. Chem.* **2018**, *90*, 11775–11779. [CrossRef]
76. Wang, H.; Chen, M.; Sun, Y.; Xu, L.; Li, F.; Han, J. Machine Learning-Assisted Pattern Recognition of Amyloid Beta Aggregates with Fluorescent Conjugated Polymers and Graphite Oxide Electrostatic Complexes. *Anal. Chem.* **2022**, *94*, 2757–2763. [CrossRef]
77. Wang, H.; Zhou, L.; Qin, J.; Chen, J.; Stewart, C.; Sun, Y.; Huang, H.; Xu, L.; Li, L.; Han, J.; et al. One-Component Multichannel Sensor Array for Rapid Identification of Bacteria. *Anal. Chem.* **2022**, *94*, 10291–10298. [CrossRef]
78. Yan, P.; Zheng, X.; Liu, S.; Dong, Y.; Fu, T.; Tian, Z.; Wu, Y. Colorimetric Sensor Array for Identification of Proteins and Classification of Metabolic Profiles under Various Osmolyte Conditions. *ACS Sens.* **2023**, *8*, 133–140. [CrossRef]
79. Izenman, A.J. Linear Discriminant Analysis. In *Modern Multivariate Statistical Techniques: Regression, Classification, and Manifold Learning*; Springer: New York, NY, USA, 2008; pp. 237–280. [CrossRef]
80. Wold, S.; Esbensen, K.; Geladi, P. Principal component analysis. *Chemom. Intell. Lab. Syst.* **1987**, *2*, 37–52. [CrossRef]
81. Bridges, C.C. Hierarchical Cluster Analysis. *Psychol. Rep.* **1966**, *18*, 851–854. [CrossRef]
82. Izenman, A.J. *Modern Multivariate Statistical Techniques: Regression, Classification, and Manifold Learning*; Springer: New York, NY, USA, 2008; pp. 1–731. [CrossRef]
83. Jolliffe, I.T.; Cadima, J. Principal component analysis: A review and recent developments. *Philos. Trans. R. Soc. A-Math. Phys. Eng. Sci.* **2016**, *374*, 20150202. [CrossRef]
84. Frades, I.; Matthiesen, R. Overview on Techniques in Cluster Analysis. In *Bioinformatics Methods in Clinical Research*; Matthiesen, R., Ed.; Humana Press: Totowa, NJ, USA, 2010; Volume 593, pp. 81–107.
85. Mitchell, L.; New, E.J.; Mahon, C.S. Macromolecular Optical Sensor Arrays. *ACS Appl. Polym. Mater.* **2021**, *3*, 506–530. [CrossRef]
86. Yang, X.; Yang, M.; Pang, B.; Vara, M.; Xia, Y. Gold Nanomaterials at Work in Biomedicine. *Chem. Rev.* **2015**, *115*, 10410–10488. [CrossRef]
87. Jiang, Y.; Shi, M.L.; Liu, Y.; Wan, S.; Cui, C.; Zhang, L.Q.; Tan, W.H. Aptamer/AuNP Biosensor for Colorimetric Profiling of Exosomal Proteins. *Angew. Chem.-Int. Ed.* **2017**, *56*, 11916–11920. [CrossRef]
88. Nejati, K.; Dadashpour, M.; Gharibi, T.; Mellatyar, H.; Akbarzadeh, A. Biomedical Applications of Functionalized Gold Nanoparticles: A Review. *J. Clust. Sci.* **2022**, *33*, 1–16. [CrossRef]
89. Miranda, O.R.; Li, X.; Garcia-Gonzalez, L.; Zhu, Z.J.; Yan, B.; Bunz, U.H.; Rotello, V.M. Colorimetric bacteria sensing using a supramolecular enzyme-nanoparticle biosensor. *J. Am. Chem. Soc.* **2011**, *133*, 9650–9653. [CrossRef]
90. Bajaj, A.; Miranda, O.R.; Kim, I.B.; Phillips, R.L.; Jerry, D.J.; Bunz, U.H.; Rotello, V.M. Detection and differentiation of normal, cancerous, and metastatic cells using nanoparticle-polymer sensor arrays. *Proc. Natl. Acad. Sci. USA* **2009**, *106*, 10912–10916. [CrossRef]
91. Bajaj, A.; Rana, S.; Miranda, O.R.; Yawe, J.C.; Jerry, D.J.; Bunz, U.H.F.; Rotello, V.M. Cell surface-based differentiation of cell types and cancer states using a gold nanoparticle-GFP based sensing array. *Chem. Sci.* **2010**, *1*, 134. [CrossRef]
92. Le, N.D.B.; Yesilbag Tonga, G.; Mout, R.; Kim, S.T.; Wille, M.E.; Rana, S.; Dunphy, K.A.; Jerry, D.J.; Yazdani, M.; Ramanathan, R.; et al. Cancer Cell Discrimination Using Host-Guest "Doubled" Arrays. *J. Am. Chem. Soc.* **2017**, *139*, 8008–8012. [CrossRef]
93. Yang, J.E.; Lu, Y.X.; Ao, L.; Wang, F.Y.; Jing, W.J.; Zhang, S.C.; Liu, Y.Y. Colorimetric sensor array for proteins discrimination based on the tunable peroxidase-like activity of AuNPs-DNA conjugates. *Sens. Actuators B-Chem.* **2017**, *245*, 66–73. [CrossRef]
94. Wang, F.Y.; Zhang, X.; Lu, Y.X.; Yang, J.O.; Jing, W.J.; Zhang, S.C.; Liu, Y.Y. Continuously evolving 'chemical tongue' biosensor for detecting proteins. *Talanta* **2017**, *165*, 182–187. [CrossRef]
95. Lin, X.; Hai, X.; Wang, N.; Chen, X.W.; Wang, J.H. Dual-signal model array sensor based on GQDs/AuNPs system for sensitive protein discrimination. *Anal. Chim. Acta* **2017**, *992*, 105–111. [CrossRef]
96. Mao, J.P.; Lu, Y.X.; Chang, N.; Yang, J.E.; Zhang, S.C.; Liu, Y.Y. Multidimensional colorimetric sensor array for discrimination of proteins. *Biosens. Bioelectron.* **2016**, *86*, 56–61. [CrossRef]
97. Jiang, M.D.; Gupta, A.; Zhang, X.Z.; Chattopadhyay, A.N.; Fedeli, S.; Huang, R.; Yang, J.; Rotello, V.M. Identification of Proteins Using Supramolecular Gold Nanoparticle-Dye Sensor Arrays. *Anal. Sens.* **2023**, *3*, e202200080. [CrossRef]
98. Bian, M.M.; Xu, M.; Yuan, Y.L.; Nie, J.F. Colorimetric Array Sensor for Discrimination of Multiple Heavy Metal Ions Based on Different Lengths of DNA-AuNPs. *Chin. J. Anal. Chem.* **2020**, *48*, 863–870. [CrossRef]
99. Li, L.; CiRen, D.; Chen, Z.B. Gold Nanoparticles-Based Dual-Channel Colorimetric Array Sensors for Discrimination of Metal Ions. *ACS Appl. Nano Mater.* **2022**, *5*, 18270–18275. [CrossRef]
100. Liu, Q.; Yeh, Y.C.; Rana, S.; Jiang, Y.; Guo, L.; Rotello, V.M. Differentiation of cancer cell type and phenotype using quantum dot-gold nanoparticle sensor arrays. *Cancer Lett.* **2013**, *334*, 196–201. [CrossRef]
101. Wei, X.C.; Chen, Z.B.; Tan, L.L.; Lou, T.H.; Zhao, Y. DNA-Catalytically Active Gold Nanoparticle Conjugates-Based Colorimetric Multidimensional Sensor Array for Protein Discrimination. *Anal. Chem.* **2017**, *89*, 556–559. [CrossRef]

102. Xu, Y.; Qian, C.; Yu, Y.; Yang, S.J.; Shi, F.F.; Gao, X.; Liu, Y.H.; Huang, H.; Stewart, C.; Li, F.; et al. Machine Learning-Assisted Nanoenzyme/Bioenzyme Dual-Coupled Array for Rapid Detection of Amyloids. *Anal. Chem.* **2023**, *95*, 4605–4611. [CrossRef]
103. Abbasi-Moayed, S.; Orouji, A.; Hormozi-Nezhad, M.R. Multiplex Detection of Biogenic Amines for Meat Freshness Monitoring Using Nanoplasmonic Colorimetric Sensor Array. *Biosensors* **2023**, *13*, 803. [CrossRef]
104. Chen, X.L.; Liang, Y. Colorimetric sensing strategy for multiplexed detection of proteins based on three DNA-gold nanoparticle conjugates sensors. *Sens. Actuators B-Chem.* **2021**, *329*, 129202. [CrossRef]
105. Wong, S.F.; Khor, S.M. Differential colorimetric nanobiosensor array as bioelectronic tongue for discrimination and quantitation of multiple foodborne carcinogens. *Food Chem.* **2021**, *357*, 129801. [CrossRef]
106. Li, Y.N.; Liu, Q.Y.; Chen, Z.B. A colorimetric sensor array for detection and discrimination of antioxidants based on Ag nanoshell deposition on gold nanoparticle surfaces. *Analyst* **2019**, *144*, 6276–6282. [CrossRef]
107. Yang, J.Y.; Yang, T.; Wang, X.Y.; Wang, Y.T.; Liu, M.X.; Chen, M.L.; Yu, Y.L.; Wang, J.H. A Novel Three-Dimensional Nanosensing Array for the Discrimination of Sulfur-Containing Species and Sulfur Bacteria. *Anal. Chem.* **2019**, *91*, 6012–6018. [CrossRef]
108. Li, D.Y.; Dong, Y.H.; Li, B.Y.; Wu, Y.Y.; Wang, K.; Zhang, S.C. Colorimetric sensor array with unmodified noble metal nanoparticles for naked-eye detection of proteins and bacteria. *Analyst* **2015**, *140*, 7672–7677. [CrossRef]
109. Jia, F.; Huang, J.; Wei, W.; Chen, Z.; Zhou, Q. Visual sensing of flavonoids based on varying degrees of gold nanoparticle aggregation via linear discriminant analysis. *Sens. Actuators B Chem.* **2021**, *348*, 130685. [CrossRef]
110. Leng, Y.; Cheng, J.; Liu, C.; Wang, D.; Lu, Z.; Ma, C.; Zhang, M.; Dong, Y.; Xing, X.; Yao, L.; et al. A rapid reduction of Au(I→0) strategy for the colorimetric detection and discrimination of proteins. *Microchim. Acta* **2021**, *188*, 249. [CrossRef]
111. Qiang, H.; Wei, X.; Liu, Q.; Chen, Z. Iodide-Responsive Cu–Au Nanoparticle-Based Colorimetric Sensor Array for Protein Discrimination. *ACS Sustain. Chem. Eng.* **2018**, *6*, 15720–15726. [CrossRef]
112. Xi, H.; He, W.; Liu, Q.; Chen, Z. Protein Discrimination Using a Colorimetric Sensor Array Based on Gold Nanoparticle Aggregation Induced by Cationic Polymer. *ACS Sustain. Chem. Eng.* **2018**, *6*, 10751–10757. [CrossRef]
113. Das Saha, N.; Sasmal, R.; Meethal, S.K.; Vats, S.; Gopinathan, P.V.; Jash, O.; Manjithaya, R.; Gagey-Eilstein, N.; Agasti, S.S. Multichannel DNA Sensor Array Fingerprints Cell States and Identifies Pharmacological Effectors of Catabolic Processes. *ACS Sens.* **2019**, *4*, 3124–3132. [CrossRef]
114. Mayilo, S.; Kloster, M.A.; Wunderlich, M.; Lutich, A.; Klar, T.A.; Nichtl, A.; Kürzinger, K.; Stefani, F.D.; Feldmann, J. Long-Range Fluorescence Quenching by Gold Nanoparticles in a Sandwich Immunoassay for Cardiac Troponin T. *Nano Lett.* **2009**, *9*, 4558–4563. [CrossRef]
115. Hung, S.Y.; Shih, Y.C.; Tseng, W.L. Tween 20-stabilized gold nanoparticles combined with adenosine triphosphate-BODIPY conjugates for the fluorescence detection of adenosine with more than 1000-fold selectivity. *Anal. Chim. Acta* **2015**, *857*, 64–70. [CrossRef]
116. Lu, S.S.; Wang, S.; Chen, C.X.; Sun, J.; Yang, X.R. Enzyme-free aptamer/AuNPs-based fluorometric and colorimetric dual-mode detection for ATP. *Sens. Actuators B-Chem.* **2018**, *265*, 67–74. [CrossRef]
117. Lv, L.; Jin, Y.D.; Kang, X.J.; Zhao, Y.Y.; Cui, C.B.; Guo, Z.J. PVP-coated gold nanoparticles for the selective determination of ochratoxin A via quenching fluorescence of the free aptamer. *Food Chem.* **2018**, *249*, 45–50. [CrossRef]
118. Saad, S.M.; Abdullah, J.; Abd Rashid, S.; Fen, Y.W.; Salam, F.; Yih, L.H. A fluorescence quenching based gene assay for *Escherichia coli* O157:H7 using graphene quantum dots and gold nanoparticles. *Microchim. Acta* **2019**, *186*, 804. [CrossRef]
119. Chen, Y.S.; Chen, Z.W.; Yuan, Y.W.; Chen, K.C.; Liu, C.P. Fluorescence Quenchers Manipulate the Peroxidase-like Activity of Gold-Based Nanomaterials. *ACS Omega* **2020**, *5*, 24487–24494. [CrossRef]
120. Saad, S.M.; Abdullah, J.; Abd Rashid, S.; Fen, Y.W.; Salam, F.; Yih, L.H. A carbon dots based fluorescence sensing for the determination of *Escherichia coli* O157:H7. *Measurement* **2020**, *160*, 107845. [CrossRef]
121. Jiang, W.; Wang, Z.; Beier, R.C.; Jiang, H.; Wu, Y.; Shen, J. Simultaneous determination of 13 fluoroquinolone and 22 sulfonamide residues in milk by a dual-colorimetric enzyme-linked immunosorbent assay. *Anal. Chem.* **2013**, *85*, 1995–1999. [CrossRef]
122. Bajaj, A.; Miranda, O.R.; Phillips, R.; Kim, I.-B.; Jerry, D.J.; Bunz, U.H.F.; Rotello, V.M. Array-Based Sensing of Normal, Cancerous, and Metastatic Cells Using Conjugated Fluorescent Polymers. *J. Am. Chem. Soc.* **2010**, *132*, 1018–1022. [CrossRef]
123. Kumar, V.V.; Anthony, S.P. Highly selective colorimetric sensing of Hg^{2+} ions by label free AuNPs in aqueous medium across wide pH range. *Sens. Actuators B-Chem.* **2016**, *225*, 413–419. [CrossRef]
124. Liu, X.H.; Wang, Y.; Chen, P.; McCadden, A.; Palaniappan, A.; Zhang, J.L.; Liedberg, B. Peptide Functionalized Gold Nanoparticles with Optimized Particle Size and Concentration for Colorimetric Assay Development: Detection of Cardiac Troponin I. *ACS Sens.* **2016**, *1*, 1416–1422. [CrossRef]
125. Yang, J.J.; Feng, L.J.; Liu, J.; Li, S.; Li, N.; Zhang, X.F. DNA-mediated charge neutralization of AuNPs for colorimetric sensing of Hg^{2+} in environmental waters and cosmetics. *Sens. Actuators B-Chem.* **2024**, *398*, 134697. [CrossRef]
126. Lu, Y.; Liu, Y.; Zhang, S.; Wang, S.; Zhang, S.; Zhang, X. Aptamer-based plasmonic sensor array for discrimination of proteins and cells with the naked eye. *Anal. Chem.* **2013**, *85*, 6571–6574. [CrossRef]
127. Sun, W.; Lu, Y.; Mao, J.; Chang, N.; Yang, J.; Liu, Y. Multidimensional sensor for pattern recognition of proteins based on DNA-gold nanoparticles conjugates. *Anal. Chem.* **2015**, *87*, 3354–3359. [CrossRef]
128. Wei, X.; Wang, Y.; Zhao, Y.; Chen, Z. Colorimetric sensor array for protein discrimination based on different DNA chain length-dependent gold nanoparticles aggregation. *Biosens. Bioelectron.* **2017**, *97*, 332–337. [CrossRef]

129. Yang, X.; Li, J.; Pei, H.; Zhao, Y.; Zuo, X.; Fan, C.; Huang, Q. DNA-gold nanoparticle conjugates-based nanoplasmonic probe for specific differentiation of cell types. *Anal. Chem.* **2014**, *86*, 3227–3231. [CrossRef]
130. Gao, X.; Li, M.; Zhao, M.; Wang, X.; Wang, S.; Liu, Y. Metabolism-Triggered Colorimetric Sensor Array for Fingerprinting and Antibiotic Susceptibility Testing of Bacteria. *Anal. Chem.* **2022**, *94*, 6957–6966. [CrossRef]
131. Tang, Z.; Chen, F.; Wang, D.; Xiong, D.; Yan, S.; Liu, S.; Tang, H. Fabrication of avidin-stabilized gold nanoclusters with dual emissions and their application in biosensing. *J. Nanobiotechnology* **2022**, *20*, 306. [CrossRef]
132. Biswas, A.; Banerjee, S.; Gart, E.V.; Nagaraja, A.T.; McShane, M.J. Gold Nanocluster Containing Polymeric Microcapsules for Intracellular Ratiometric Fluorescence Biosensing. *ACS Omega* **2017**, *2*, 2499–2506. [CrossRef]
133. Niazi, S.; Khan, I.M.; Akhtar, W.; ul Haq, F.; Pasha, I.; Khan, M.K.I.; Mohsin, A.; Ahmad, S.; Zhang, Y.; Wang, Z. Aptamer functionalized gold nanoclusters as an emerging nanoprobe in biosensing, diagnostic, catalysis and bioimaging. *Talanta* **2024**, *268*, 125270. [CrossRef]
134. Zhang, Y.; Zhang, C.; Xu, C.; Wang, X.; Liu, C.; Waterhouse, G.I.N.; Wang, Y.; Yin, H. Ultrasmall Au nanoclusters for biomedical and biosensing applications: A mini-review. *Talanta* **2019**, *200*, 432–442. [CrossRef]
135. Gan, Z.B.; Lin, Y.J.; Luo, L.; Han, G.M.; Liu, W.; Liu, Z.J.; Yao, C.H.; Weng, L.H.; Liao, L.W.; Chen, J.S.; et al. Fluorescent Gold Nanoclusters with Interlocked Staples and a Fully Thiolate-Bound Kernel. *Angew. Chem.-Int. Ed.* **2016**, *55*, 11567–11571. [CrossRef] [PubMed]
136. Guo, Y.H.; Amunyela, H.; Cheng, Y.L.; Xie, Y.F.; Yu, H.; Yao, W.R.; Li, H.W.; Qian, H. Natural protein-templated fluorescent gold nanoclusters: Syntheses and applications. *Food Chem.* **2021**, *335*, 127657. [CrossRef]
137. Sonia; Komal; Kukreti, S.; Kaushik, M. Gold nanoclusters: An ultrasmall platform for multifaceted applications. *Talanta* **2021**, *234*, 122623. [CrossRef]
138. Sun, Y.F.; Zhou, Z.P.; Shu, T.; Qian, L.S.; Su, L.; Zhang, X.J. Multicolor Luminescent Gold Nanoclusters: From Structure to Biosensing and Bioimaging. *Prog. Chem.* **2021**, *33*, 179–187. [CrossRef]
139. Chen, S.Q.; Li, S.S.; Wang, Y.L.; Chen, Z.H.; Wang, H.; Zhang, X.D. Gold Nanoclusters for Tumor Diagnosis and Treatment. *Adv. Nanobiomed Res.* **2023**, *3*, 2300082. [CrossRef]
140. Ivanova, N.K.; Karpushkin, E.A.; Lopatina, L.I.; Sergeyev, V.G. DNA as a template for synthesis of fluorescent gold nanoclusters. *Mendeleev Commun.* **2023**, *33*, 346–348. [CrossRef]
141. Liu, Z.Y.; Luo, L.S.; Jin, R.C. Visible to NIR-II Photoluminescence of Atomically Precise Gold Nanoclusters. *Adv. Mater.* **2023**, *36*, e2309073. [CrossRef] [PubMed]
142. Zhou, S.C.; Gustavsson, L.; Beaune, G.; Chandra, S.; Niskanen, J.; Ruokolainen, J.; Timonen, J.V.I.; Ikkala, O.; Peng, B.; Ras, R.H.A. pH-Responsive Near-Infrared Emitting Gold Nanoclusters. *Angew. Chem.-Int. Ed.* **2023**, *62*, e202312679. [CrossRef]
143. Sun, M.; Wu, L.; Ren, H.; Chen, X.; Ouyang, J.; Na, N. Radical-Mediated Spin-Transfer on Gold Nanoclusters Driven an Unexpected Luminescence for Protein Discrimination. *Anal. Chem.* **2017**, *89*, 11183–11188. [CrossRef]
144. Xu, S.; Gao, T.; Feng, X.; Fan, X.; Liu, G.; Mao, Y.; Yu, X.; Lin, J.; Luo, X. Near infrared fluorescent dual ligand functionalized Au NCs based multidimensional sensor array for pattern recognition of multiple proteins and serum discrimination. *Biosens. Bioelectron.* **2017**, *97*, 203–207. [CrossRef]
145. Xu, S.; Li, W.; Zhao, X.; Wu, T.; Cui, Y.; Fan, X.; Wang, W.; Luo, X. Ultrahighly Efficient and Stable Fluorescent Gold Nanoclusters Coated with Screened Peptides of Unique Sequences for Effective Protein and Serum Discrimination. *Anal. Chem.* **2019**, *91*, 13947–13952. [CrossRef] [PubMed]
146. Wu, Y.; Wang, B.; Wang, K.; Yan, P. Identification of proteins and bacteria based on a metal ion–gold nanocluster sensor array. *Anal. Methods* **2018**, *10*, 3939–3944. [CrossRef]
147. Lu, H.; Lu, Q.; Sun, H.; Wang, Z.; Shi, X.; Ding, Y.; Ran, X.; Pei, J.; Pan, Y.; Zhang, Q. ROS-Responsive Fluorescent Sensor Array for Precise Diagnosis of Cancer via pH-Controlled Multicolor Gold Nanoclusters. *ACS Appl. Mater. Interfaces* **2023**, *15*, 38381–38390. [CrossRef] [PubMed]
148. Xiao, Y.; Cheng, P.; Zhu, X.; Xu, M.; Liu, M.; Li, H.; Zhang, Y.; Yao, S. Antimicrobial Agent Functional Gold Nanocluster-Mediated Multichannel Sensor Array for Bacteria Sensing. *Langmuir* **2024**, *40*, 2369–2376. [CrossRef]
149. Li, X.; Kong, H.; Mout, R.; Saha, K.; Moyano, D.F.; Robinson, S.M.; Rana, S.; Zhang, X.; Riley, M.A.; Rotello, V.M. Rapid Identification of Bacterial Biofilms and Biofilm Wound Models Using a Multichannel Nanosensor. *ACS Nano* **2014**, *8*, 12014–12019. [CrossRef] [PubMed]
150. Tomita, S.; Matsuda, A.; Nishinami, S.; Kurita, R.; Shiraki, K. One-Step Identification of Antibody Degradation Pathways Using Fluorescence Signatures Generated by Cross-Reactive DNA-Based Arrays. *Anal. Chem.* **2017**, *89*, 7818–7822. [CrossRef] [PubMed]
151. Xu, L.Q.; Wang, L.; Zhang, B.; Lim, C.H.; Chen, Y.; Neoh, K.-G.; Kang, E.-T.; Fu, G.D. Functionalization of reduced graphene oxide nanosheets via stacking interactions with the fluorescent and water-soluble perylene bisimide-containing polymers. *Polymer* **2011**, *52*, 2376–2383. [CrossRef]
152. Suguna, S.; David, C.I.; Prabhu, J.; Nandhakumar, R. Functionalized graphene oxide materials for the fluorometric sensing of various analytes: A mini review. *Mater. Adv.* **2021**, *2*, 6197–6212. [CrossRef]
153. Tao, Y.; Auguste, D.T. Array-based identification of triple-negative breast cancer cells using fluorescent nanodot-graphene oxide complexes. *Biosens. Bioelectron.* **2016**, *81*, 431–437. [CrossRef]
154. Hizir, M.S.; Robertson, N.M.; Balcioglu, M.; Alp, E.; Rana, M.; Yigit, M.V. Universal sensor array for highly selective system identification using two-dimensional nanoparticles. *Chem. Sci.* **2017**, *8*, 5735–5745. [CrossRef]

155. Park, J.S.; Goo, N.I.; Kim, D.E. Mechanism of DNA Adsorption and Desorption on Graphene Oxide. *Langmuir* **2014**, *30*, 12587–12595. [CrossRef]
156. Lu, C.; Huang, P.J.J.; Liu, B.W.; Ying, Y.B.; Liu, J.W. Comparison of Graphene Oxide and Reduced Graphene Oxide for DNA Adsorption and Sensing. *Langmuir* **2016**, *32*, 10776–10783. [CrossRef]
157. Lu, C.; Liu, Y.B.; Ying, Y.B.; Liu, J.W. Comparison of MoS_2, WS_2, and Graphene Oxide for DNA Adsorption and Sensing. *Langmuir* **2017**, *33*, 630–637. [CrossRef]
158. Lu, Z.J.; Wang, P.; Xiong, W.W.; Qi, B.P.; Shi, R.J.; Xiang, D.S.; Zhai, K. Simultaneous detection of mercury (II), lead (II) and silver (I) based on fluorescently labelled aptamer probes and graphene oxide. *Environ. Technol.* **2021**, *42*, 3065–3072. [CrossRef]
159. Morales-Narvaez, E.; Merkoci, A. Graphene Oxide as an Optical Biosensing Platform: A Progress Report. *Adv. Mater.* **2019**, *31*, e1805043. [CrossRef]
160. Yu, W.; Sisi, L.; Haiyan, Y.; Jie, L. Progress in the functional modification of graphene/graphene oxide: A review. *RSC Adv.* **2020**, *10*, 15328–15345. [CrossRef]
161. Behera, P.; De, M. Nano-Graphene Oxide Based Multichannel Sensor Arrays towards Sensing of Protein Mixtures. *Chem.-Asian J.* **2019**, *14*, 553–560. [CrossRef]
162. Zhu, Q.Y.; Zhang, F.R.; Du, Y.; Zhang, X.X.; Lu, J.Y.; Yao, Q.F.; Huang, W.T.; Ding, X.Z.; Xia, L.Q. Graphene-Based Steganographically Aptasensing System for Information Computing, Encryption and Hiding, Fluorescence Sensing and in Vivo Imaging of Fish Pathogens. *ACS Appl. Mater. Interfaces* **2019**, *11*, 8904–8914. [CrossRef]
163. Tomita, S.; Ishihara, S.; Kurita, R. A Multi-Fluorescent DNA/Graphene Oxide Conjugate Sensor for Signature-Based Protein Discrimination. *Sensors* **2017**, *17*, 2194. [CrossRef]
164. Lin, M.; Li, W.S.; Wang, Y.L.; Yang, X.H.; Wang, K.M.; Wang, Q.; Wang, P.; Chang, Y.J.; Tan, Y.Y. Discrimination of hemoglobins with subtle differences using an aptamer based sensing array. *Chem. Commun.* **2015**, *51*, 8304–8306. [CrossRef]
165. Fu, M.Q.; Wang, X.C.; Dou, W.T.; Chen, G.R.; James, T.D.; Zhou, D.M.; He, X.P. Supramolecular fluorogenic peptide sensor array based on graphene oxide for the differential sensing of ebola virus. *Chem. Commun.* **2020**, *56*, 5735–5738. [CrossRef] [PubMed]
166. Nandu, N.; Smith, C.W.; Uyar, T.B.; Chen, Y.S.; Kachwala, M.J.; He, M.H.; Yigit, M.V. Machine-Learning Single-Stranded DNA Nanoparticles for Bacterial Analysis. *ACS Applied. Nano Mater.* **2020**, *3*, 11709–11714. [CrossRef] [PubMed]
167. Shen, J.; Hu, R.; Zhou, T.; Wang, Z.; Zhang, Y.; Li, S.; Gui, C.; Jiang, M.; Qin, A.; Tang, B.Z. Fluorescent Sensor Array for Highly Efficient Microbial Lysate Identification through Competitive Interactions. *ACS Sens.* **2018**, *3*, 2218–2222. [CrossRef]
168. Martynenko, I.V.; Litvin, A.P.; Purcell-Milton, F.; Baranov, A.V.; Fedorov, A.V.; Gun'ko, Y.K. Application of semiconductor quantum dots in bioimaging and biosensing. *J. Mater. Chem. B* **2017**, *5*, 6701–6727. [CrossRef]
169. Freire, R.M.; Le, N.D.B.; Jiang, Z.W.; Kim, C.S.; Rotello, V.M.; Fechine, P.B.A. NH_2-rich Carbon Quantum Dots: A protein-responsive probe for detection and identification. *Sens. Actuators B-Chem.* **2018**, *255*, 2725–2732. [CrossRef]
170. Zheng, L.B.; Qi, P.; Zhang, D. Identification of bacteria by a fluorescence sensor array based on three kinds of receptors functionalized carbon dots. *Sens. Actuators B-Chem.* **2019**, *286*, 206–213. [CrossRef]
171. Wang, M.; Ye, H.; You, L.; Chen, X. A Supramolecular Sensor Array Using Lanthanide-Doped Nanoparticles for Sensitive Detection of Glyphosate and Proteins. *ACS Appl. Mater. Interfaces* **2016**, *8*, 574–581. [CrossRef]
172. Wu, Y.; Tan, Y.; Wu, J.; Chen, S.; Chen, Y.Z.; Zhou, X.; Jiang, Y.; Tan, C. Fluorescence array-based sensing of metal ions using conjugated polyelectrolytes. *ACS Appl. Mater. Interfaces* **2015**, *7*, 6882–6888. [CrossRef]
173. Wu, P.; Miao, L.N.; Wang, H.F.; Shao, X.G.; Yan, X.P. A multidimensional sensing device for the discrimination of proteins based on manganese-doped ZnS quantum dots. *Angew. Chem. Int. Ed. Engl.* **2011**, *50*, 8118–8121. [CrossRef]
174. Wang, K.; Dong, Y.; Li, B.; Li, D.; Zhang, S.; Wu, Y. Differentiation of proteins and cancer cells using metal oxide and metal nanoparticles-quantum dots sensor array. *Sens. Actuators B Chem.* **2017**, *250*, 69–75. [CrossRef]
175. Chen, S.; Wei, L.; Chen, X.W.; Wang, J.H. Suspension Array of Ionic Liquid or Ionic Liquid-Quantum Dots Conjugates for the Discrimination of Proteins and Bacteria. *Anal. Chem.* **2015**, *87*, 10902–10909. [CrossRef]
176. Lu, Z.; Lu, N.; Xiao, Y.; Zhang, Y.; Tang, Z.; Zhang, M. Metal-Nanoparticle-Supported Nanozyme-Based Colorimetric Sensor Array for Precise Identification of Proteins and Oral Bacteria. *ACS Appl. Mater. Interfaces* **2022**, *14*, 11156–11166. [CrossRef]
177. Zhang, L.; Qi, Z.; Yang, Y.; Lu, N.; Tang, Z. Enhanced "Electronic Tongue" for Dental Bacterial Discrimination and Elimination Based on a DNA-Encoded Nanozyme Sensor Array. *ACS Appl. Mater. Interfaces* **2024**, *16*, 11228–11238. [CrossRef]
178. Chen, H.; Guo, S.; Zhuang, Z.; Ouyang, S.; Lin, P.; Zheng, Z.; You, Y.; Zhou, X.; Li, Y.; Lu, J.; et al. Intelligent Identification of Cerebrospinal Fluid for the Diagnosis of Parkinson's Disease. *Anal. Chem.* **2024**, *96*, 2534–2542. [CrossRef]
179. Yang, J.; Lu, S.; Chen, B.; Hu, F.; Li, C.; Guo, C. Machine learning-assisted optical nano-sensor arrays in microorganism analysis. *TrAC Trends Anal. Chem.* **2023**, *159*, 116945. [CrossRef]
180. Behera, P.; Singh, K.K.; Pandit, S.; Saha, D.; Saini, D.K.; De, M. Machine Learning-Assisted Array-Based Detection of Proteins in Serum Using Functionalized MoS_2 Nanosheets and Green Fluorescent Protein Conjugates. *ACS Appl. Nano Mater.* **2021**, *4*, 3843–3851. [CrossRef]
181. Yan, P.; Ding, Z.; Li, X.; Dong, Y.; Fu, T.; Wu, Y. Colorimetric Sensor Array Based on Wulff-Type Boronate Functionalized AgNPs at Various pH for Bacteria Identification. *Anal. Chem.* **2019**, *91*, 12134–12137. [CrossRef]
182. Yang, H.M.; Jie, X.; Wang, L.; Zhang, Y.; Wang, M.; Wei, W.L. An array consisting of glycosylated quantum dots conjugated to MoS_2 nanosheets for fluorometric identification and quantitation of lectins and bacteria. *Microchimica Acta* **2018**, *185*, 512. [CrossRef]

183. Behera, P.; Kumar Singh, K.; Kumar Saini, D.; De, M. Rapid Discrimination of Bacterial Drug Resistivity by Array-Based Cross-Validation Using 2D MoS$_2$. *Chem.—A Eur. J.* **2022**, *28*, e202201386. [CrossRef]
184. He, Y.; He, X.; Liu, X.; Gao, L.; Cui, H. Dynamically tunable chemiluminescence of luminol-functionalized silver nanoparticles and its application to protein sensing arrays. *Anal. Chem.* **2014**, *86*, 12166–12171. [CrossRef]
185. Pu, F.; Ran, X.; Guan, M.; Huang, Y.; Ren, J.; Qu, X. Biomolecule-templated photochemical synthesis of silver nanoparticles: Multiple readouts of localized surface plasmon resonance for pattern recognition. *Nano Res.* **2017**, *11*, 3213–3221. [CrossRef]
186. Ran, X.; Pu, F.; Ren, J.; Qu, X. A CuS-based chemical tongue chip for pattern recognition of proteins and antibiotic-resistant bacteria. *Chem. Commun.* **2015**, *51*, 2675–2678. [CrossRef] [PubMed]
187. Wan, Y.; Sun, Y.; Qi, P.; Wang, P.; Zhang, D. Quaternized magnetic nanoparticles-fluorescent polymer system for detection and identification of bacteria. *Biosens. Bioelectron.* **2014**, *55*, 289–293. [CrossRef] [PubMed]
188. Sabela, M.; Balme, S.; Bechelany, M.; Janot, J.M.; Bisetty, K. A Review of Gold and Silver Nanoparticle-Based Colorimetric Sensing Assays. *Adv. Eng. Mater.* **2017**, *19*, 1700270. [CrossRef]
189. Geng, Y.; Peveler, W.J.; Rotello, V.M. Array-based "Chemical Nose" Sensing in Diagnostics and Drug Discovery. *Angew. Chem.-Int. Ed.* **2019**, *58*, 5190–5200. [CrossRef]
190. Medrano-Lopez, J.A.; Villalpando, I.; Salazar, M.I.; Torres-Torres, C. Hierarchical Nanobiosensors at the End of the SARS-CoV-2 Pandemic. *Biosensors* **2024**, *14*, 108. [CrossRef]
191. Behera, P.; De, M.M. Surface-Engineered Nanomaterials for Optical Array Based Sensing. *Chempluschem* **2023**, e202300610. [CrossRef]

Disclaimer/Publisher's Note: The statements, opinions and data contained in all publications are solely those of the individual author(s) and contributor(s) and not of MDPI and/or the editor(s). MDPI and/or the editor(s) disclaim responsibility for any injury to people or property resulting from any ideas, methods, instructions or products referred to in the content.

Review

PCR Independent Strategy-Based Biosensors for RNA Detection

Xinran Li [1,†], Haoqian Wang [2,†], Xin Qi [1], Yi Ji [3], Fukai Li [1], Xiaoyun Chen [3,*], Kai Li [1,*] and Liang Li [1,*]

1. Institute of Quality Standard and Testing Technology for Agro-Products, Chinese Academy of Agricultural Sciences, Beijing 100081, China; lllxinran_05@163.com (X.L.); qixin_0908@163.com (X.Q.); lifukai@caas.cn (F.L.)
2. Development Center of Science and Technology, Ministry of Agriculture and Rural Affairs, Beijing 100176, China; wanghaoqian@agri.gov.cn
3. State Key Laboratory for Managing Biotic and Chemical Threats to the Quality and Safety of Agro-Products, Zhejiang Academy of Agricultural Sciences, Hangzhou 310021, China; jymemory12138@163.com
* Correspondence: xiaoyunchen_2016@163.com (X.C.); likaij@163.com (K.L.); liliang@caas.cn (L.L.)
† These authors contributed equally to this work.

Abstract: RNA is an important information and functional molecule. It can respond to the regulation of life processes and is also a key molecule in gene expression and regulation. Therefore, RNA detection technology has been widely used in many fields, especially in disease diagnosis, medical research, genetic engineering and other fields. However, the current RT-qPCR for RNA detection is complex, costly and requires the support of professional technicians, resulting in it not having great potential for rapid application in the field. PCR-free techniques are the most attractive alternative. They are a low-cost, simple operation method and do not require the support of large instruments, providing a new concept for the development of new RNA detection methods. This article reviews current PCR-free methods, overviews reported RNA biosensors based on electrochemistry, SPR, microfluidics, nanomaterials and CRISPR, and discusses their challenges and future research prospects in RNA detection.

Keywords: RNA detection; PCR-free; biosensor; nanomaterial

Citation: Li, X.; Wang, H.; Qi, X.; Ji, Y.; Li, F.; Chen, X.; Li, K.; Li, L. PCR Independent Strategy-Based Biosensors for RNA Detection. *Biosensors* **2024**, *14*, 200. https://doi.org/10.3390/bios14040200

Received: 13 March 2024
Revised: 11 April 2024
Accepted: 15 April 2024
Published: 18 April 2024

Copyright: © 2024 by the authors. Licensee MDPI, Basel, Switzerland. This article is an open access article distributed under the terms and conditions of the Creative Commons Attribution (CC BY) license (https://creativecommons.org/licenses/by/4.0/).

1. Introduction

Ribonucleic acid (RNA) is made up of phosphoric acid, ribose and base. It is usually found in biological cells as well as some viruses and viroids, and a small number of viruses are based on RNA as genetic material. RNA plays an essential role in the cells and is involved in biological processes such as protein synthesis, the regulation of gene expression and the transmission of genetic information [1–3]. Because RNA has many functions and meanings in biology, the detection of RNA has become particularly important. RNA detection has important applications in various fields. For example, RNA detection technology can be used in tumor diagnosis and viral infection, and the occurrence and prognosis of some diseases can be predicted by detection methods such as microRNAs (miRNAs). At the same time, RNA detection technology also plays a vital role in medical research, such as RNA sequence analysis, which can be used to study the mechanism of gene expression and transcription regulation. In addition, RNA modifications can also be detected by RNA detection techniques [4,5].

At present, the gold standard for RNA detection is still reverse transcription quantitative polymerase chain reaction (RT-qPCR). Although this method is very reliable and reasonably analytical, it requires the support of technicians and expensive instruments [6]. These problems make the nucleic acid testing technology unsuitable for its integration into miniaturized devices for clinical applications. Therefore, in order to avoid dependence on professional technicians and instruments in the process of nucleic acid testing, developing new diagnostic methods to achieve accurate, rapid and portable nucleic acid testing and quantification is urgently needed.

Among them, PCR-free methods are the most attractive solutions because they can perform molecular detection without the complexity of PCR, thereby reducing the cost of RNA detection and improving the application of nucleic acid testing [7]. Although the PCR-free-based nucleic acid testing methods address the limitations of traditional PCR methods, such as high cost, cumbersome operation and large instrument support, some limitations still need to be addressed with the transition from conventional diagnostic laboratories to portable bedside devices. So far, most of the reported reviews of RNA detection focus on direct detection without amplification [8–10], but there is a certain lack of sensitivity in the direct detection of RNA. Compared with direct detection without amplification, readers pay more attention to detection sensitivity. With the researchers' continuous efforts, the limitations of traditional PCR methods have been solved in electrochemistry, surface plasma resonance (SPR), microfluidics, nanomaterials and CRISPR, and there have been successful cases; we will review these aspects (Figure 1).

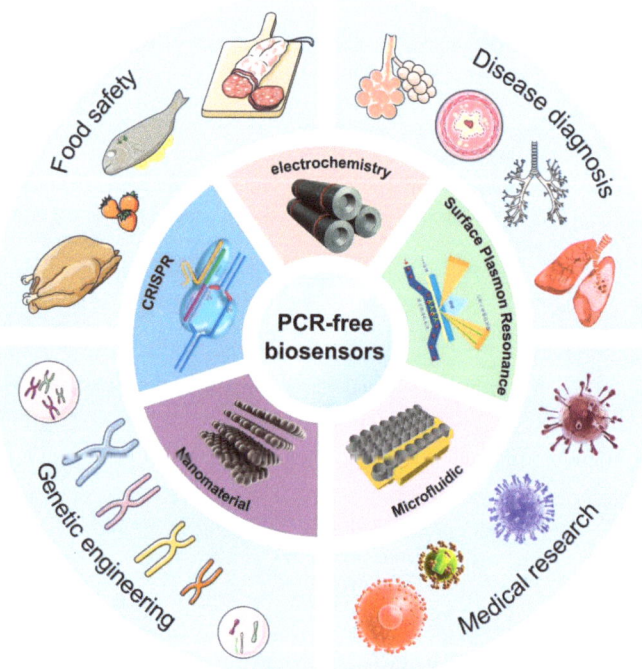

Figure 1. PCR-free biosensors for RNA detection. This figure provides an overview of RNA biosensors based on electrochemistry, SPR, microfluidics, nanomaterials and CRISPR, and applications in different fields such as disease diagnosis, medical research, genetic engineering, and food safety.

2. Electrochemical-Based RNA Biosensors

Recently, electrochemical methods have made significant advances in the detection of clinically relevant RNA [11,12]. Most of these methods are based on the hybridization of the target RNA sequence to complementary probes (mainly DNA oligonucleotides) bound on the electrode surface. The hybridization of RNA with the probe produces a measurable electrochemical signal. Here, signal transduction depends on various factors, including the inherent electrical activity of the nucleobase, the presence of redox indicators (e.g., ferrocene, methylene blue), covalently bound redox markers (e.g., nanoparticles) or reporter enzymes (e.g., phosphatase, peroxidase) [13]. Finally, RNA is detected mainly by voltammetry, amperometry and impedance methods [14]. Electrochemical methods are promising for RNA detection due to their high sensitivity, rapid detection, cost-effectiveness and compatibility with small portability.

2.1. Electrochemical Biosensors for the Ultra-Sensitive Detection of RNA

One of the earlier RNA electrochemical detection methods was developed by Kosuke Mukumoto [15]. The method used ferrocenylcarbodiimide (I) to directly label messenger RNA (mRNA), which was coupled to an electrode immobilized with a DNA probe. The observed peak charge had a good correlation with the concentration of mRNA, as measured by Osteryoung Square Wave Voltammetry (SWV), which had successfully achieved electrochemical detection of labeled mRNAs, with a limit of detection (LOD) at the sub-nanogram level.

However, one of the biggest challenges faced by RNA biomarker detection in clinical applications is the simultaneous screening of minimal amounts of readily available RNA biomarkers in complex heterogeneous samples that may contain many non-specific targets. To address this challenge, several new approaches have been developed for high-sensitivity analysis of RNA by amplifying or using novel nanostructured electrochemical sensors. For example, Yang et al. [16] developed a triple signal amplification strategy technique combining gold nanoparticles (AuNPs), reverse transcription loop-mediated isothermal amplification (RT-LAMP), and a high-affinity biotin-affinity system to detect HPV E6/E7 mRNA. This novel signal amplification strategy exhibited a 0.1 fM (~100 copies) detection limit for HPV0 E08/E100 mRNA detection (Figure 2A). Thanyarat Chaibun et al. [17] designed a multiplex rolling circle amplification (RCA)-based electrochemical biosensor for rapid detection of nucleocapsid phosphoprotein (N gene) and Spike protein (S gene) of SARS-CoV-2 in clinical samples. Combining the high amplification capacity of RCAs with the sensitivity of electrochemical detection methods, viral N or S genes as low as 1 copy/μL could be detected within two hours, resulting in detection performance comparable to RT-qPCR in clinical samples. Zhang et al. [18] proposed a novel polymerase-assisted cyclic electrochemiluminescent aptamer biosensor for ultrasensitive leukemia marker gene mRNA detection. Combining polymerase-assisted signal amplification with AuNPs, the detection limit was 4.3×10^{-17} mol/L, which led to a much higher detection sensitivity. Peng et al. [19] developed an electrochemical biosensor that combined the signal amplification ability of catalytic hairpin assembly (CHA) [20–23] and terminal transferase (TDT) [23]. The electrochemical signal was significantly amplified by the electrostatic adsorption of a large number of negatively charged long single stranded DNA (ssDNA) and a large number of positively charged $Ru(NH_3)_6^{3+}$, and the detection limit was as low as 26 fM. At the same time, it has been applied in complex matrices and highly stable clinical patient samples, showing great clinical application prospects (Figure 2B). Recently, an apurinic/apyrimidinic endonuclease 1 (APE1) mediated target-responsive Structure Switching Electrochemical (SSE) biosensor was developed by the Li' group for Strawberry Mottle Virus (SMoV) RNA detection. The essence of the proposed SSE biosensor relied on the structure switching that caused the position conversion of the AP site within dsDNA and ssDNA. They used an SSE biosensor to detect target RNA, achieving a limit detection at the fM level, and successfully verified its performance in detecting SMoV in strawberry leaf-like varieties [24].

Most conventional electrochemical strategies for targeting nucleotides face tedious interfacial manipulation and washing procedures, as well as stringent reaction conditions for tool enzymes, thus limiting their potential applications. To address this problem, a series of enzyme-free electrochemical biosensors has been developed. For example, Cheng et al. [25] and Zhao et al. [26] proposed a non-enzymatic, ultrasensitive electrochemical biosensor using a hybridization chain reaction (HCR) strategy for signal amplification. For sensitive signal amplification and highly specific detection of target mRNA, ideal sensitivities with detection limits of 3 fM and 30 fM were achieved, respectively. Atie Roohizadeh et al. [27] developed an ultrasensitive label-free nano biosensor for the detection of hepatitis C virus (HCV) RNA without target denaturation. Copper oxide (CuO) and AuNPs were utilized to increase the electron transfer conductivity and reaction kinetics and improve the biosensor conductance; this strategy achieved a low detection limit of 1 fM. Emily Kerr et al. [28] studied a sensitive, rapid and portable electrochemiluminescence (ECL)-based biosensor for detecting miRNA-21. The biosensor combined turned-on ECL molecular beacons (MBs) with magnetic bead-based extraction of miRNA target sequences without the need for complex signal amplification strategies using enzymes or hairpin probes, resulting in a limit of detection up to 500 amol, which could be easily applied to point-of-care (POC) applications [29]. Overall, these methods avoid the steps of mostly enzymatic amplification of target RNA and address the problems of sample manipulation, amplification bias and longer detection time caused by the enzymatic amplification step.

2.2. Electrochemical Biosensors for the Rapid Detection of RNA

In addition, based on the ultra-sensitivity to RNA electrochemical detection in pursuit of rapid detection, researchers have combined nanomaterials with simple electrical readout methods. For example, Maha Alafeef et al. [30] invented a fast (less than 2 min), low-cost, quantitative paper-based electrochemical biosensor chip using AuNPs covered with highly specific antisense oligonucleotides (ssDNA) targeting the viral N gene. The device, which imparted a sensing probe to a paper-based electrochemical platform to generate nucleic acid tests, was relatively portable and fast, and its readings could be recorded with a simple handheld reader to enable digital detection of SARS-CoV-2 genetic material (Figure 2C). Ye et al. [31] designed a rapid and sensitive detection method of RNA using composite screen-printed carbon electrodes (SPCEs) modified with multi-walled carbon nanotubes (MWNTs). MWNTs displayed the catalytic properties of direct electrochemical oxidation of the adenine residues of RNA, resulting in indicator-free detection of RNA concentration. Within 5 min, the proposed method allowed for the rapid detection of yeast transfer RNA (tRNA) ranging from 8.2 $\mu g\ mL^{-1}$ to 4.1 $mg\ mL^{-1}$.

2.3. Electrochemical Biosensors in RNA POCT

Moreover, in order to solve the cumbersome biosensor manufacturing steps in the process of using electrochemical detection of RNA, some miniaturized and portable electrochemical biosensors have been developed successively; these generally replace traditional bulky electrodes by easy to-manufacture and miniaturized electrodes or use portable devices such as smartphones instead of traditional machines for reading. For example, Md. Nazmul Islam et al. [32] developed an amplification-free electrochemical method using screen-printed gold electrodes (SPE-Au) for the sensitive and selective detection of mRNA. Target mRNA was selectively isolated by magnetic separation and directly adsorbed onto the unmodified SPE-Au. In addition to not requiring any prior enzymatic amplification of the mRNA, it used mRNA adsorbing directly to the surface of the unmodified SPE-Au electrode, thus avoiding the cumbersome manufacturing steps of traditional biosensors. In addition, researchers have developed a simple yet fast and sensitive electrocatalytic assay for bacterial ribosomal RNA (rRNA), exploiting DNA and rRNA hybridization to the hairpin DNA probe, immobilized on the SPE-Au surface, and DNA-mediated electrocatalysis for signal amplification. The detection limit of the developed method for *E. coli* rRNA was as low as the fM level [33]. Fu et al. [34] constructed a portable and smartphone-controlled biosensing platform based on disposable organic electrochemical transistors for ultrasensitive analyses of miRNA biomarkers in less than 1 h, opening a window for low-cost mobile diagnostics of various diseases (Figure 2D). Li et al. [35] designed and prepared a portable electrochemical isothermal nucleic acid amplification test (E-INAAT) device integrating real-time monitoring and labeling-free electrochemical detection functions and a supporting plug-and-play disposable pH-sensitive potential sensor. The device, integrated with a Bluetooth module, could be implemented in smartphones for real-time monitoring of isothermal nucleic acid amplification tests (INAATs), rather than relying on heavy instruments, in the home for SARS-CoV-2 pathogens. Ultra-rapid self-inspection provides a simple, efficient and low-cost method for the development of portable, fully integrated medical detection equipment against infectious diseases.

Overall, these electrochemical biosensors are widely used in the field of nucleic acid testing due to their outstanding advantages, such as high sensitivity, simplicity of equipment, cost-effectiveness and miniaturized portability. Despite significant advances in these electrochemical methods, most of these biosensors are highly dependent on a series of optimization steps in a well-equipped laboratory setup as they are only proof-of-concept demonstrations. Several obstacles to translating these laboratory-based proof-of-concept demonstrations into real-world clinical applications exist. At this stage, our main efforts should focus on improving blocking biosensor surfaces with variously designed self-assembled monolayers or the co-immobilization of blocking molecules.

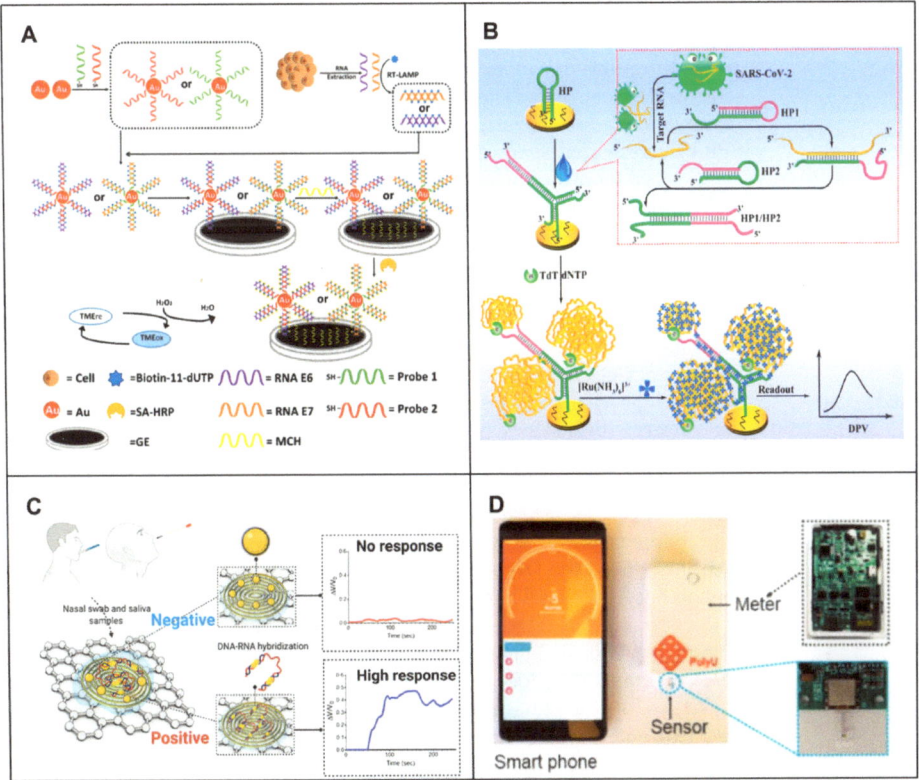

Figure 2. Principle of electrochemical-based RNA biosensors. (**A**) Schematic of the principle of an electrochemical biosensor based on a triple signal amplification strategy combining AuNPs, RT-LAMPs and a high-affinity biotin-affinity system for sensitive detection of mRNA [16]. (**B**) Schematic principle of an electrochemical biosensor based on CHA and TDT signal amplification for sensitive SARS-CoV-2 RNA detection [19]. (**C**) Schematic diagram of the working principle of the COVID-19 electrochemical biosensing platform [30]. (**D**) Schematic of the design of a portable biosensing platform based on organic thin film transistors. The OECT miRNA sensor is inserted into a portable meter and a smartphone is used to communicate with the portable meter via Bluetooth [34].

3. SPR-Based RNA Biosensors

Surface Plasmon Resonance (SPR), known as a label-free optical biosensor, is a direct detection method that utilizes a specific mode (surface plasmon) of a metal-dielectric waveguide to measure changes in the refractive index caused by biomolecular interactions occurring on the surface of the SPR biosensor. SPR is a highly sensitive method with many advantages, such as excellent reliability, selectivity and reproducibility. It has a wide range of applications in the real-time monitoring of biomolecular interactions and the detection of biological and chemical analytes based on labeled or unlabeled forms. Recently, the use of SPR biosensors for RNA detection studies has been reported.

3.1. Nanomaterial-Enhanced SPR Biosensors in RNA Detection

Based on a variety of signal amplification methods, including nanoparticle enhancement, super-sandwich assembly, streptavidin/biotin complex, antibody amplification, enzymatic reaction, triple structure formation and CHA, the limitations of SPR methods in detecting low-concentration biomolecules can be overcome, making them suitable for clinical diagnosis [36].

Among these methods, nanotechnology has enhanced the performance and sensitivity of SPR in development. Nanoparticles can provide numerous biosensing functions and applications due to their excellent biocompatibility, large specific surface area, wide structural diversity, and significant biological simulation characteristics. As early as 2008, researchers had developed a highly sensitive detection of 16S rRNA in *E. coli* using an SPR biosensor combined with AuNPs. In this method, a cationic gold nanoparticle was synthesized by using the neutral skeleton characteristics of a peptide acid probe (PAN), and the signal was amplified by ion interaction with the 16S rRNA hybridized on the SPR biosensor chip immobilized with a PNA probe. The detection limit of this method for *E. coli* rRNA was 58.2 ± 1.37 pg mL^{-1}, and *Staphylococcus aureus* could be detected without the purification of rRNA using this method [37]. Subsequently, Zhang et al. [38] constructed a highly sensitive SPR RNA biosensor using a two-dimensional metallic material called GeP$_5$ nanosheets as the sensing material. Theoretical evaluations have shown that the presence of GeP$_5$ nanosheets can significantly enhance the plasma electric field of Au films, thereby improving sensing sensitivity. The functionalization of GeP$_5$ enabled GeP to realize nanosheets with specific complementary DNA (cDNA) probes for detecting SARS-CoV-2 RNA sequences with high sensitivity down to 10 aM and excellent selectivity. Mansoureh Z. Mousavi et al. [39] demonstrated an ultrasensitive assay for the detection of mRNA biomarkers based on SPR on functionalized magnetic nanoparticles (MNPs) intercalated with gold nanoscale. The assay used MNP to capture biomarker target molecules and then introduced the target-carrying MNP into the SPR chip to hybridize with a probe immobilized on a gold nanoslit to enhance the signal, which enabled the measurement of target molecules as low as 7 fM (equivalent to 1.26×10^5 molecules) in a 30 µL sample (Figure 3A). Li et al. [40] developed a novel, sensitive and multifunctional SPR biosensor based on graphene oxide (GO)-AuNPs composites. In this biosensor, by using two layers of GO-AuNPs for signal amplification, the GO-AuNP composite was not only used as a sensing substrate but also as a signal amplification element because the AuNPs have a large specific surface area, to the extent that they can immobilize more captured DNA molecules, which amplifies the SPR response and enables the SPR biosensor to exhibit excellent sensitivity (Figure 3B). Xue et al. [41] designed an SPR biosensor based on antimony alkene two-dimensional nanomaterials to amplify the SPR signal by gold nanorods (AuNR)-conjugating ssDNA, which achieved an extremely low detection limit (amol), exceeding existing sensing methods, and quantified miRNA molecules at trace attomole levels (Figure 3C). Zhang et al. [42] presented a newly designed SPR biosensor for cytomegalovirus (CMV)-specific miRNA, utilizing the unmodified method of polyadenine [poly(A)]-Au interactions exhibiting a high affinity comparable to that of gold-sulfur (Au-S) interactions. In addition, MNPs are used for analyte separation, thus avoiding non-specific adsorption. Currently, the SPR biosensing platform has been successfully used for the multiplexed detection of CMV-related miRNA, UL22A-5p and UL112-3p, with detection limits of 112 fM for UL108A-24p and 3 fM for UL22-5p.

In addition, metal nanoparticles, such as AuNPs and AgNPs, have remarkable optical properties because the visible region has resonant surface plasma with resonant wavelengths, allowing them to display different colors depending on the wavelength, resulting in optical detection of [43] by anti-SPR biosensors. For example, G et al. [44] developed a bi-functional plasma biosensor that combined plasma photothermal (PPT) effects and local surface plasmon resonance (LSPR) sensing transduction for the clinical diagnosis of SARS-CoV-2. In this study, DNA targets were detected by nucleic acid hybridization using a two-dimensional gold nanoisland (AuNI) functionalized with cDNA receptors modified by mercapto. SARS-CoV-2's RNA-dependent RNA polymerase (RdRp) sequence LOD was approximately 0.22 ± 0.08 pM using this LSPR-based biosensor. In a study by Yasaman-Sadat Borghei et al. [45], a dual-mode sensing system based on fluorescent DNA-modified silver nanoclusters and AuNPs was presented, which allowed naked-eye visualization of miRNA and provided rapid FRET detection. Using nanoclusters and AuNPs to transfer energy between them, the team identified and quantified miRNA in biological samples without using expensive and sophisticated instruments (Figure 3D).

3.2. Signal Amplification Strategy-Enhanced SPR Biosensors in RNA Detection

In addition to the high sensitivity detection of label-free optical biosensors using nanoparticles, several other methods to amplify signals have been developed, including super-sandwich assembly, streptavidin/biotin complex, antibody amplification, enzymatic reactions, triplex structure formation and catalytic hairpin assembly. For example, Wang et al. [46] developed an enzyme-free sensitive SPR biosensor based on AuNPs and DNA super-sandwich for miRNA detection using amplification of AuNPs coupled to DNA super-sandwich with a detection limit of 21 fM. Ding et al. [47] reported an SPR biosensor for nucleic acid testing. Through signal amplification-enabled sensitive nucleic acid analysis without enzyme assistance based on DNA super-sandwich assembly and the biotin/streptavidin system, this strategy was highly sensitive, and the selective detection of miRNA could detect target miRNA as low as 30 pM in 9 min and could be applied to the determination of miRNA in real samples (Figure 3E). Li et al. [48] developed an SPR biosensor coupling mismatch CHA amplification with programmable streptavidin aptamer (SA-aptamer) for the specific and highly sensitive detection of target miRNAs. Under optimal conditions, this design strategy could detect target miRNAs as low as 1 pM and was successfully applied to the determination of spiked miRNAs in human total RNA samples (Figure 3F). Li et al. [49] have developed an ultra-sensitive multiplex SPR biosensor for the quantification of a standard-free miRNA. This approach introduced a mass transfer restriction (MTL) strategy for absolute miRNA quantification. By evaluating the factors affecting the probe/target interaction (including length and structure), the MTL and quantitative detection of the miRNA were achieved with an LOD of 500 fM without any signal amplification.

It can be seen that the SPR-based detection methods only need to capture the RNA at the sensor site, and the methods are simple and highly sensitive. However, in order to achieve a high sensitivity, some signal amplification strategies must be performed. Furthermore, SPR-based sensors have made advances in reusability and miniaturization, as they usually require only one light source and one detector as a device configuration. Combined with these elements, the social implementation of SPR detection sensors will enable workers in the field to perform rapid virus detection in minutes using a combination of smartphones and simple detection kits.

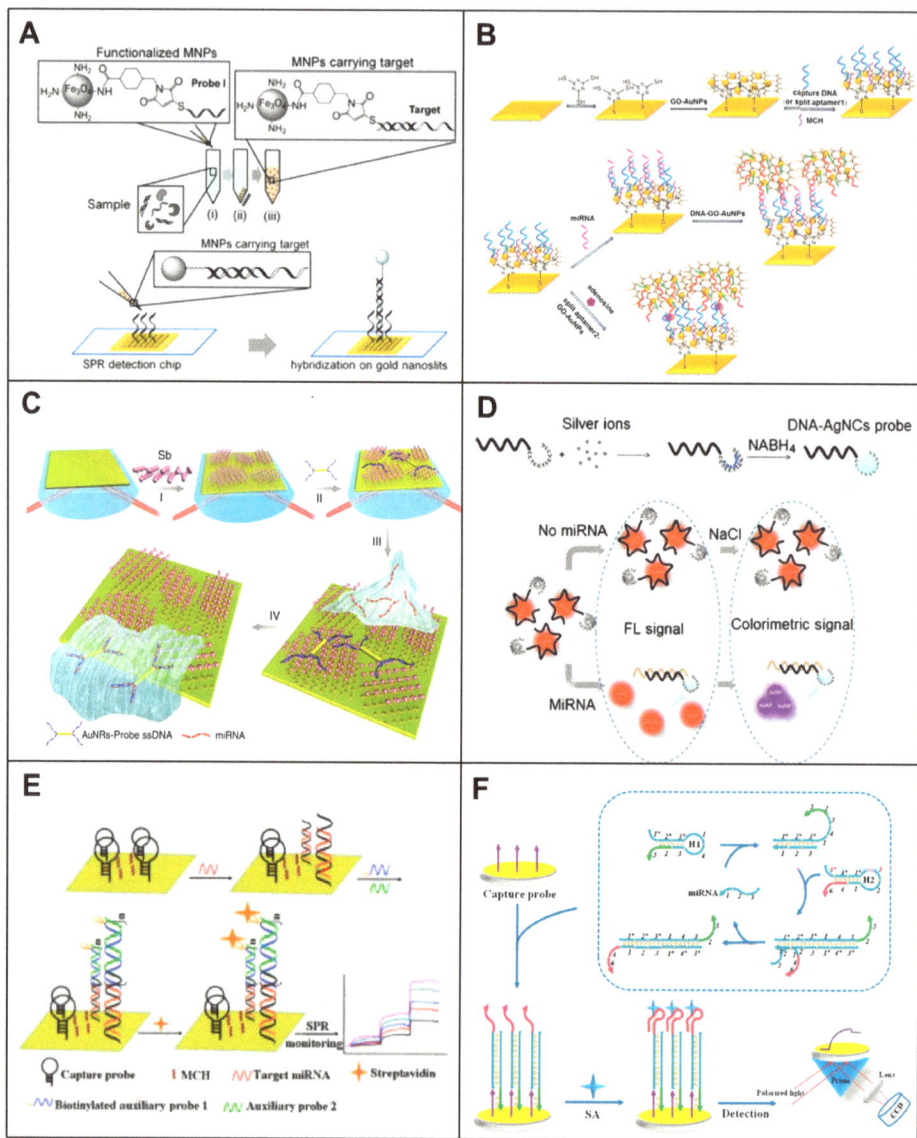

Figure 3. Principle of SPR-based RNA biosensors. (**A**) Schematic diagram of SPR biosensor based on functionalized MNP with gold nanoslit for mRNA detection [39]. (**B**) Schematic of SPR biosensor based on GO-AuNP composites. GO-AuNP composites were used as sensing substrate and signal amplification element [40]. (**C**) Schematic diagram of the strategy used to test the principle of antimonene-miRNA hybridization [41]. (**D**) Construction of DNA templated AgNC (DNA/AgNC) fluorescent probe for the detection of microRNA-155 and the schematic illustration of detection procedure by the FRET-based nano-biosensor [45]. (**E**) Schematic representation of miRNA detection assay using SPR biosensor based on DNA super-sandwich assemblies and streptavidin amplification [47]. (**F**) Schematic representation of miRNA SPR biosensor based on mismatched catalytic hairpin assembly amplification coupled with streptavidin aptamer [48].

4. Microfluidic-Based RNA Biosensors

4.1. Paper-Based Microfluidic RNA Biosensors

The application scenario for PCR-free RNA detection is mostly point-of-care testing (POCT) [50]. It is a rapidly growing field that involves using paper-based microfluidic devices as a tool to conduct POCTs, particularly since they come with many inherent advantages, including low cost, folding ability, ideal biocompatibility and disposability, and they are rapidly becoming more popular. Their analyzing capabilities are also attractive, including capillary force-driven sample transfer [51–53]. As a result of the porous structure of paper, fluid can flow through it capillarily, which is suitable for storage, mixing, flow control and the multi-analysis of reagents [54]. There is a diversity of applications and biological targets that can be studied using the type of paper, the geometry of the device and the coating of the paper used in PADs.

The lateral flow assay (LFA) is one of the main techniques used in PAD devices. Among the main components of the LFA device are a sample pad, a conjugate pad, a nitrocellulose membrane (NC) and an absorbent pad. Using strips cut into strips, these four parts are interconnected to form a one-dimensional flow [55]. After being introduced into the sample pad, the sample contains the target, which reacts with the recognition probe after passing through the conjugate pad. After passing through the NC membrane, the target-coupled probe complex passes over a control line and a test line, each containing an immobilized antibody specific to the target. Excess samples and unreacted probes are removed from the sample as it flows into the absorbent pad. Positive samples produce both the test and control lines, whereas negative samples only produce the control line. Liu's team [56] developed a sandwich-type nucleic acid hybridization reaction using DNA probes labeled with AuNPs for the detection of miRNAs using an LFA-based paper-based microfluidic system. Upon accumulating AuNPs on the test line, a colorimetric signal was produced which could be compared with the control line visually or quantitatively using a portable strip reader. This same approach was used by Zheng et al. [57] to develop a microfluidic device for the simultaneous measurement of three miRNAs, miRNA-21, miRNA-210 and miRNA-155, using NC membranes. Bhagwan S. Batule et al. [58] demonstrated a two-step strategy for extracting and detecting viral RNA from infectious diseases within one hour. A ready-to-use device for viral RNA extraction and detection was successfully prepared using paper as a substrate. The strategy used a handheld RNA extraction paper strip device to capture and elute viral RNA (e.g., Zika, Dengue and Chikungunya), followed by an RT-LAMP assay using another paper microarray device. The entire process (extraction to detection of viral RNA) was completed in less than 1 h and was simple, sensitive and cost-effective (Figure 4A). Natalia M. Rodriguez et al. [59] developed a test strip-based assay that used a rapid, isothermal, RT-LAMP assay without the need for a thermal cycler. Sample-to-result testing could be completed in as little as 45 min at the POC and had a clinically relevant viral load LOD of 106 copies/mL, a 10-fold improvement in performance over current rapid immunoassays. The method is, therefore, suitable for rapid diagnosis, providing a simple and inexpensive platform for immediate test development.

In addition to this, there are many other fabrication techniques for μPADs. For forming hydrophobic microfluidic channels on paper substrates, many methods have been proposed for fabricating PADs. They can usually be categorized into printed and non-printed methods. Two different groups [60] demonstrated the fabrication of PADs using a solid wax printer and a hot plate back in 2009. To create a hydrophobic barrier on paper, a wax pattern is printed on the surface and melted into the paper. In a similar manner, Ashok Mulchandani et al. [61] and Kattika Kaarj et al. [62] used wax paper printers to fabricate PADs targeting the sensitive detection of miRNAs and ZIKV, respectively (Figure 4D). Even today, wax printing is the most widely used printing method for PAD fabrication due to its simplicity and low cost. It was recently demonstrated that invasive fungi can be visible and quantifiable at the point of time with a hydrogel-integrated paper-based analysis device (ReaCH-PAD) with a microfluidic scale readout and CRISPR Cas12a response. A series of enzymatic reactions is used to amplify and transduce signals using DNA hydrogels combined with a series of enzymatic reactions, as well as a paper-based microfluidic chip for visual quantitative analysis [63]. Its detection targets are 18S rRNA generic conserved fragments linked with the CRISPR Cas12a system. For non-printed fabrication methods, fluid-constrained barriers are created on the substrate by means of masters, stamps or masks instead of printing a hydrophobic barrier layout onto a paper substrate. Using this method, the Whiteside group demonstrated the simultaneous detection of glucose and proteins in 2007 [52]. A hydrophobic barrier pattern was created on the paper substrate by irradiating photoresist-impregnated paper with ultraviolet (UV) rays before baking and developing (Figure 4B). The photolithography process has high resolution and dimensional stability, but it is susceptible to lateral spreading after the hot plate heating step, resulting in a loss of resolution.

4.2. Microchip-Based RNA Biosensors

As mentioned above, µPADs enable rapid, low-cost and sensitive nucleic acid testing analysis, which is promising for POC disease diagnosis and on-site molecular testing. However, because one of the challenges of paper-based devices is usually analytical sensitivity, and because they also bring disadvantages, such as cross-reactions, false positive signals and even environmental pollution, researchers have developed other more accurate and environmentally friendly device-chip microfluidic devices.

For example, Han's team [64] invented a microfluidic biochip for the rapid and ultrasensitive detection of SARS-CoV-2 by taking advantage of the specific SARS-CoV-2 RNA and probe DNA reactions in the microfluidic channel and fluorescence signaling modulation by nanomaterials, which enabled the ultrasensitive optical detection of SARS-CoV-2 RNA without the need for a molecular amplification step (Figure 4C). Qin et al. [65] proposed an NoV digital isothermal detection (NoV-DID) chip based on a gas-driven microfluidic chamber which uses a simple monolayer of polydimethylsiloxane (PDMS) for the detection of NoV GII.4. Combined with reverse transcription recombinase-assisted amplification (RT-RAA), it overcomes the limitations of the NoV detection technology and effectively reduces time, cost and dependence on instruments. In contrast to methods using reverse transcription, Zhang et al. [66] reported a new microfluidic RNA microarray (MIRC, a prototype of microchips) strategy based on the genomic replication of DNA polymerase-extended RNA primers on DNA templates with dNTP [67,68], which allowed the direct detection of RNAs without the need for reverse transcription, thus overcoming the tedious reverse transcription process. The method is characterized by rapid detection (within 20 min), high sensitivity, automation and high throughput [69,70]. In addition, the introduction of a microfluidic chip reduces reaction time, reagent usage and assay complexity.

Therefore, compared with RNA detection by PCR, highly miniaturized microfluidic technologies can integrate complex nucleic acid detection processes on a piece of paper or a chip [71,72], thus reducing the complexity of the operation and helping to build an automated and efficient diagnostic system [73–76]. Especially in the last two years, with the huge demand for POCT for COVID-19 testing, highly miniaturized microfluidic devices have provided essential tools for integrating complex nucleic acid testing processes and will increasingly become the trend and backbone of pandemic disease response.

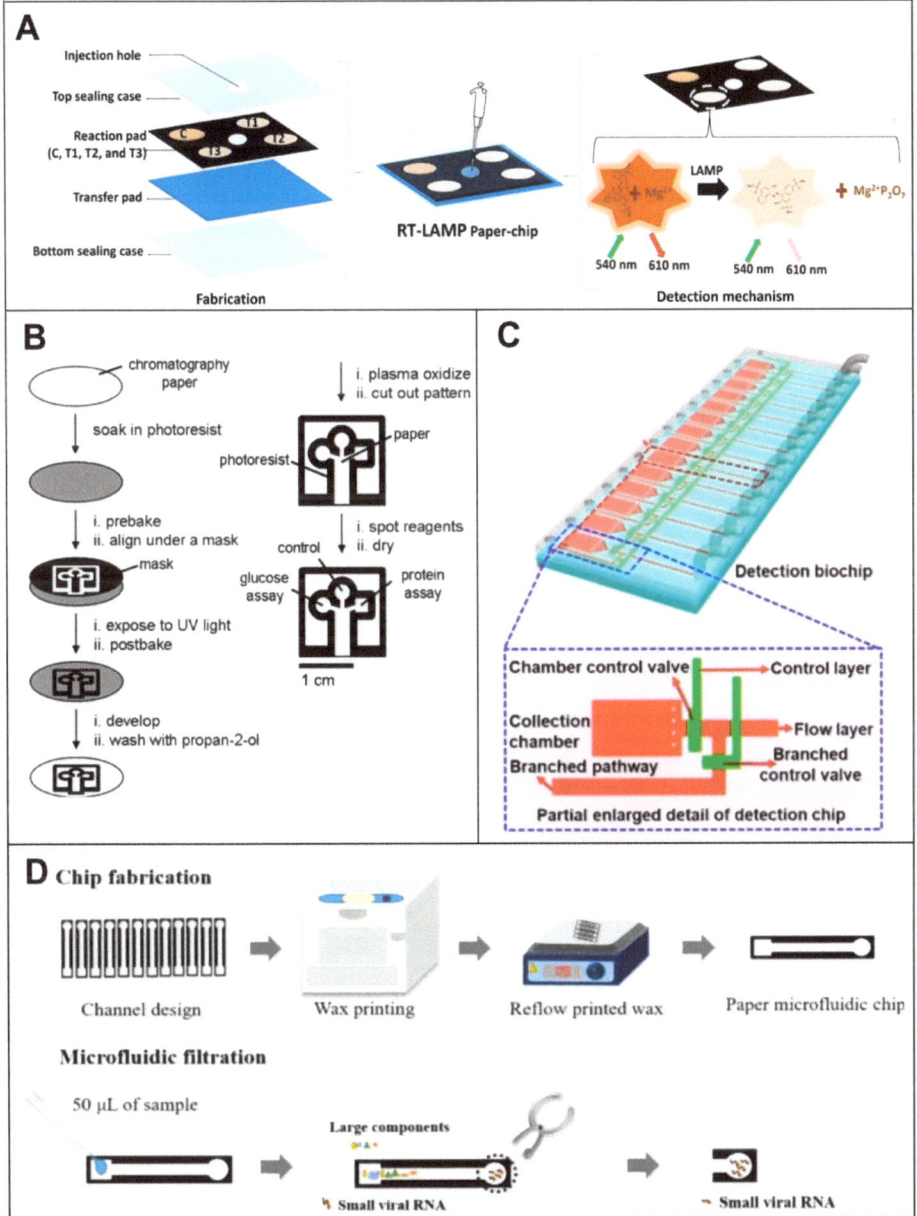

Figure 4. Principle of microfluidic-based RNA biosensors. (**A**) Fabrication of a paper-chip device for viral RNA amplification and detection [58]. (**B**) Principles of fabricating PADs using photolithography [52]. (**C**) Schematic representation of the microfluidic biochip structure for the detection of SARS-CoV-2 [64]. (**D**) Principle of wax printing for manufacturing PADs [62].

5. Nanomaterial-Based RNA Biosensors

Since viral RNA detection and identification involves longer operating times and greater device complexity, there is a great need to identify alternative viral detection targets and procedures for a simpler and more rapid diagnosis. Generally speaking, nanomaterials are usually used in vitro in combination with other methods to amplify signals and improve sensitivity. However, due to their own properties, such as the quenching of graphene oxide and the fluorescence of carbon nanotubes themselves, nanomaterials can be used to detect RNA [77,78], but individual detection is usually achieved in the cell. Based on this, nanomaterials as a separate analytical tool also provide a feasible alternative to RT-PCR for rapid and accurate virus detection.

5.1. Graphene Oxide-Based RNA Biosensors

Graphene oxide (GO) is a single-atom-thick two-dimensional carbon nanomaterial with properties such as large specific surface area, biocompatibility and effective fluorescence burst. Taking advantage of these promising properties, some methods for the direct detection of RNA based on GO have been developed. For example, Jiang et al. [79] reported a multiplexed GO fluorescent nanoprobe for the intracellular detection and quantification of mRNA in living cells by utilizing the fluorescence bursting property of GO. The detection limit of this GO-based nanoprobe was as low as 0.26 nM for mRNA mimics, and the nanoprobe was able to simultaneously perform relative quantification and intracellular detection of multiple target mRNAs in living cells compared to conventional mRNA detection methods. Do Won Hwang et al. [80] developed a robust nanoprobe platform for the simultaneous quantification and intracellular detection of multiple target mRNAs in living cells using GO quenching and fluorescence in situ hybridization (G-FISH). They also explored in situ hybridization recovery for sensitive RNA detection in formalin-fixed paraffin-embedded (FFPE) tissues (Figure 5A). Li et al. [81] presented a novel GO-based CHA and HCR signal dual amplification system (GO-CHA-HCR, or GO-AR) for circ-Foxo3 imaging detection in living cells. This method enabled the detection limit of circ-Foxo3 to be as low as 15 pM with excellent sensitivity and selectivity (Figure 5B). Yang et al. [82] developed a highly sensitive strategy for live cell and in vivo miRNA fluorescence imaging detection based on MB with GO enhanced signaling molecular bursts. The detection limit was as low as 30 pM in the presence of miRNA. This simple and effective strategy provided a new sensing platform for highly sensitive detection and simultaneous imaging analysis of multiple low-level biomarkers in living cells and in vivo.

5.2. Carbon Nanotube-Based RNA Biosensors

Carbon nanotubes (CNTs) are a widely used nanomaterial. In addition to properties such as a large specific surface area and carrying an abundance of electrons, the high accessibility of CNTs and easy-to-use fluorescence analyses allow them to be used as a material for the direct detection of RNA [83,84]. For example, Shrute Kannappan et al. [85] reported a fluorine-based CNT-DNA biosensor by introducing short complementary sequences (SCSs) that could regulate the binding strength of the probe sequence to CNT, thereby enhancing its reactivity to target oligonucleotides. The introduction of SCSs significantly increased the LOD of the biosensor, and this strategy could also be used to multiplex a set of miRNAs for a range of other pathological states by redesigning the probe sequence and measuring the corresponding fluorescence in a very short time (~1 h) (Figure 5C). Ma et al. [86] developed a sensitive sensing platform for the detection of a potential marker of breast cancer miRNA-155 [87,88] based on multiwall carbon nanotube-gold nanocomposites (MWCNT/AuNCs) as a new platform of fluorescence quenching coupled with DSN-assisted recovery signal amplification (Figure 5D).

Nanotechnology is likely to play an important role in the continued development of PCR-free methods for RNA detection. PCR-free methods are especially valuable in developing countries and resource-constrained settings. In order to progress in this field, cutting-edge innovations in nanotechnology will be essential, including nanoparticles, GO and DNA nanostructures, as well as nanomaterials such as CNTs, nanowires and quantum dots. This innovation and development will benefit PCR-free nucleic acid testing methods.

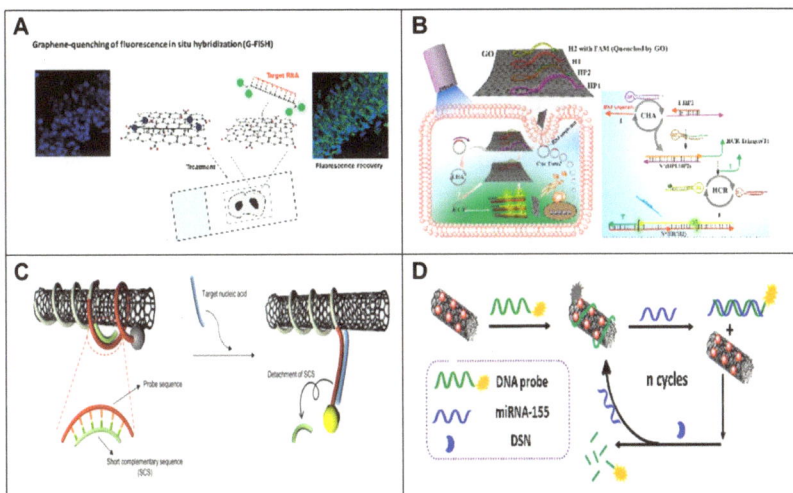

Figure 5. Principle of nanomaterial-based RNA biosensors. (**A**) GO nanoplatform for non-coding RNA detection in FFPE tissue specimens. Schematic of G-FISH [80]. (**B**) Schematic representation of a GO-based CHA and HCR signal dual amplification system for circ-Foxo3 imaging in living cells [81]. (**C**) A fluorescence-based CNT-DNA biosensor using a SCS. The schematic diagram of the improved CNT-DNA biosensor with an SCS improves the sensitivity of detecting the target miRNA [85]. (**D**) Schematic diagram of MWCNT/AuNCs used as a novel fluorescence bursting platform for miRNA-155 detection [86].

6. CRISPR-Based RNA Biosensors

CRISPR (clusters of regularly interspaced short palindromic repeats) was first discovered in the 1980s and is thought to be an adaptive immune system coupled to Cas proteins which uses RNA-directed nucleases to cleave invading nucleic acids [89], thereby allowing resistance to invading exogenous DNA and viruses in bacteria and archaea. The system is divided into two categories [90], with the most studied Cas9, Cas12 and Cas13 all belonging to the second category. In addition to being used for gene editing, Cas12 and Cas13 are involved in non-specific ssDNA or RNA cleavage after specific recognition of the target (trans-cleavage activity), making CRISPR-Cas a promising tool for the detection of nucleic acids [91,92]. Abnormal expression and mutation of RNA may be harmful to cells and cause disease, so the detection of abnormally expressed RNA or disease-related RNA mutations provides an avenue for disease diagnosis [93,94]. However, the detection of most disease-associated RNAs is demanding in terms of accuracy and detection limit due to low abundance and high sequence similarity among family members. The low tolerance of the CRISPR-Cas system to base mismatches in target nucleic acid sequences gives it excellent recognition of single-base mismatches. Therefore, CRISPR-Cas-based biosensors have a broad application prospect in RNA detection.

6.1. CRISPR-Cas9-Based RNA Biosensors

In recent years, some researchers have been trying to expand the application of CRISPR-Cas9 to the field of RNA detection. Cas9 has been used as a tool to detect miRNAs by converting RNA targets into substrates capable of triggering CRISPR-Cas9 responses. For example, Qiu et al. [95] were the first to perform miRNA detection using CRISPR-Cas9. The assay system incorporated isothermal amplification, detection and reporting based on RCA, CRISPR-Cas9 and split-horseradish peroxidase technologies. First, the miRNA was sequentially converted into a large DNA fragment containing multiple repeating complementary sequences and random neck loop structures of the dCas9 target by RCA amplification reaction. The Split-HRP-dCas9 protein recognized and localized to the RCA product under the guidance of specific small guide RNAs (sgRNAs), which subsequently led to the formation of active horseradish peroxidase (HRP), catalyzing the oxidation of tetramethylbenzidine (TMB). This method enabled the detection of trace miRNA in samples with single base specificity (Figure 6A). Wang et al. [96] developed an miRNA biosensor consisting of dCas9, miRNA-mediated sgRNA and red fluorescent protein. The biosensor provided an example for measuring miRNA activity and tracking cell state transitions in order to allow timely monitoring of miRNA activity in stem cell differentiation and cancer progression. Moe Hirosawa et al. [97] designed an miRNA-responsive AcrIIA4 switch based on the expression of endogenous miRNA activity-controlled *S. pyogenes* Cas4 inhibitor AcrIIA9, which, together with Cas9 or dCas9-VPR guide RNA complex, indirectly activates Cas9, enabling multiple intracellular miRNA sensing. By sensing intracellular miRNAs, this system could provide a powerful tool for future therapeutic applications and genome engineering.

In addition to detecting miRNA, there are some reports of the CRISPR-CAS system in the detection of mRNA and viral RNA. For example, Li et al. [98] designed an mRNA CRISPR biosensor that activated the cleavage function of Cas9 by switching the blocked sgRNA with the target RNA. In this strategy, mRNA-sensing CRISPR was constructed by guide RNA (gRNA) reconstitution and toe-mediated strand shift, in which each target site could be controlled independently. Experiments have shown that the switch could be embedded into the gRNA and used as an RNA biosensor, which could orthogonally detect multiple mRNA inputs and provide CRISPR/Cas9 response outputs. Bonhan Koo et al. [99] developed an improved molecular diagnostic tool that utilized a CRISPR/dCas9-mediated biosensor to couple dCas9 and a single micro-ring resonator biosensor for label-free and real-time detection of pathogenic RNA, achieving single-molecule sensitivity for RNA detection and 100-fold more sensitivity than RT-PCR detection. It improved the sensitivity and specificity of pathogen diagnosis in clinical samples (Figure 6B). Tin Marsic et al. [100] devised a method utilizing CRISPR/Cas9 enzymes for DNA scanning and recognition and VirD2 release covalently binding to ssDNA probes for LFA conjugation for SARS-CoV-2 viral RNA detection. The method employed a chimeric fusion between dCas9 and VirD2 in combination with an ssDNA reporter as a detection complex. A sensitive, specific and low-cost detection method was realized. In addition, CRISPR/cas9-based tools have been used as antiviral drugs for the treatment of HIV infections and for the detection of Zika virus and methicillin-resistant Staphylococcus aureus infections.

6.2. CRISPR-Cas12-Based RNA Biosensors

The Cas12 protein is a member of the CRISPR family and can be programmed with CRISPR-deriver RNA (crRNA) to specifically bind to complementary ssDNA and double stranded DNA (dsDNA) targets [101,102]. Cas12 is an alternative to Cas9 due to its unique properties, such as the ability to target T-rich motifs and the absence of the need for trans-activation crRNA. Along with specific double strand breaks (DSBs), Cas12a also undergoes non-specific cleavage on other ssDNA molecules. These non-specific tendencies are triggered only when the crRNA binds to its complementary target, known as trans-cleavage activity [103]. However, Cas12a has weak trans-cleavage activity, making nucleic acid testing less sensitive [91,104]. When combined with preamplification, cas12a-mediated detection can detect concentrations as low as 2 aM [104,105]. For example, James P. Broughton et al. [106] developed a CRISPR-cas12-based lateral flow assay for the detection of viral infection [107] called SARS-CoV-2 DNA Endonuclease Targeting CRISPR Trans Reporter (DETECTR). This method relied on the trans-cleavage activity of Cas12a proteins activated after Cas12a recognition of the target RNA. In addition, the LAMP step was combined with the DETECTR technology of CRISPR-Cas12 to enrich the target sequences. The purpose of rapid (30–40 min) detection, easy implementation and high accuracy of SARS-CoV-2 in clinical samples was achieved, providing a visual and faster alternative for RT-PCR detection (Figure 6C). Shi-Yuan Li et al., using the characteristic [108] of non-targeting ssDNA during the formation of Cas12a/crRNA/target DNA ternary complex, developed a low-cost multi-purpose efficient detection system HOLMES (one-hour low-cost multi-purpose highly efficient system), which could be used for the rapid and low-cost detection of target RNA [109]. At the same time, Cas12a-based HOLMES can also detect nucleic acids with aM sensitivity. Compared to Cas12a, Cas12b exhibited higher activity against the exes. Subsequently, Liang et al. developed an updated version of HOLMESV2 that combined Cas2b and isothermal amplification to detect nucleotides, distinguish single nucleotide polymorphisms (SNPs), and quantify dsRNA and RNA methylation [110]. Unlike the DETECTR, which is only used for qualitative measurements, HOLMES can be used for quantitative detection. A successful attempt was made to create an active Cas12a nanocomposite that could be used as a biosensor without shell deconstruction or enzyme release.

Instead of using preamplification to achieve signal amplification, Zhi Run Ji et al. [111] proposed a strategy to use metal-organic frameworks (MOFs) to protect Cas12a from harsh environments and successfully constructed an active Cas12a nanocomposite, Cas12a-on-MAF-7 (COM), as a biosensor for the first time, without the need of isothermal amplification. This strategy achieved an ultra-sensitive SARS-CoV-2 RNA detection with a detection limit of one copy and solved the problem of poor stability of CRISPR-Cas12a (Figure 6D).

6.3. CRISPR-Cas13-Based RNA Biosensors

CRISPR-Cas13a is the only CRISPR effector that targets RNA and has the ability to go beyond signal amplification [112]. One of the milestones in using CRISPR-Cas for RNA detection was the discovery by Zhang's team in 2016 of the trans-cleavage activity of Cas13a (also known as C2c2) [92]. The target RNA is specifically recognized by crRNA and subsequently cleaved by Cas13a [113]. When used together, activated Cas13a utilizes its unique trans-cleavage activity, the RNA provided by the lateral cleavage RNA probe, and the fluorophore quencher labeled RNA reporter with trans-cleavage activity [103], which could be used together to determine specific target sequences. The CRISPR-Cas13 system has great potential for detecting viral RNA due to its reliability, high sensitivity and ease of implementation [114]. RNA can be directly detected by the side branch cleavage of CRISPR-Cas13a. For example, Alexandra East-Seletsky et al. [115] described a method for the direct detection of RNA using Cas13a trans-cleavage. By designing fluorophore quencher labeled RNA reporters, known as reporter genes, they achieved efficient signal amplification detection of target RNAs with pM sensitivity, as each activated Cas13a is capable of cleaving thousands of reporter genes. Subsequently, Hajime Shinoda et al. developed a platform that enables the accurate and rapid detection of single stranded RNA (ssRNA) at the single-molecule level, the CRISPR-based Amplification-Free Digital RNA Assay (SATORI) platform. The combination of CRISPR-Cas13-based RNA detection technology and microchamber array technology avoids the long detection time and false negative or false positive results due to amplification errors caused by the preamplification process in CRISPR-based methods, resulting in a maximum sensitivity of 10 fM for the detection of ssRNA targets with high specificity and a short detection time (less than 5 min) [116] (Figure 6E).

Despite the high specificity and simplicity of direct detection of RNA with Cas13a, the abundance of RNA in organisms is particularly low. To further improve sensitivity, incorporating nucleic acid amplification is an effective strategy. Max J. Kellner et al. recently established a CRISPR-based diagnostic platform that combines nucleic acid preamplification with the CRISPR-Cas13 system for specific recognition of the desired RNA sequence. The platform, known as Specific High Sensitivity Enzyme Reporter Unlock (SHERLOCK), allows for multiplexed, portable and ultra-sensitive RNA detection from clinically relevant samples. However, a drawback of SHERLOCK made it unsuitable for RNA quantitative detection [105]. In the following year, Gootenberg et al. upgraded SHERLOCK version 2 (SHERLOCKv2) [105] to allow simultaneous detection of multiple targets. The SHERLOCKv2 is a powerful tool for nucleic acid testing because of its increased sensitivity and quantitative and visual readouts. CRISPR-Cas13a cascade-based viral RNA detection in clinical samples was reported by Yuxi Wang et al. as a label-free, isothermal method [117]. SARS-CoV-2 RNA was directly detected using Cas13a/crRNA to activate transcriptional amplification for light-up RNA aptamer output [118–120]. This assay achieved high sensitivity for SARS-CoV-2 RNA detection at a detection limit of 0.216 fM by integrating Cas13a/crRNA's RNA-specific recognition capability and trans-cleavage activity into cascade amplification.

As a whole, the CRISPR-Cas system has the potential to be a valuable tool for nucleic acid testing [121–124]. A molecular diagnosis of SARS-CoV-2 is mainly performed using RT-qPCR, which is time-consuming and expensive. Further, patients may experience anxiety, irritability and fear as a result of false positives produced by RT-qPCR. A CRISPR-based approach, on the other hand, eliminates expensive probes (e.g., quenchers and fluorescent-modified RNA probes) and expensive equipment (e.g., thermal circulators), thus reducing detection complexity and equipment costs [125–128]. Furthermore, the CRISPR-Cas system can detect pathogens more rapidly in food due to its highly accurate and efficient characteristics. This significantly improves the efficiency of the food safety detection process. Additionally, CRISPR technology can identify various pollutants in food and monitor food safety conditions [129,130].

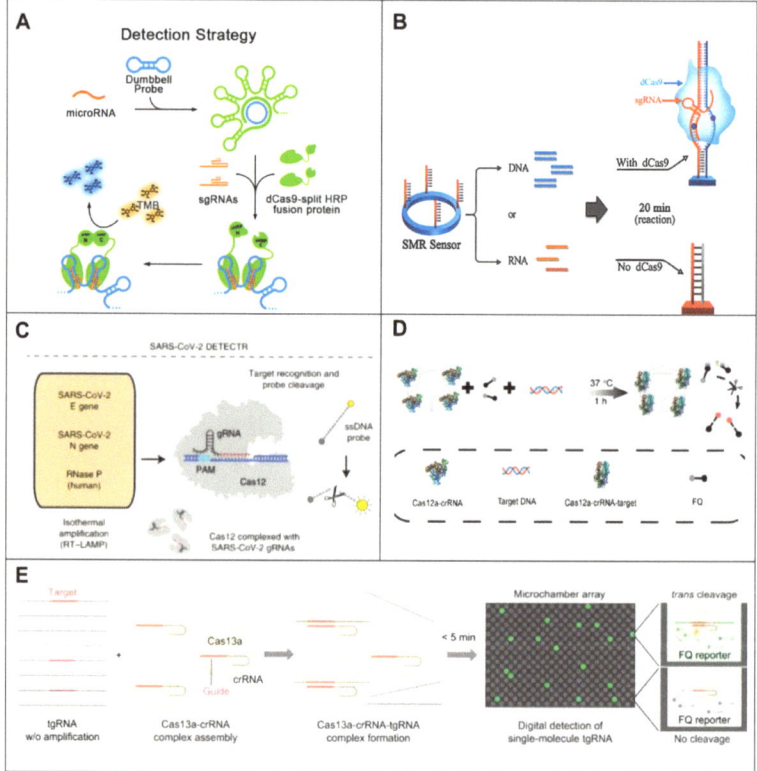

Figure 6. Principle of CRISPR-based RNA biosensors. (**A**) Schematic representation of the RCA-CRISPR-split-HRP (RCH) method based on RCA, CRISPR-Cas9 and cleaved-root peroxidase technologies for miRNA detection [95]. (**B**) Schematic of CRISPR/dCas9-mediated biosensor. SMR biosensor, silicon mirroring resonator biosensor [99]. (**C**) Schematic of SARS-CoV-2 DETECTR workflow [106]. (**D**) Cas12a assembled with MAF-7 to form nanobiocomposites with higher stability and trans-cleavage activity for nucleic acid testing [111]. (**E**) Schematic illustration of SATORI. LwaCas13a–crRNA–tgRNA cleaves FQ reporters, leading to fluorescence increases in a microchamber array device [116].

7. Discussion

RNA detection has important applications in various fields. In particular, the COVID-19 pandemic has introduced the need for new accurate and efficient diagnostic tools for SARS-CoV-2 RNA detection. Therefore, it is essential to choose the proper detection method. Most detection techniques are still based on PCR and RT-PCR. While PCR-based techniques are highly sensitive and specific, their analysis requires a variety of equipment and technicians and can only be performed in laboratories. To overcome this limitation, considerable efforts have been made to perform RNA detection, and various improved or innovative PCR-free methods have been developed, such as electrochemical methods, SPR, microfluidic devices, nanotechnology and CRISPR-based detection techniques, to name a few. They overcome the limitations of PCR-based detection of RNA, and since they do not require expensive reagents and instruments, the application of PCR-free detection methods may help reduce the cost of RNA detection and thus improve the applicability of RNA detection. However, there are also some problems in the existing technologies (Table 1). For example, PCR-free-based biosensors are faced with technical problems such as miniaturization, portability, high precision and low energy consumption, which limit their popularization and development. At the same time, the PCR-free-based biosensors also have limitations in their reusability and signal interference. In addition, PCR-free-based biosensors involve multidisciplinary cross-integration, such as materials science, computers, communications, bioinformatics, biochips, etc., and require more innovation and cooperation to achieve technological breakthroughs. In conclusion, with the rapid development of new technologies and methods, we believe that more excellent and efficient detection methods will be developed in the future, which will provide scientists and clinicians with more choices. At the same time, the most economical and optimal choice can only be obtained by weighing the advantages and disadvantages of various detection methods according to the specific purpose.

Table 1. Summary of PCR-free-based biosensors.

System	Combination	Sensitivity	Time	Target	Ref.
Electrochemical-based RNA biosensors	AuNPs/RT-LAMP/high affinity biotin-avidin system	0.1 fmol·L^{-1}	~1 h	mRNA	[6]
	RCA	1 copy/µL	<2 h	viral N or S genes	[7]
	AuNPs/polymerase-assisted signal amplification	4.3×10^{-17} mol/L	<1 h	mRNA	[8]
	CHA/TDT	26 fmol·L^{-1}	<1 h	SARS-CoV-2 RNA	[9]
	HCR	3 fmol·L^{-1}	<1 h	mRNA	[14]
	CuO/AuNPs	1 fmol·L^{-1}	~1 h	HCV RNA	[16]
	MWNTs/SPE	8.2 µg mL^{-1}	<5 min	tRNA	[19]
	SPE-Au	fmol·L^{-1}	<1 h	mRNA	[20]
SPR-based RNA biosensors	GeP5	10 amol·L^{-1}	<1 h	SARS-CoV-2 RNA	[24]
	MNPs/AuNPs	7 fmol·L^{-1}	~1 h	mRNA	[25]
	Antimonene two-dimensional nanomaterials/AuNR	amol·L^{-1}	~1 h	miRNA	[27]
	MNP	3 fmol·L^{-1}	~2 h	miRNA	[28]
	DNA-AgNCs/AuNPs	fmol·L^{-1}	<2 h	miRNA	[31]
	AuNPs/DNA super-sandwich	21 fmol·L^{-1}	~1 h	miRNA	[32]
	DNA super-sandwich/biotin-streptavidin system	30 pmol·L^{-1}	<9 min	miRNA	[33]
	CHA/streptavidin aptamer	1 pmol·L^{-1}	<1 h	miRNA	[34]
	MTL	500 fmol·L^{-1}	~1 h	miRNA	[35]

Table 1. *Cont.*

System	Combination	Sensitivity	Time	Target	Ref.
Microfluidic-based RNA biosensors	AuNPs	fmol·L^{-1}	~2 h	miRNA	[40]
	RT-LAMP	fmol·L^{-1}	<1 h	Viral RNA	[42]
	RT-LAMP	160 copies/μL	<45 min	Viral RNA	[43]
	NoV-DID/PDMS/RT-RAA	fmol·L^{-1}	~1 h	Viral RNA	[48]
Nanomaterial-based RNA biosensors	GO	0.26 nmol·L^{-1}	<2 h	mRNA	[61]
	GO/CHA/HCR	15 pmol·L^{-1}	<2 h	circRNA	[63]
	GO/MB	30 pmol·L^{-1}	~1 h	miRNA	[64]
	MWCNT/AuNCs	33.4 fmol·L^{-1}	~1 h	miRNA	[68]
CRISPR-based RNA biosensors	RCA/CRISPR-Cas9	fmol·L^{-1}	<1 h	miRNA	[77]
	DETECTR	fmol·L^{-1}	30–40 min	SARS-CoV-2 RNA	[88]
	HOLMES	amol·L^{-1}	~1 h	Viral RNA	[90]
	CRISPR-Cas12/MoFs	1 copy	~1 h	SARS-CoV-2 RNA	[93]
	SATORI	10 fmol·L^{-1}	<5 min	SARS-CoV-2 RNA	[97]
	SHERLOCK	fmol·L^{-1}	~1 h	SARS-CoV-2 RNA	[86]
	SHERLOCKv2	fmol·L^{-1}	~1 h	SARS-CoV-2 RNA	[86]
	CRISPR-Cas13	0.216 fmol·L^{-1}	~1 h	SARS-CoV-2 RNA	[99]

Author Contributions: Conceptualization, X.L.; resources and data curation, H.W., X.Q., Y.J. and F.L.; writing—original draft preparation, X.L.; writing—review and editing, X.C., K.L. and L.L.; visualization, X.L.; supervision, X.C., K.L. and L.L.; funding acquisition, X.C., K.L. and L.L. All authors have read and agreed to the published version of the manuscript.

Funding: This research was funded by the Science and Technology Innovation (2030) Agricultural Biological Breeding Major Project (2022ZD0402013); the Key Laboratory of Traceability for Agricultural GMOs of MARA (2022KF02); the IQSTAP project (1610072023101); and the Agricultural Science and Technology Innovation Program of CAAS (CAAS-ASTIP-IQSTAP-2024).

Institutional Review Board Statement: Not applicable.

Informed Consent Statement: Not applicable.

Data Availability Statement: The data presented in this study are available on request from the corresponding authors.

Conflicts of Interest: The authors declare no conflicts of interest.

Abbreviations

AgNPs	Silver nanoparticles
AuNIs	gold nanoisland
AuNPs	gold nanoparticles
AuNR	gold nanorods
Au-S	gold-sulfur
cDNA	complementary DNA
CHA	catalytic hairpin assembly
CMV	cytomegalovirus
CNTs	carbon nanotubes
COM	Cas12a-on-MAF-7
CRISPR	clusters of regularly interspaced short palindromic repeats
crRNA	CRISPR-deriver RNA
CuO	copper oxide
DETECTR	DNA Endonuclease Targeting CRISPR Trans Reporter
DNA-AgNC	DNA-silver nanocluster
DSBs	double strand breaks
dsDNA	double stranded DNA
E-INAATs	electrochemical isothermal nucleic acid amplification tests

ELC	electrochemiluminescence
FFPE	formalin-fixed paraffin-embedded
G-FISH	graphene oxide-fluorescence in situ hybridization
GO	graphene oxide
gRNA	Guide RNA
HCR	hybridization chain reaction
HCV	hepatitis C virus
HOLMES	one-hour low-cost multipurpose highly efficient system
HRP	horseradish peroxidase
INAATs	isothermal nucleic acid amplification tests
LFA	lateral flow assay
LOD	limit of detection
LSPR	local surface plasmon resonance
MB	molecular beacons
miRNAs	microRNAs
MNPs	magnetic nanoparticles
MOFs	metal-organic frameworks
mRNA	messenger RNA
MTL	mass transfer restriction
MWCNT/AuNCs	multiwall carbon nanotube-gold nanocomposites
MWNTs	multi-walled carbon nanotubes
NC	nitrocellulose
N gene	nucleocapsid phosphoprotein
PAD	Paper-based microfludics
PDMS	polydimethylsiloxane
POC	point-of-care
POCT	point-of-care testing
Poly(A)	polyadenine
PPT	plasma photothermal
PAN	Peptide acid probe
rRNA	ribosomal RNA
RCA	rolling circle amplification
RdRp	RNA-dependent RNA polymerase
RT-LAMP	reverse transcription loop-mediated isothermal amplification
RT-qPCR	reverse transcription quantitative polymerase chain reaction
RT-RAA	reverse-transcription recombinase-assisted amplification
SA-aptamer	streptavidin aptamer
SCS	short complementary sequences
S gene	Spike protein
sgRNA	Small guide RNA
SHERLOCK	Specific High Sensitivity Enzyme Reporter Unlock
SNPs	single nucleotide polymorphisms
SPCEs	screen-printed carbon electrodes
SPE-Au	screen-printed gold electrodes
SPR	surface plasma resonance
ssRNA	single stranded RNA
SWV	Square Wave Voltammetry
TDT	terminal transferase
TMB	tetramethylbenzidine
tRNA	transfer RNA
UV	ultraviolet rays
µPADs	Microfluidic paper-based analytical devices

References

1. Mortimer, S.A.; Kidwell, M.A.; Doudna, J.A. Insights into RNA Structure and Function from Genome-Wide Studies. *Nat. Rev. Genet.* **2014**, *15*, 469–479. [CrossRef] [PubMed]
2. Cui, L.; Ma, R.; Cai, J.; Guo, C.; Chen, Z.; Yao, L.; Wang, Y.; Fan, R.; Wang, X.; Shi, Y. RNA Modifications: Importance in Immune Cell Biology and Related Diseases. *Signal Transduct. Target. Ther.* **2022**, *7*, 334. [CrossRef] [PubMed]
3. Laurent, G.S.; Shtokalo, D.; Tackett, M.R.; Yang, Z.; Vyatkin, Y.; Milos, P.M.; Seilheimer, B.; McCaffrey, T.A.; Kapranov, P. On the Importance of Small Changes in RNA Expression. *Methods* **2013**, *63*, 18–24. [CrossRef] [PubMed]
4. Kellner, S.; Burhenne, J.; Helm, M. Detection of RNA Modifications. *RNA Biol.* **2010**, *7*, 237–247. [CrossRef] [PubMed]
5. Motorin, Y.; Lyko, F.; Helm, M. 5-Methylcytosine in RNA: Detection, Enzymatic Formation and Biological Functions. *Nucleic Acids Res.* **2010**, *38*, 1415–1430. [CrossRef]
6. Ma, X.; Xu, J.; Zhou, F.; Ye, J.; Yang, D.; Wang, H.; Wang, P.; Li, M. Recent Advances in PCR-Free Nucleic Acid Detection for SARS-CoV-2. *Front. Bioeng. Biotechnol.* **2022**, *10*, 999358. [CrossRef]
7. Calorenni, P.; Leonardi, A.A.; Sciuto, E.L.; Rizzo, M.G.; Lo Faro, M.J.; Fazio, B.; Irrera, A.; Conoci, S. PCR-Free Innovative Strategies for SARS-CoV-2 Detection. *Adv. Healthc. Mater.* **2023**, *12*, 2300512. [CrossRef]
8. Shinoda, H.; Lida, T.; Makino, A.; Yoshimura, M.; Lshikawa, J.; Ando, J.; Murai, K.; Sugiyama, K.; Muramoto, Y.; Nakano, M. Automated Amplification-Free Digital RNA Detection Platform for Rapid and Sensitive SARS-CoV-2 Diagnosis. *Commun. Biol.* **2022**, *5*, 473. [CrossRef] [PubMed]
9. Ueda, T.; Shinoda, H.; Makino, A.; Yoshimure, M.; Lida, T.; Watanabe, R. Purification/Amplification-Free RNA Detection Platform for Rapid and Multiplex Diagnosis of Plant Viral Infections. *Anal. Chem.* **2023**, *95*, 9680–9686. [CrossRef] [PubMed]
10. Wang, D.; Wang, X.; Ye, F.; Zou, J.; Qu, J.; Jiang, X. An Integrated Amolification-Free Digital Crispr/Cas-Assisted Assay for Single Molecule Detection of RNA. *ACS Nano* **2023**, *17*, 7250–7256. [CrossRef]
11. Tavallaie, R.; Darwish, N.; Hibbert, D.B.; Gooding, J.J. Nucleic-Acid Recognition Interfaces: How the Greater Ability of RNA Duplexes to Bend towards the Surface Influences Electrochemical Sensor Performance. *Chem. Commun.* **2015**, *51*, 16526–16529. [CrossRef] [PubMed]
12. Labib, M.; Sargent, E.H.; Kelley, S.O. Electrochemical Methods for the Analysis of Clinically Relevant Biomolecules. *Chem. Rev.* **2016**, *116*, 9001–9090. [CrossRef] [PubMed]
13. Hartman, M.R.; Ruiz, R.C.H.; Hamada, S.; Xu, C.; Yancey, K.G.; Yu, Y.; Han, W.; Luo, D. Point-of-Care Nucleic Acid Detection Using Nanotechnology. *Nanoscale* **2013**, *5*, 10141–10154. [CrossRef]
14. Johnson, B.N.; Mutharasan, R. Biosensor-Based Microrna Detection: Techniques, Design, Performance, and Challenges. *Analyst* **2014**, *139*, 1576–1588. [CrossRef] [PubMed]
15. Mukumoto, K.; Nojima, T.; Sato, S.; Waki, M.; Takenaka, S. Direct Modification of mRNA by Ferrocenyl Carbodiimide and Its Application to Electrochemical Detection of mRNA. *Anal. Sci.* **2007**, *23*, 115–119. [CrossRef] [PubMed]
16. Yang, N.; Liu, P.; Cai, C.; Zhang, R.; Sang, K.; Shen, P.; Huang, Y.; Lu, Y. Triple Signal Amplification Strategy for the Ultrasensitive Electrochemical Detection of Human Papillomavirus 16 E6/E7 mRNA. *Enzym. Microb. Technol.* **2021**, *149*, 109855. [CrossRef]
17. Chaibun, T.; Puenpa, J.; Ngamdee, T.; Boonapatcharoen, N.; Athamanolap, P.; O'Mullane, A.P.; Vongpunsawad, S.; Poovorawan, Y.; Lee, S.Y.; Lertanantawong, B. Rapid Electrochemical Detection of Coronavirus SARS-CoV-2. *Nat. Commun.* **2021**, *12*, 802. [CrossRef]
18. Zhang, M.; Zhou, F.; Zhou, D.; Chen, D.; Hai, H.; Li, J. An Aptamer Biosensor for Leukemia Marker mRNA Detection Based on Polymerase-Assisted Signal Amplification and Aggregation of Illuminator. *Anal. Bioanal. Chem.* **2019**, *411*, 139–146. [CrossRef] [PubMed]
19. Peng, Y.; Pan, Y.; Sun, Z.; Li, J.; Yi, Y.; Yang, J.; Li, G. An Electrochemical Biosensor for Sensitive Analysis of the SARS-CoV-2 RNA. *Biosens. Bioelectron.* **2021**, *186*, 113309. [CrossRef]
20. Dai, W.; Zhang, J.; Meng, X.; He, J.; Zhang, K.; Cao, Y.; Wang, D.; Dong, H.; Zhang, X. Catalytic Hairpin Assembly Gel Assay for Multiple and Sensitive Microrna Detection. *Theranostics* **2018**, *8*, 2646. [CrossRef]
21. Li, P.; Wei, M.; Zhang, F.; Su, J.; Wei, W.; Zhang, Y.; Liu, S. Novel Fluorescence Switch for Microrna Imaging in Living Cells Based on Dnazyme Amplification Strategy. *ACS Appl. Mater. Interfaces* **2018**, *10*, 43405–43410. [CrossRef]
22. Li, P.; Wei, M.; Zhang, F.; Su, J.; Wei, W.; Zhang, Y.; Liu, S. Application of Spectral Crosstalk Correction for Improving Multiplexed Microrna Detection Using a Single Excitation Wavelength. *Anal. Chem.* **2017**, *89*, 3430–3436. [CrossRef] [PubMed]
23. Xu, E.; Feng, Y.; Yang, H.; Li, P.; Kong, L.; Wei, W.; Liu, S. Ultrasensitive and Specific Multi-Mirna Detection Based on Dual Signal Amplification. *Sens. Actuators B Chem.* **2021**, *337*, 129745. [CrossRef]
24. Li, K.; Chen, T.; Wang, M.; Li, F.; Qi, X.; Song, X.; Fan, L.; Li, L. Ape1 Mediated Target-Responsive Structure Switching Electrochemical (SSE) Biosensor for RNA Detection. *Sens. Actuators B Chem.* **2024**, *398*, 134782. [CrossRef]
25. Cheng, Y.-H.; Liu, S.-J.; Jiang, J.-H. Enzyme-Free Electrochemical Biosensor Based on Amplification of Proximity-Dependent Surface Hybridization Chain Reaction for Ultrasensitive mRNA Detection. *Talanta* **2021**, *222*, 121536. [CrossRef]
26. Zhao, T.; Zhang, H.-S.; Tang, H.; Jiang, J.-H. Nanopore Biosensor for Sensitive and Label-Free Nucleic Acid Detection Based on Hybridization Chain Reaction Amplification. *Talanta* **2017**, *175*, 121–126. [CrossRef] [PubMed]

27. Roohizadeh, A.; Ghaffarinejad, A.; Salahandish, R.; Omidinia, E. Label-Free Rna-Based Electrochemical Nanobiosensor for Detection of Hepatitis C. *Curr. Res. Biotechnol.* **2020**, *2*, 187–192. [CrossRef]
28. Kerr, E.; Farr, R.; Doeven, E.H.; Nai, Y.H.; Alexander, R.; Guijt, R.M.; Prieto-Simon, B.; Francis, P.S.; Dearnley, M.; Hayne, D.J. Amplification-Free Electrochemiluminescence Molecular Beacon-Based Microrna Sensing Using a Mobile Phone for Detection. *Sens. Actuators B Chem.* **2021**, *330*, 129261. [CrossRef]
29. Bhaiyya, M.; Pattnaik, P.K.; Goel, S.D. A Brief Review on Miniaturized Electrochemiluminescence Devices: From Fabrication to Applications. *Curr. Opin. Electrochem.* **2021**, *30*, 100800. [CrossRef]
30. Alafeef, M.; Dighe, K.; Moitra, P.; Pan, D. Rapid, Ultrasensitive, and Quantitative Detection of SARS-CoV-2 Using Antisense Oligonucleotides Directed Electrochemical Biosensor Chip. *ACS Nano* **2020**, *14*, 17028–17045. [CrossRef]
31. Ye, Y.; Ju, H. Rapid Detection of Ssdna and RNA Using Multi-Walled Carbon Nanotubes Modified Screen-Printed Carbon Electrode. *Biosens. Bioelectron.* **2005**, *21*, 735–741. [CrossRef] [PubMed]
32. Islam, N.; Gopalan, V.; Haque, H.; Masud, M.K.; Al Hossain, S.; Yamauchi, Y.; Nguyen, N.-T.; Lam, A.K.-Y.; Shiddiky, M.J. A Pcr-Free Electrochemical Method for Messenger RNA Detection in Cancer Tissue Samples. *Biosens. Bioelectron.* **2017**, *98*, 227–233. [CrossRef] [PubMed]
33. Jamal, R.B.; Vitasovic, T.; Gosewinkel, U.; Ferapontova, E.E. Detection of E. Coli 23s Rrna by Electrocatalytic "Off-on" DNA Beacon Assay with Femtomolar Sensitivity. *Biosens. Bioelectron.* **2023**, *228*, 115214. [CrossRef] [PubMed]
34. Fu, Y.; Wang, N.; Yang, A.; Xu, Z.; Zhang, W.; Liu, H.; Law, H.K.-W.; Yan, F. Ultrasensitive Detection of Ribonucleic Acid Biomarkers Using Portable Sensing Platforms Based on Organic Electrochemical Transistors. *Anal. Chem.* **2021**, *93*, 14359–14364. [CrossRef] [PubMed]
35. Li, Q.; Li, Y.; Gao, Q.; Jiang, C.; Tian, Q.; Ma, C.; Shi, C. Real-Time Monitoring of Isothermal Nucleic Acid Amplification on a Smartphone by Using a Portable Electrochemical Device for Home-Testing of SARS-CoV-2. *Anal. Chim. Acta* **2022**, *1229*, 340343. [CrossRef] [PubMed]
36. Jebelli, A.; Oroojalian, F.; Fathi, F.; Mokhtarzadeh, A.; de la Guardia, M. Recent Advances in Surface Plasmon Resonance Biosensors for Micrornas Detection. *Biosens. Bioelectron.* **2020**, *169*, 112599. [CrossRef] [PubMed]
37. Joung, H.-A.; Lee, N.-R.; Lee, S.K.; Ahn, J.; Shin, Y.B.; Choi, H.-S.; Lee, C.-S.; Kim, S.; Kim, M.-G. High Sensitivity Detection o 16s Rrna Using Peptide Nucleic Acid Probes and a Surface Plasmon Resonance Biosensor. *Anal. Chim. Acta* **2008**, *630*, 168–173. [CrossRef]
38. Chang, S.; Liu, L.; Mu, C.; Wen, F.; Xiang, J.; Zhai, K.; Wang, B.; Wu, L.; Nie, A.; Shu, Y.; et al. An Ultrasensitive Spr Biosensor for RNA Detection Based on Robust Gep5 Nanosheets. *J. Colloid Interface Sci.* **2023**, *651*, 938–947. [CrossRef]
39. Mousavi, M.Z.; Chen, H.-Y.; Wu, S.-H.; Peng, S.-W.; Lee, K.-L.; Wei, P.-K.; Cheng, J.-Y. Magnetic Nanoparticle-Enhanced Spr on Gold Nanoslits for Ultra-Sensitive, Label-Free Detection of Nucleic Acid Biomarkers. *Analyst* **2013**, *138*, 2740–2748. [CrossRef]
40. Li, Q.; Wang, Q.; Yang, X.; Wang, K.; Zhang, H.; Nie, W. High Sensitivity Surface Plasmon Resonance Biosensor for Detection of Microrna and Small Molecule Based on Graphene Oxide-Gold Nanoparticles Composites. *Talanta* **2017**, *174*, 521–526. [CrossRef]
41. Xue, T.; Liang, W.; Li, Y.; Sun, Y.; Xiang, Y.; Zhang, Y.; Dai, Z.; Duo, Y.; Wu, L.; Qi, K.; et al. Ultrasensitive Detection of Mirna with an Antimonene-Based Surface Plasmon Resonance Sensor. *Nat. Commun.* **2019**, *10*, 28. [CrossRef] [PubMed]
42. Chang, Y.-F.; Chou, Y.-T.; Cheng, C.-Y.; Hsu, J.-F.; Su, L.-C.; Ho, J.-A.A. Amplification-Free Detection of Cytomegalovirus Mirna Using a Modification-Free Surface Plasmon Resonance Biosensor. *Anal. Chem.* **2021**, *93*, 8002–8009. [CrossRef] [PubMed]
43. Camarca, A.; Varriale, A.; Capo, A.; Pennacchio, A.; Calabrese, A.; Giannattasio, C.; Almuzara, C.M.; D'Auria, S.; Staiano, M. Emergent Biosensing Technologies Based on Fluorescence Spectroscopy and Surface Plasmon Resonance. *Sensors* **2021**, *21*, 906. [CrossRef] [PubMed]
44. Qiu, G.; Gai, Z.; Tao, Y.; Schmitt, J.; Kullak-Ublick, G.A.; Wang, J. Dual-Functional Plasmonic Photothermal Biosensors for Highly Accurate Severe Acute Respiratory Syndrome Coronavirus 2 Detection. *ACS Nano* **2020**, *14*, 5268–5277. [CrossRef] [PubMed]
45. Borghei, Y.-S.; Hosseini, M.; Ganjali, M.R.; Ju, H. Colorimetric and Energy Transfer Based Fluorometric Turn-on Method for Determination of Microrna Using Silver Nanoclusters and Gold Nanoparticles. *Microchim. Acta* **2018**, *185*, 286. [CrossRef] [PubMed]
46. Wang, Q.; Liu, R.; Yang, X.; Wang, K.; Zhu, J.; He, L.; Li, Q. Surface Plasmon Resonance Biosensor for Enzyme-Free Amplified Microrna Detection Based on Gold Nanoparticles and DNA Supersandwich. *Sens. Actuators B Chem.* **2016**, *223*, 613–620. [CrossRef]
47. Ding, X.; Yan, Y.; Li, S.; Zhang, Y.; Cheng, W.; Cheng, Q.; Ding, S. Surface Plasmon Resonance Biosensor for Highly Sensitive Detection of Microrna Based on DNA Super-Sandwich Assemblies and Streptavidin Signal Amplification. *Anal. Chim. Acta* **2015**, *874*, 59–65. [CrossRef]
48. Li, J.; Lei, P.; Ding, S.; Zhang, Y.; Yang, J.; Cheng, Q.; Yan, Y. An Enzyme-Free Surface Plasmon Resonance Biosensor for Real-Time Detecting Microrna Based on Allosteric Effect of Mismatched Catalytic Hairpin Assembly. *Biosens. Bioelectron.* **2016**, *77*, 435–441. [CrossRef]
49. Li, K.; An, N.; Wu, L.; Wang, M.; Li, F.; Li, L. Absolute Quantification of Micrornas Based on Mass Transport Limitation under a Laminar Flow Spr System. *Biosens. Bioelectron.* **2024**, *244*, 115776. [CrossRef]

50. Kulkarni, M.B.; Goel, S. Mini-Thermal Platform Integrated with Microfluidic Device with on-Site Detection for Real-Time DNA Amplification. *Biotechniques* **2023**, *74*, 158–171. [CrossRef]
51. Martinez, A.W.; Phillips, S.T.; Whitesides, G.M. Devices (UPADS)-Are a New Platform Designed for Assured. *Anal. Chem.* **2010**, *82*, 3–10. [CrossRef] [PubMed]
52. Martinez, A.W.; Phillips, S.T.; Butte, M.J.; Whitesides, G.M. Patterned Paper as a Platform for Inexpensive, Low-Volume, Portable Bioassays. *Angew. Chem.* **2007**, *119*, 1340–1342. [CrossRef]
53. Kumari, M.; Gupta, V.; Kumar, N.; Arun, R.K. Microfluidics-Based Nanobiosensors for Healthcare Monitoring. *Mol. Biotechnol.* **2024**, *66*, 378–401. [CrossRef] [PubMed]
54. Yamada, K.; Shibata, H.; Suzuki, K.; Citterio, D. Toward Practical Application of Paper-Based Microfluidics for Medical Diagnostics: State-of-the-Art and Challenges. *Lab Chip* **2017**, *17*, 1206–1249. [CrossRef] [PubMed]
55. Noviana, E.; Ozer, T.; Carrell, C.S.; Link, J.S.; McMahon, C.; Jang, I.; Henry, C.S. Microfluidic Paper-Based Analytical Devices: From Design to Applications. *Chem. Rev.* **2021**, *121*, 11835–11885. [CrossRef] [PubMed]
56. Gao, X.; Xu, H.; Baloda, M.; Gurung, A.S.; Xu, L.-P.; Wang, T.; Zhang, X.; Liu, G. Visual Detection of Microrna with Lateral Flow Nucleic Acid Biosensor. *Biosens. Bioelectron.* **2014**, *54*, 578–584. [CrossRef] [PubMed]
57. Zheng, W.; Yao, L.; Teng, J.; Yan, C.; Qin, P.; Liu, G.; Chen, W. Lateral Flow Test for Visual Detection of Multiple Micrornas. *Sens. Actuators B Chem.* **2018**, *264*, 320–326. [CrossRef] [PubMed]
58. Batule, B.S.; Seok, Y.; Kim, M.-G. Based Nucleic Acid Testing System for Simple and Early Diagnosis of Mosquito-Borne RNA Viruses from Human Serum. *Biosens. Bioelectron.* **2020**, *151*, 111998. [CrossRef] [PubMed]
59. Rodriguez, N.M.; Linnes, J.C.; Fan, A.; Ellenson, C.K.; Pollock, N.R.; Klapperich, C.M. Based RNA Extraction, in Situ Isothermal Amplification, and Lateral Flow Detection for Low-Cost, Rapid Diagnosis of Influenza a (H1N1) from Clinical Specimens. *Anal. Chem.* **2015**, *87*, 7872–7879. [CrossRef]
60. Carrilho, E.; Martinez, A.W.; Whitesides, G.M. Understanding Wax Printing: A Simple Micropatterning Process for Paper-Based Microfluidics. *Anal. Chem.* **2009**, *81*, 7091–7095. [CrossRef]
61. Shen, Y.; Mulchandani, A. Affordable Paper-Based Swnts Field-Effect Transistor Biosensors for Nucleic Acid Amplification-Free and Label-Free Detection of Micro RNAs. *Biosens. Bioelectron. X* **2023**, *14*, 100364. [CrossRef]
62. Kaarj, K.; Akarapipad, P.; Yoon, J.-Y. Simpler, Faster, and Sensitive Zika Virus Assay Using Smartphone Detection of Loop-Mediated Isothermal Amplification on Paper Microfluidic Chips. *Sci. Rep.* **2018**, *8*, 12438. [CrossRef] [PubMed]
63. Huang, D.; Ni, D.; Fang, M.; Shi, Z.; Xu, Z. Microfluidic Ruler-Readout and Crispr Cas12a-Responded Hydrogel-Integrated Paper-Based Analytical Devices (Mreach-Pad) for Visible Quantitative Point-of-Care Testing of Invasive Fungi. *Anal. Chem.* **2021**, *93*, 16965–16973. [CrossRef] [PubMed]
64. Chu, Y.; Qiu, J.; Wang, Y.; Wang, M.; Zhang, Y.; Han, L. Rapid and High-Throughput SARS-CoV-2 RNA Detection without RNA Extraction and Amplification by Using a Microfluidic Biochip. *Chem.—A Eur. J.* **2022**, *28*, e202104054. [CrossRef] [PubMed]
65. Qin, Z.; Xiang, X.; Xue, L.; Cai, W.; Gao, J.; Yang, J.; Liang, Y.; Wang, L.; Chen, M.; Pang, R. Development of a Novel Raa-Based Microfluidic Chip for Absolute Quantitative Detection of Human Norovirus. *Microchem. J.* **2021**, *164*, 106050. [CrossRef]
66. Zhang, S.; Chen, J.; Liu, D.; Hu, B.; Luo, G.; Huang, Z. A Novel Microfluidic RNA Chip for Direct, Single-Nucleotide Specific, Rapid and Partially-Degraded RNA Detection. *Talanta* **2022**, *239*, 122974. [CrossRef] [PubMed]
67. Burgers, P.M.J.; Kunkel, T.A. Eukaryotic DNA Replication Fork. *Annu. Rev. Biochem.* **2017**, *86*, 417–438. [CrossRef] [PubMed]
68. McHenry, C.S. DNA Replicases from a Bacterial Perspective. *Annu. Rev. Biochem.* **2011**, *80*, 403–436. [CrossRef] [PubMed]
69. Tian, T.; Bi, Y.; Xu, X.; Zhu, Z.; Yang, C. Integrated Paper-Based Microfluidic Devices for Point-of-Care Testing. *Anal. Methods* **2018**, *10*, 3567–3581. [CrossRef]
70. Basiri, A.; Heidari, A.; Nadi, M.F.; Fallahy, M.T.P.; Nezamabadi, S.S.; Sedighi, M.; Saghazadeh, A.; Rezaei, N. Microfluidic Devices for Detection of RNA Viruses. *Rev. Med. Virol.* **2021**, *31*, 1–11. [CrossRef]
71. Gao, D.; Ma, Z.; Jiang, Y. Recent Advances in Microfluidic Devices for Foodborne Pathogens Detection. *TrAC Trends Anal. Chem.* **2022**, *157*, 116788. [CrossRef]
72. Chen, Y.; Liu, Y.; Shi, Y.; Ping, J.; Wu, J.; Chen, H. Magnetic Particles for Integrated Nucleic Acid Purification, Amplification and Detection without Pipetting. *TrAC Trends Anal. Chem.* **2020**, *127*, 115912. [CrossRef] [PubMed]
73. Chen, Y.; Zong, N.; Ye, F.; Mei, Y.; Qu, J.; Jiang, X. Dual-Crispr/Cas12a-Assisted Rt-Raa for Ultrasensitive SARS-CoV-2 Detection on Automated Centrifugal Microfluidics. *Anal. Chem.* **2022**, *94*, 9603–9609. [CrossRef] [PubMed]
74. Chen, Y.; Mei, Y.; Jiang, X. Universal and High-Fidelity DNA Single Nucleotide Polymorphism Detection Based on a Crispr/Cas12a Biochip. *Chem. Sci.* **2021**, *12*, 4455–4462. [CrossRef] [PubMed]
75. Zong, N.; Gao, X.; Chen, Y.; Luo, X.; Jiang, X. Automated Centrifugal Microfluidic Chip Integrating Pretreatment and Molecular Diagnosis for Hepatitis B Virus Genotyping from Whole Blood. *Anal. Chem.* **2022**, *94*, 5196–5203. [CrossRef] [PubMed]
76. Chen, Y.; Mei, Y.; Zhao, X.; Jiang, X. Reagents-Loaded, Automated Assay That Integrates Recombinase-Aided Amplification and Cas12a Nucleic Acid Detection for a Point-of-Care Test. *Anal. Chem.* **2020**, *92*, 14846–14852. [CrossRef] [PubMed]
77. Pinals, R.L.; Ledesma, F.; Yang, D.; Navarro, N.; Jeong, S.; Pak, J.E.; Kuo, L.; Chuang, Y.-C.; Cheng, Y.-W.; Sun, H.-Y. Rapid SARS-CoV-2 Spike Protein Detection by Carbon Nanotube-Based near-Infrared Nanosensors. *Nano Lett.* **2021**, *21*, 2272–2280. [CrossRef]

78. Kumar, S.; Singh, H.; Feder-Kubis, J.; Nguyen, D.D. Recent Advances in Nanobiosensors for Sustainable Healthcare Applications: A Systematic Literature Review. *Environ. Res.* **2023**, *238*, 117177. [CrossRef]
79. Jiang, H.; Li, F.-R.; Li, W.; Lu, X.; Ling, K. Multiplexed Determination of Intracellular Messenger RNA by Using a Graphene Oxide Nanoprobe Modified with Target-Recognizing Fluorescent Oligonucleotides. *Microchim. Acta* **2018**, *185*, 552. [CrossRef]
80. Hwang, D.W.; Choi, Y.R.; Kim, H.; Park, H.Y.; Kim, K.W.; Kim, M.Y.; Park, C.-K.; Lee, D. Graphene Oxide-Quenching-Based Fluorescence in Situ Hybridization (G-Fish) to Detect RNA in Tissue: Simple and Fast Tissue RNA Diagnostics. *Nanomed. Nanotechnol. Biol. Med.* **2019**, *16*, 162–172. [CrossRef]
81. Li, H.; Zhang, B.; He, X.; Zhu, L.; Zhu, L.; Yang, M.; Huang, K.; Luo, H.; Xu, W. Intracellular Circrna Imaging and Signal Amplification Strategy Based on the Graphene Oxide-DNA System. *Anal. Chim. Acta* **2021**, *1183*, 338966. [CrossRef] [PubMed]
82. Yang, L.; Liu, B.; Wang, M.; Li, J.; Pan, W.; Gao, X.; Li, N.; Tang, B. A Highly Sensitive Strategy for Fluorescence Imaging of Microrna in Living Cells and in Vivo Based on Graphene Oxide-Enhanced Signal Molecules Quenching of Molecular Beacon. *ACS Appl. Mater. Interfaces* **2018**, *10*, 6982–6990. [CrossRef]
83. Harvey, J.D.; Jena, P.V.; Baker, H.A.; Zerze, G.H.; Williams, R.M.; Galassi, T.V.; Roxbury, D.; Mittal, J.; Heller, D.A. A Carbon Nanotube Reporter of Microrna Hybridization Events in Vivo. *Nat. Biomed. Eng.* **2017**, *1*, 0041. [CrossRef] [PubMed]
84. Kim, K.; Yoon, S.; Chang, J.; Lee, S.; Cho, H.H.; Jeong, S.H.; Jo, K.; Lee, J.H. Multifunctional Heterogeneous Carbon Nanotube Nanocomposites Assembled by DNA-Binding Peptide Anchors. *Small* **2020**, *16*, 1905821. [CrossRef] [PubMed]
85. Kannappan, S.; Chang, J.; Sundharbaabu, P.R.; Heo, J.H.; Sung, W.-K.; Ro, J.C.; Kim, K.K.; Rayappan, J.B.B.; Lee, J.H. DNA-Wrapped Cnt Sensor for Small Nucleic Acid Detection: Influence of Short Complementary Sequence. *BioChip J.* **2022**, *16*, 490–500. [CrossRef]
86. Ma, H.; Xue, N.; Li, Z.; Xing, K.; Miao, X. Ultrasensitive Detection of Mirna-155 Using Multi-Walled Carbon Nanotube-Gold Nanocomposites as a Novel Fluorescence Quenching Platform. *Sens. Actuators B Chem.* **2018**, *266*, 221–227. [CrossRef]
87. Zhu, W.; Qin, W.; Atasoy, U.; Sauter, E.R. Circulating Micrornas in Breast Cancer and Healthy Subjects. *BMC Res. Notes* **2009**, *2*, 89. [CrossRef] [PubMed]
88. Mattiske, S.; Suetani, R.J.; Neilsen, P.M.; Callen, D.F. The Oncogenic Role of Mir-155 in Breast Cancer. *Cancer Epidemiol. Biomark. Prev.* **2012**, *21*, 1236–1243. [CrossRef] [PubMed]
89. Wu, H.; Chen, X.; Zhang, M.; Wang, X.; Chen, Y.; Qian, C.; Wu, J.; Xu, J. Versatile Detection with Crispr/Cas System from Applications to Challenges. *TrAC Trends Anal. Chem.* **2021**, *135*, 116150. [CrossRef]
90. Safari, F.; Hatam, G.; Behbahani, A.B.; Rezaei, V.; Barekati-Mowahed, M.; Petramfar, P.; Khademi, F. Crispr System: A High-Throughput Toolbox for Research and Treatment of Parkinson's Disease. *Cell. Mol. Neurobiol.* **2020**, *40*, 477–493. [CrossRef]
91. Chen, J.S.; Ma, E.; Harrington, L.B.; Da Costa, M.; Tian, X.; Palefsky, J.M.; Doudna, J.A. Crispr-Cas12a Target Binding Unleashes Indiscriminate Single-Stranded Dnase Activity. *Science* **2018**, *360*, 436–439. [CrossRef]
92. Abudayyeh, O.O.; Gootenberg, J.S.; Konermann, S.; Joung, J.; Slaymaker, I.M.; Cox, D.B.T.; Shmakov, S.; Makarova, K.S.; Semenova, E.; Minakhin, L.; et al. C2c2 Is a Single-Component Programmable Rna-Guided Rna-Targeting Crispr Effector. *Science* **2016**, *353*, aaf5573. [CrossRef]
93. Wang, S.; Wei, S.; Wang, S.; Zhu, X.; Lei, C.; Huang, Y.; Nie, Z.; Yao, S. Chimeric DNA-Functionalized Titanium Carbide Mxenes for Simultaneous Mapping of Dual Cancer Biomarkers in Living Cells. *Anal. Chem.* **2018**, *91*, 1651–1658. [CrossRef]
94. Wang, S.; Song, W.; Wei, S.; Zeng, S.; Yang, S.; Lei, C.; Huang, Y.; Nie, Z.; Yao, S. Functional Titanium Carbide Mxenes-Loaded Entropy-Driven RNA Explorer for Long Noncoding RNA Pca3 Imaging in Live Cells. *Anal. Chem.* **2019**, *91*, 8622–8629. [CrossRef] [PubMed]
95. Qiu, X.-Y.; Zhu, L.-Y.; Zhu, C.-S.; Ma, J.-X.; Hou, T.; Wu, X.-M.; Xie, S.-S.; Min, L.; Tan, D.-A.; Zhang, D.-Y.; et al. Highly Effective and Low-Cost Microrna Detection with Crispr-Cas9. *ACS Synth. Biol.* **2018**, *7*, 807–813. [CrossRef]
96. Wang, X.-W.; Hu, L.-F.; Hao, J.; Liao, L.-Q.; Chiu, Y.-T.; Shi, M.; Wang, Y. A Microrna-Inducible Crispr–Cas9 Platform Serves as a Microrna Sensor and Cell-Type-Specific Genome Regulation Tool. *Nat. Cell Biol.* **2019**, *21*, 522–530. [CrossRef] [PubMed]
97. Hirosawa, M.; Fujita, Y.; Saito, H. Cell-Type-Specific Crispr Activation with Microrna-Responsive Acrlla4 Switch. *ACS Synth. Biol.* **2019**, *8*, 1575–1582. [CrossRef] [PubMed]
98. Li, Y.; Teng, X.; Zhang, K.; Deng, R.; Li, J. RNA Strand Displacement Responsive Crispr/Cas9 System for mRNA Sensing. *Anal. Chem.* **2019**, *91*, 3989–3996. [CrossRef]
99. Koo, B.; Kim, D.-E.; Kweon, J.; Jin, C.E.; Kim, S.-H.; Kim, Y.; Shin, Y. Crispr/Dcas9-Mediated Biosensor for Detection of Tick-Borne Diseases. *Sens. Actuators B Chem.* **2018**, *273*, 316–321. [CrossRef]
100. Marsic, T.; Ali, Z.; Tehseen, M.; Mahas, A.; Hamdan, S.; Mahfouz, M. Vigilant: An Engineered Vird2-Cas9 Complex for Lateral Flow Assay-Based Detection of SARS-CoV2. *Nano Lett.* **2021**, *21*, 3596–3603. [CrossRef]
101. Aman, R.; Mahas, A.; Mahfouz, M. Nucleic Acid Detection Using Crispr/Cas Biosensing Technologies. *ACS Synth. Biol.* **2020**, *9*, 1226–1233. [CrossRef] [PubMed]
102. Pickar-Oliver, A.; Gersbach, C.A. The Next Generation of Crispr–Cas Technologies and Applications. *Nat. Rev. Mol. Cell Biol.* **2019**, *20*, 490–507. [CrossRef]
103. Li, Y.; Li, S.; Wang, J.; Liu, G. Crispr/Cas Systems Towards Next-Generation Biosensing. *Trends Biotechnol.* **2019**, *37*, 730–743. [CrossRef]

104. Gootenberg, J.S.; Abudayyeh, O.O.; Kellner, M.J.; Joung, J.; Collins, J.J.; Zhang, F. Multiplexed and Portable Nucleic Acid Detection Platform with Cas13, Cas12a, and Csm6. *Science* **2018**, *360*, 439–444. [CrossRef] [PubMed]
105. Gootenberg, J.S.; Abudayyeh, O.O.; Lee, J.W.; Essletzbichler, P.; Dy, A.J.; Joung, J.; Verdine, V.; Donghia, N.; Daringer, N.M.; Freije, C.A.; et al. Nucleic Acid Detection with Crispr-Cas13a/C2c2. *Science* **2017**, *356*, 438–442. [CrossRef]
106. Broughton, J.P.; Deng, X.; Yu, G.; Fasching, C.L.; Servellita, V.; Singh, J.; Miao, X.; Streithorst, J.A.; Granados, A.; Sotomayor-Gonzalez, A. Crispr-Cas12-Based Detection of SARS-CoV-2. *Nat. Biotechnol.* **2020**, *38*, 870–874. [CrossRef]
107. Mustafa, M.I.; Makhawi, A.M. Sherlock and Detectr: Crispr-Cas Systems as Potential Rapid Diagnostic Tools for Emerging Infectious Diseases. *J. Clin. Microbiol.* **2021**, *59*, 10-1128. [CrossRef]
108. Li, S.-Y.; Cheng, Q.-X.; Liu, J.-K.; Nie, X.-Q.; Zhao, G.-P.; Wang, J. Crispr-Cas12a Has Both Cis-and Trans-Cleavage Activities on Single-Stranded DNA. *Cell Res.* **2018**, *28*, 491–493. [CrossRef]
109. Li, S.-Y.; Cheng, Q.-X.; Wang, J.-M.; Li, X.-Y.; Zhang, Z.-L.; Gao, S.; Cao, R.-B.; Zhao, G.-P.; Wang, J. Crispr-Cas12a-Assisted Nucleic Acid Detection. *Cell Discov.* **2018**, *4*, 20. [CrossRef] [PubMed]
110. Liang, M.; Li, Z.; Wang, W.; Liu, J.; Liu, L.; Zhu, G.; Karthik, L.; Wang, M.; Wang, K.-F.; Wang, Z.; et al. A Crispr-Cas12a-Derived Biosensing Platform for the Highly Sensitive Detection of Diverse Small Molecules. *Nat. Commun.* **2019**, *10*, 3672. [CrossRef] [PubMed]
111. Ji, Z.; Zhou, B.; Shang, Z.; Liu, S.; Li, X.; Zhang, X.; Li, B. Active Crispr-Cas12a on Hydrophilic Metal–Organic Frameworks: A Nanobiocomposite with High Stability and Activity for Nucleic Acid Detection. *Anal. Chem.* **2023**, *95*, 10580–10587. [CrossRef] [PubMed]
112. Tian, T.; Shu, B.; Jiang, Y.; Ye, M.; Liu, L.; Guo, Z.; Han, Z.; Wang, Z.; Zhou, X. An Ultralocalized Cas13a Assay Enables Universal and Nucleic Acid Amplification-Free Single-Molecule RNA Diagnostics. *ACS Nano* **2020**, *15*, 1167–1178. [CrossRef] [PubMed]
113. Gao, G.; Zhu, X.; Lu, B. Development and Application of Sensitive, Specific, and Rapid Crispr-Cas13-Based Diagnosis. *J. Med. Virol.* **2021**, *93*, 4198–4204.
114. Shan, Y.; Zhou, X.; Huang, R.; Xing, D. High-Fidelity and Rapid Quantification of Mirna Combining Crrna Programmability and Crispr/Cas13a Trans-Cleavage Activity. *Anal. Chem.* **2019**, *91*, 5278–5285. [CrossRef] [PubMed]
115. East-Seletsky, A.; O'Connell, M.R.; Knight, S.C.; Burstein, D.; Cate, J.H.D.; Tjian, R.; Doudna, J.A. Two Distinct Rnase Activities of Crispr-C2c2 Enable Guide-RNA Processing and RNA Detection. *Nature* **2016**, *538*, 270–273. [CrossRef] [PubMed]
116. Shinoda, H.; Taguchi, Y.; Nakagawa, R.; Makino, A.; Okazaki, S.; Nakano, M.; Muramoto, Y.; Takahashi, C.; Takahashi, I.; Ando, J.; et al. Amplification-Free RNA Detection with Crispr–Cas13. *Commun. Biol.* **2021**, *4*, 476. [CrossRef] [PubMed]
117. Wang, Y.; Xue, T.; Wang, M.; Ledesma-Amaro, R.; Lu, Y.; Hu, X.; Zhang, T.; Yang, M.; Li, Y.; Xiang, J. Crispr-Cas13a Cascade-Based Viral RNA Assay for Detecting SARS-CoV-2 and Its Mutations in Clinical Samples. *Sens. Actuators B Chem.* **2022**, *362*, 131765. [CrossRef] [PubMed]
118. Paige, J.S.; Nguyen-Duc, T.; Song, W.; Jaffrey, S.R. Fluorescence Imaging of Cellular Metabolites with RNA. *Science* **2012**, *335*, 1194. [CrossRef] [PubMed]
119. Paige, J.S.; Wu, K.Y.; Jaffrey, S.R. RNA Mimics of Green Fluorescent Protein. *Science* **2011**, *333*, 642–646. [CrossRef]
120. Ying, Z.-M.; Wu, Z.; Tu, B.; Tan, W.; Jiang, J.-H. Genetically Encoded Fluorescent RNA Sensor for Ratiometric Imaging of Microrna in Living Tumor Cells. *J. Am. Chem. Soc.* **2017**, *139*, 9779–9782. [CrossRef]
121. Fozouni, P.; Son, S.; de León Derby, M.D.; Knott, G.J.; Gray, C.N.; D'Ambrosio, M.V.; Zhao, C.; Switz, N.A.; Kumar, G.R.; Stephens, S.I. Amplification-Free Detection of SARS-CoV-2 with Crispr-Cas13a and Mobile Phone Microscopy. *Cell* **2021**, *184*, 323–333.e9. [CrossRef] [PubMed]
122. Arizti-Sanz, J.; Freije, C.A.; Stanton, A.C.; Petros, B.A.; Boehm, C.K.; Siddiqui, S.; Shaw, B.M.; Adams, G.; Kosoko-Thoroddsen, T.S.-F.; Kemball, M.E. Streamlined Inactivation, Amplification, and Cas13-Based Detection of SARS-CoV-2. *Nat. Commun.* **2020**, *11*, 5921. [CrossRef] [PubMed]
123. Patchsung, M.; Jantarug, K.; Pattama, A.; Aphicho, K.; Suraritdechachai, S.; Meesawat, P.; Sappakhaw, K.; Leelahakorn, N.; Ruenkam, T.; Wongsatit, T. Clinical Validation of a Cas13-Based Assay for the Detection of SARS-CoV-2 RNA. *Nat. Biomed. Eng.* **2020**, *4*, 1140–1149. [CrossRef] [PubMed]
124. Ke, Y.; Huang, S.; Ghalandari, B.; Li, S.; Warden, A.R.; Dang, J.; Kang, L.; Zhang, Y.; Wang, Y.; Sun, Y. Hairpin-Spacer Crrna-Enhanced Crispr/Cas13a System Promotes the Specificity of Single Nucleotide Polymorphism (SNP) Identification. *Adv. Sci.* **2021**, *8*, 2003611. [CrossRef] [PubMed]
125. World Health Organization. *Laboratory Testing for Coronavirus Disease 2019 (COVID-19) in Suspected Human Cases: Interim Guidance, 2 March 2020*; World Health Organization: Geneva, Switzerland, 2020.
126. Feng, W.; Newbigging, A.M.; Le, C.; Pang, B.; Peng, H.; Cao, Y.; Wu, J.; Abbas, G.; Song, J.; Wang, D.-B.; et al. Molecular Diagnosis of COVID-19: Challenges and Research Needs. *Anal. Chem.* **2020**, *92*, 10196–10209. [CrossRef] [PubMed]
127. Yan, S.; Ahmad, K.Z.; Warden, A.R.; Ke, Y.; Maboyi, N.; Zhi, X.; Ding, X. One-Pot Pre-Coated Interface Proximity Extension Assay for Ultrasensitive Co-Detection of Anti-SARS-CoV-2 Antibodies and Viral RNA. *Biosens. Bioelectron.* **2021**, *193*, 113535. [CrossRef] [PubMed]

128. Li, S.; Huang, S.; Ke, Y.; Chen, H.; Dang, J.; Huang, C.; Liu, W.; Cui, D.; Wang, J.; Zhi, X. A Hipad Integrated with Rgo/Mwcnts Nano-Circuit Heater for Visual Point-of-Care Testing of SARS-CoV-2. *Adv. Funct. Mater.* **2021**, *31*, 2100801. [CrossRef] [PubMed]
129. Wei, Y.; Tao, Z.; Wan, L.; Zong, C.; Wu, J.; Tan, X.; Wang, B.; Guo, Z.; Zhang, L.; Yuan, H.; et al. Aptamer-Based Cas14a1 Biosensor for Amplification-Free Live Pathogenic Detection. *Biosens. Bioelectron.* **2022**, *211*, 114282. [CrossRef]
130. Ge, H.; Wang, X.; Xu, J.; Lin, H.; Zhou, H.; Hao, T.; Wu, Y.; Guo, Z. A Crispr/Cas12a-Mediated Dual-Mode Electrochemical Biosensor for Polymerase Chain Reaction-Free Detection of Genetically Modified Soybean. *Anal. Chem.* **2021**, *93*, 14885–14891. [CrossRef]

Disclaimer/Publisher's Note: The statements, opinions and data contained in all publications are solely those of the individual author(s) and contributor(s) and not of MDPI and/or the editor(s). MDPI and/or the editor(s) disclaim responsibility for any injury to people or property resulting from any ideas, methods, instructions or products referred to in the content.

Review

Biosensors for Cancer Biomarkers Based on Mesoporous Silica Nanoparticles

Minja Mladenović [†], Stefan Jarić [†], Mirjana Mundžić, Aleksandra Pavlović, Ivan Bobrinetskiy and Nikola Ž. Knežević *

BioSense Institute, University of Novi Sad, Dr Zorana Djindjica 1, 21000 Novi Sad, Serbia; minja.mladenovic@biosense.rs (M.M.); sjaric@biosense.rs (S.J.); mirjana.mundzic@biosense.rs (M.M.); aleksandra.pavlovic@biosense.rs (A.P.)
* Correspondence: nknezevic@biosense.rs
[†] These authors contributed equally to this work.

Abstract: Mesoporous silica nanoparticles (MSNs) exhibit highly beneficial characteristics for devising efficient biosensors for different analytes. Their unique properties, such as capabilities for stable covalent binding to recognition groups (e.g., antibodies or aptamers) and sensing surfaces, open a plethora of opportunities for biosensor construction. In addition, their structured porosity offers capabilities for entrapping signaling molecules (dyes or electroactive species), which could be released efficiently in response to a desired analyte for effective optical or electrochemical detection. This work offers an overview of recent research studies (in the last five years) that contain MSNs in their optical and electrochemical sensing platforms for the detection of cancer biomarkers, classified by cancer type. In addition, this study provides an overview of cancer biomarkers, as well as electrochemical and optical detection methods in general.

Keywords: MSNs; biosensors; cancer biomarkers; electrochemical detection; optical detection

1. Introduction

Mesoporous silica nanoparticles (MSNs) are a mesostructured porous network of silicon oxide produced through the hydrolytic sol–gel process involving the hydrolysis and condensation of silicon alkoxide precursors under acidic or basic conditions in the presence of a surfactant template [1]. The formation of mesoporous silica typically involves a silica–surfactant micelle templating process, in which condensation of the silicon alkoxide precursor takes place around the surfactant acting as a structure-directing agent [2]. In this manner, hexagonally ordered spherical particles are obtained. MSNs possess precisely controllable physicochemical characteristics, such as particle size, morphology, pore dimensions and volume, surface area, and surface properties [3]. Furthermore, facile surface modification offers the possibility of tailoring the chemical and physical properties of MSNs to achieve specific characteristics or functionalities [4]. The adaptability of MSNs in terms of size, shape, and composition has led to their widespread utilization across various fields [5–8].

Bioanalytical devices that integrate nanotechnology with biological recognition elements, along with physicochemical transducers to detect and quantify specific biological and chemical substances, are considered nanobiosensors [9]. A transducer is an element that enables the conversion of the target–bioreceptor interaction into a measurable signal. On the other hand, a bioreceptor is a molecule that can interact with a specific analyte and gives the biosensor its specificity [10]. Selectivity, sensitivity, and stability are some of the main characteristics to consider when developing biosensors [11].

Integrating nanomaterials (NMs) in diagnostics opens a potential for increased sensitivity, reduced processing times, and improved cost-effectiveness [12] due to favorable NM characteristics such as increased relative surface area, high surface-to-volume ratio,

Citation: Mladenović, M.; Jarić, S.; Mundžić, M.; Pavlović, A.; Bobrinetskiy, I.; Knežević, N.Ž. Biosensors for Cancer Biomarkers Based on Mesoporous Silica Nanoparticles. *Biosensors* **2024**, *14*, 326. https://doi.org/10.3390/bios14070326

Received: 12 June 2024
Revised: 25 June 2024
Accepted: 28 June 2024
Published: 30 June 2024

Copyright: © 2024 by the authors. Licensee MDPI, Basel, Switzerland. This article is an open access article distributed under the terms and conditions of the Creative Commons Attribution (CC BY) license (https://creativecommons.org/licenses/by/4.0/).

and quantum confinement effects [13]. Among the variety of NMs, MSNs are of significant relevance in medical diagnostics and sensing applications, presenting a promising tool for the development of advanced nanobiosensors [14–16]. However, the successful deployment of MSN-based nanobiosensors depends on a comprehensive understanding of how synthesis methods and post-synthesis modifications influence their performance, considering that the presence of functional groups as well as particle morphology and size is of vital influence on their behavior [13].

On the other hand, cancer remains a leading cause of death worldwide. In 2022, there were an estimated 9.7 million cancer-related deaths globally [17]. The early diagnosis of cancer is crucial for improving patient outcomes. Detecting cancer at an early stage allows for more effective treatment options and potentially better prognosis, increasing the chances of successful therapy and resulting in higher survival rates. Cancer arises from disruptions in normal cell signaling pathways, leading to the emergence of cancer cells with a significant growth advantage [18]. These changes result from a variety of genetic and epigenetic alterations, activating oncogenes and deactivating tumor suppressor genes [19]. However, there is not a single universal gene mutation found across all cancers, and patterns of genetic changes vary not only by tumor location but also within tumors from the same location. With over 200 different cancer types affecting various parts of the body, clinical testing becomes intricate. Given the complexity of, and variability in, cancer-related changes, selecting specific biomarkers for diagnosis is challenging [20]. The National Cancer Institute defines a biomarker as "a biological molecule found in blood, other body fluids, or tissues that is a sign of a normal or abnormal process, or of a condition or disease". They are produced as the body's immune response or by the cancerous cells themselves [21]. Determination of cancer biomarker levels in biofluids can be used to detect cancer at different stages or to monitor the outcome of therapy [22]. An important step in ensuring good sensitivity and selectivity for early-stage cancer detection is the identification of appropriate biomarkers as well as the type of biofluid [23]. In addition, the recognition interaction between the biorecognition molecule and the biomarker may also dictate sensitivity [24]. Furthermore, during disease diagnosis, a range of biomarkers is typically analyzed. This means that reliable non-invasive cancer diagnosis often requires simultaneous determination of multiple biomarkers found in different body fluids using different techniques, which are still looking for standardization and validation [25], although there are some techniques that are used in clinical diagnosis as a gold standard. For example, protein biomarkers are principally quantified by the enzyme-linked immunosorbent assay (ELISA), where the target molecule is specifically bound to its natural counterpart—antibody—while the secondary antibody labeled with a certain enzyme acts as a messenger providing a colorimetric signal when the appropriate substrate is added to the reaction well [26]. While the ELISA method can be highly sensitive and selective in complex matrices, it is limited by the moderate risk of false-positive signal production due to its colorimetric nature of detection or nonspecific binding to the reaction well and the high cost of production. To detect DNA/RNA biomarkers, the amplification of nucleic acid is performed to increase the probability of signal detection. Among different methods, the polymerase chain reaction (PCR) method stood out as a golden standard in nucleic acid amplification tools, where using the specific set of oligonucleotides (primers) and enzyme polymerases, the amplification of the DNA segment previously denatured through multiple temperature variation cycles is achieved [27]. However, PCR methods often require a specific laboratory environment and equipment since it is very sensitive to contamination. Despite challenges originating from these and other standard methods, their application is still the first choice due to the lack of appropriate options in clinical use. Additionally, there is a demand for the (RE)ASSURED ((Real-time connectivity, Ease of specimen collection, and environmental friendliness), Affordable, Sensitive, Specific, User-friendly, Rapid and robust, Equipment-free, and Deliverable to end-users) criteria proposed by the World Health Organization to describe and develop an ideal point-of-care testing (POCT) system, primarily in medical applications [28].

There are different strategies in which MSNs are incorporated into biosensors to close the ASSURED circuit. Based on the transduction signal, the most common biosensors for cancer biomarker detection are electrochemical, optical, and colorimetric [29].

2. Cancer Biomarkers

2.1. General Discussion on Cancer Biomarkers and Relevant Biosensors

Current clinical practice in oncology emphasizes the importance of early diagnosis, proper prognostication, and screening for malignancy in its pre-invasive stage (before metastasis). The important role of biomarkers is widely recognized in research, medicine, and pharmacology. Apart from replacing clinical endpoints and reducing the time and the costs for Phase I and Phase II clinical trials, their levels are measured to diagnose a disease or to monitor treatment efficacy and disease progression [30].

Markers usually differentiate an affected patient from a healthy person. Upon tumor formation, levels of tumor markers rise accordingly, stressing the importance of limits of detection (LOD) for early screening stages. In the case of carcinoembryonic antigen (CEA), the threshold is as low as 3 µg/L of the sample or it can go up to 12.5 mg/L in the case of neuron-specific enolase (NSE) [31]. Apart from sensitivity, specificity is another important perspective. Tumor markers can be associated with different tumors, and most of the tumors have more than one marker associated with their onset and growth. The specificity and sensitivity of a lot of markers are being evaluated for clinical use [32]. Unfortunately, none of the currently described biomarkers achieve 100% sensitivity or specificity. For example, the sensitivity of the prostate-specific antigen (PSA), a serum biomarker for prostate cancer, is greater than 90%, but it has a specificity of only around 25%, resulting in patients needing to undergo a biopsy for the final confirmation of disease [33].

Cancer biomarkers can be different types of molecules such as proteins, DNA, RNA, micro-RNA, peptides, hormones, oncofetal antigens, cytokeratins, exosomes, and carbohydrates. One of the first discovered tumor antigens was carcinoembryonic antigen CEA, a glycoprotein molecule isolated in 1965 [34]. Biomarkers can be intra or extracellular. In cases in which biomarkers are intracellular, cells need to be lysed to collect them. Biomarkers are detectable in tissues and/or biological fluids like blood (whole blood, serum, or plasma) and secretions (stool, urine, sputum, or nipple discharge), and thus can be collected non-invasively [35].

There are a few traditional cancer screening methods such as mammography and the fecal occult blood test followed by colonoscopy. Nowadays, scientists are creating tools at the molecular level to measure molecular alterations in the process of tumor growth. Genetic alterations can be inherited, confirmed as sequence variations in isolated DNA, or somatic, identified as mutations in isolated DNA [36]. Although DNA methylation can be studied by Southern blotting, DNA sequencing, DNA microarrays, and PCR, these genomic methods are complex and time-consuming, and genetic markers do not give information on post-translational modifications on proteins. Therefore, protein-based biomarkers are often referred to as the "classic" ones in the literature.

Some of the common biomarkers for different cancer types are listed in Table 1.

Table 1. Basic cancer biomarkers and associated cancers.

Cancer Biomarker	Cancer Type	Reference
PSA	Prostate	[37]
IgG	Prostate	[38]
PAP, PSA	Prostate	[39]
Peptide fragments	Colorectal	[40]
MMP	Colorectal	[41]
CEA, CA 19-9, CA A24-2	Colorectal and pancreatic	[42]

Table 1. *Cont.*

Cancer Biomarker	Cancer Type	Reference
P53 gene	Colorectal	[43]
CYFRA 21-1	Lung	[44]
CEA, CA 19-9, SCC antigen, NSE	Lung	[45]
EVOM	Breast	[46]
EGFR, HER2, transmembrane glycoproteins CD44 and CD24	Breast	[47]
Sialic acid	Breast and liver	[48]
AFP	Liver	[49]
CA 125, HE4	Ovarian	[50]
TRP-2, NY-ESO-1 melanoma Antigen	Melanoma	[51]

Abbreviations: IgG—immunoglobulin G; PAP—prostatic acid phosphatase; MMP—matrix metalloproteinase; CA—cancer antigen; CYFRA—cytokeratin fragment; SCC—squamous cell carcinoma; EVOM—endogenous volatile organic metabolites; EGFR—epidermal growth factor receptor; HER—human epidermal growth factor receptor; SA—sialic acid; AFR—α-fetoprotein; HE—human epididymis protein; TRP—tyrosinase-related protein.

As can be seen from Table 1, CEA and CA 19-9 are common cancer biomarkers for several different tumors. Because of that, better than relying on one single biomarker, a panel of biomarkers is more promising as disease predictors [52].

In this review, nanobiosensors that involve MSNs for the detection of biomarkers that are useful in the diagnosis of cancer are discussed.

2.2. Specific Cancer Biomarkers Targeted by MSN-Based Biosensors

Recent research studies (in the last five years) involving MSN-based biosensors have been focused on several cancer biomarkers, as detailed in this section and further elaborated with MSN-based biosensor assemblies in the following sections.

Glutathione is a tripeptide (consisting of glutamate, cysteine, and glycine) containing a sulfhydryl group by means of which it is conjugated to other molecules [53]. It is distributed in most mammalian cells and is present in intracellular concentrations from 0.1 to 10 mM [54]. It is an important non-enzymatic antioxidant with a central role in the regulation of reactive oxygen species (ROS) [55]. GSH has been found in increased levels of breast, ovarian, head and neck, and lung cancers [56].

Cytokeratins are polypeptides expressed by all epithelial cells [57]. There are 20 cytokeratins, distinguished by their molecular weight and isoelectric points, which are classified into two groups: acidic and basic–neutral [58]. The fragment of cytokeratin subunit 19, known as CYFRA 21-1, has been recognized as an accurate and specific tumor marker for detecting non-small-cell lung cancer (NSCLC), particularly the squamous cell subtype [59]. CYFRA 21-1 is frequently investigated as a lung cancer biomarker utilizing integrated sensing platforms based on MSNs.

CA 15-3, also known as MUC1, is the most used serum marker for breast cancer. It is a large transmembrane glycoprotein that is often overexpressed and abnormally glycosylated in cancerous cells. Under normal conditions, it is involved in cell adhesion, but its elevated levels in cancer may contribute to metastasis [60].

The human epidermal growth factor receptor 2 (HER2) is part of the epidermal growth factor (EGF) family. HER2 positivity is observed in approximately 15–20% of breast cancers, characterized by the overexpression of the HER2 protein. The HER2 protein promotes cell growth. However, when HER2 is overexpressed, it can lead to aggressive growth of cancer cells [61].

The epithelial cell adhesion molecule (EpCAM) is a type I transmembrane protein consisting of 314 amino acids that are involved in cell signaling and carcinogenesis [62]. EpCAM is notably expressed in most human epithelial cancers, including colorectal, breast, gastric, prostate, ovarian, and lung cancers [63].

Total serum acid phosphatase was the first clinically useful prostate tumor marker to be discovered [64]. ACP is a lysosomal enzyme that breaks down organic phosphates in an acidic environment [65]. ACP is found to be present in the prostate in 100 times higher quantities than in any other tissue type [66]. PSA, or prostate-specific antigen, is a serine protease primarily produced by prostate cells and released into the ejaculate to help liquefy semen [67]. Normally, PSA levels in the blood are low. However, changes in the normal structure of the prostate, such as those caused by cancer, can result in increased levels of PSA in the blood [68].

The presence of high-risk human papillomavirus (HPV) is the most important cervical biomarker as it is strongly associated with the development of this type of cancer [69]. One of the high-risk types, HPV16, is closely associated with invasive cervical cancer. It is well known that the expression of HPV16 E6 oncoprotein is essential for transforming normal cells into cancerous ones [70]. Moreover, it has been proved that the HPV E6 oncogene induces the functional suppression of the tumor suppressor gene p53 [71].

Kato and Tarigoe first identified the squamous cell carcinoma (SCC) antigen in 1997 [72], which was found to be present in both neutral and acidic sub-fractions of tumor antigen 4 (TA-4). Squamous cell carcinomas account for 85–90% of all cervical cancers, while elevated serum levels of SCC have been observed in 28–88% of cases of cervical squamous cell carcinoma [73].

CA 19-9 is the most common diagnostic biomarker for pancreatic cancer, which is used for prognosis and prediction of treatment outcomes. Among other carbohydrate antigens, CEA is also being extensively investigated as a pancreatic tumor marker, but it has not been found to be more sensitive than CA 19-9 [74].

Glypican 1 (GPC1) is a membrane-anchoring protein, which is highly expressed in pancreatic cancer tissue compared to normal tissue [75].

Current strategies for screening ovarian cancer involve using a combination of blood biomarkers, CA 125, and HE-4, along with transvaginal ultrasound imaging. CA 125, also known as MUC16, has been found in up to 80% of women diagnosed with late-stage epithelial ovarian cancer at elevated levels. HE4, also known as WAP 4-disulfide core domain 2, is linked to cancer cell adhesion, migration, and tumor growth [76].

3. Detection Methods

3.1. Electrochemical Detection Methods

Concerning electrochemical biosensors, the transduction element is often denoted as an electrochemical cell predominantly consisting of three electrodes—working (WE), reference (RE), and counter electrode (CE)—fabricated on a substrate in a co-planar configuration [77]. Various materials are used for the electrode production, from metallic (gold, platinum, etc.), metal-oxides (TiO_2, ZnO, etc.), carbon-based (glassy carbon, carbon nanomaterials), to polymeric, and different microfabrication methods such as screen-printing, ink-jet printing, photolithography, and others are deployed [78]. In the electrochemical approach, a signal is obtained as a result of electron transfer between the working electrode (transducer) and electrolyte (sample) and it can be measured as current, potential, or impedance of the electrochemical cell. Direct or indirect transduction can be applied, the former is associated with enzymatic biosensors and the latter with mediator-based biosensors [79]. Mediators are small molecules with a low molecular weight, which transfer electrons from the reaction site to the electrode [80] and are usually referred to as redox probes. With the advancement of nanotechnology, in-house or commercial electrode systems have been modified with distinct nanomaterials to improve the overall electrochemical biosensor performance [81]; these systems rely on nanomaterial-enabled signal amplification by exploiting the improved electrochemical properties [82]. Today, many electrochemical techniques are used to study the biosensing mechanism at the surface of the WE and they can be grouped into potentiometric, amperometric, voltammetric, and impedimetric techniques [83]. In potentiometric measurements, the potential between two electrodes, usually WE and RE, is recorded and it can provide definite information about the target analyte presence. The working principle of amperometric biosensors relies on the current amplitude resulting from the redox processes of electroactive species on the WE at the applied potential. Voltammetric techniques are widely used in electrochemical biosensors since the current between WE and CE is monitored during a predetermined potential sweep applied between WE and RE. Depending on the potential type, there are several methods used in biosensing: cyclic voltammetry (CV), differential pulse voltammetry (DPV), square wave voltammetry (SWV), or anodic/cathodic stripping voltammetry. In impedimetric measurements, the total impedance of the electrode/electrolyte electrical circuit is monitored at applied voltage. In Table 2, an overview of the electrochemical techniques applied to MSN-based biosensing of cancer biomarkers is given.

Table 2. An overview of electrochemical techniques used for the biosensing of cancer biomarkers with MSNs applied to enable or enhance the signal of detection.

Technique	Method	MSN Role	Target Biomarker	Key Performances	Reference
Potentiometry	Commercial glucometer	Release of glucose upon target cDNA hybridization	miRNA-21	50 pM–5 nM [1] 19 pM [2]	[84]
		Release of glucose upon target binding to antibody	CYFRA 21-1	1.3–160 ng/mL [1]	[85]
	Open circuit voltage	Release of $[Fe(CN)_6]^{3-}$ upon target cDNA hybridization	miRNA-21	10 aM–1 pM [1]	[86]
	Chrono-potentiometry	MIP performance improvement	Sarcosine	10 nM–10 µM [1] 7.8 Nm [2]	[87]
Amperometry	Chrono-amperometry	Lactate oxidase immobilization	Lactic acid	40–500 µM [1]	[88]
Voltammetry	Cyclic voltammetry	Antibody immobilization with AgNP for electron transfer improvement	PSA	50 pg/mL–50 ng/mL [1] 15 pg/mL [2]	[89]
	Differential pulse voltammetry	Release of glucose from target-bound MSNs	CA 19-9	0.01–100 U/mL [1] 0.0005 U/mL [2]	[90]
		Dual-labeled MSNs with AuNRs and HRP for signal enhancement	CEA	0.1–5 pg/mL [1] 5.25 fg/mL [2]	[91]
		Sandwich-type immunoassay with MB@MSNs for signal enhancement	HPV16 E6 oncoprotein	50 fg/mL–4 ng/mL [1]	[92]
		Amino-MSNs in composite with Amino-rGO and IL for signal enhancement	Lysozyme	20 fM–50 nM [1]	[93]
		SNA-loaded MSNs for improved capture of target	MCF-7 cancer cells	1–1.0 × 10⁷ cells/mL [1] 4 cells/mL [2]	[94]
		Sandwich-type immunoassay with MMSN@AuNP-Ab2 for signal enhancement	CYFRA 21-1	0.01–1.0 pg/mL [1] 2 fg/mL [2]	[95]
		Sandwich-type immunoassay with thionine-loaded MSNs for signal enhancement	SCCA	0.01–120 ng/mL [1] 0.33 pg/mL [2]	[96]
	Square wave voltammetry	Sandwich-type immunoassay with MB-loaded MSNs for signal production by controlled MB release	PSA	10 fg/mL–100 ng/mL [1] 1.25 fg/mL [2]	[97]
		Release of MB from programmed target-enabled CHA for HCR signal amplification	miRNA-21	0.1 fM–5 pM [1]	[98]
		Sensitivity improvement by MSNs/PtNPs	CD133	5–20 cells/5 µL [1]	[99]
	Square wave anodic/cathodic stripping voltammetry	Nanocomposites for signal development and enhancement: PbS-QD@MSNs, CdTe-QD@MSNs, and AuNPs@MSNs	HE4, CA-125, and AFP	HE4: 0.02–20 pM [1]; LOD 5.07 pM CA-125: 0.45–450 IU/L [1]; LOD 3.1 IU/L AFP: 0.1–500 ng/L [1]; LOD 2.44 pg/L	[100]

Table 2. Cont.

Technique	Method	MSN Role	Target Biomarker	Key Performances	Reference
Impedimetry	Electrochemical impedance spectroscopy	Amino-MSNs in composite with Amino-rGO and IL for signal enhancement	Lysozyme	10 fM–200 nM [1]	[93]
Photoelectrochemical method	Chrono-amperometry	CD@MSB for improved sensitivity	Glutathione	34.9 nM [2]	[101]

[1] Linear range. [2] LOD. Abbreviations: MIP—molecularly imprinted polymer; cDNA—complementary DNA; Ag NPs—silver nanoparticles; AuNRs—gold nanorods; HRP—horseradish peroxidase; MB—methylene blue; HPV16—human papillomavirus 16; rGO—reduced graphene oxide; IL—ionic liquid; SNA—sambucus nigra agglutinin; MMSN—magnetic MSN; AuNPs—gold nanoparticles; CHA—catalytic hairpin assembly; HCR—hybridization chain reaction; PtNPs—platinum nanoparticles; PbS QDs—lead sulfide quantum dots; CdTe QDs—cadmium telluride quantum dots; and CDs—carbon dots.

3.2. Optical Detection Methods

Optical biosensors use optical properties of the transducer and the optical signal is detected, e.g., electromagnetic radiation in the optical range. From the perspective of signal origin, optical biosensors are usually divided into two groups: label-based and label-free biosensors [102]. Labeled optical biosensors use specific molecules (labels) responsible for signal generation; these can be colorimetric or fluorescent biosensors [103,104]. On the contrary, label-free optical biosensors utilize the change in optical radiation properties of the transducer upon biochemical interaction, such as amplitude, frequency, phase, and polarization [105], but can also be manifested by a measurable physical property of the biosensing interface, such as refractive index.

Colorimetric biosensors are an attractive sub-field because they provide a simple setup and fast analysis, but since the signal is based on a visible color change, such an approach may be insufficient for analyte quantification, complicating the colorimetric scheme of detection. Nevertheless, colorimetric biosensors offer the most criteria for the development of POCT devices precisely because of the visible signal that is easily read by the end user. The quantification of colorimetric data is usually performed using a spectrometric technique to determine the signal, such as UV-Vis absorption spectra [106]. Principally, colorimetric biosensors can be denoted as both label and label-free devices. Owing to the nanoparticle's unique properties, which can be expressed in the optical signal, their role in colorimetric biosensors is of a transducing nature, producing an optical signal detected by the above-mentioned spectroscopy. Based on the mechanism of color change, colorimetric biosensors can be divided into metal nanoparticle aggregation, enzyme catalytic activity, and chromatic transitions of conjugated polymers [107].

Luminescence is the ability of a material to emit light upon absorption of energy coming from different sources. In biosensor technology, the most significant luminescent phenomena used are fluorescence, chemiluminescence, and electrochemiluminescence. A fluorescent signal is produced by a fluorophore tag or dye anchored to the biological element and may be regulated by a quencher in the so-called "turn-on/turn-off" mechanism. There are different physical principles enabling this strategy for fluorescent-based biosensing, such as Förster resonance energy transfer (FRET) [107], fluorescence inner filter effect (IFE) [108], and others. Upon a chemical reaction, certain materials emit light, which is called chemiluminescence (CL). Furthermore, if an electric field is applied to induce electron transfer between luminescent material and electrochemical probes, the principle is called electrochemiluminescence (ECL) [109].

Considering the application of label-free biosensors for cancer biomarker detection, the most used optical principles are surface plasmon resonance (SPR) and surface-enhanced Raman spectroscopy (SERS) [110]. In SPR, a unique mode of electromagnetic field, surface plasmons (SPs), are excited by the external light in the thin metal film and propagate at the metal–dielectric interface. Due to the biosensing binding event, a change in the refractive index of the medium causes a change in SP velocity, which is measured as a change in external light properties [111]. The SERS principle is based on a Raman scattering enhancement coming from multiple physical principles, like surface plasmon resonance, substrate–molecule charge-transfer resonance at the Fermi energy level, and allowed molecular resonance. Although based on spectroscopic signals [112], SERS biosensors can also be label-based using molecules with distinct Raman signals.

The role of MSNs in optical biosensing is mostly connected to label loading for its controlled release or specific molecular/nanomaterial-based signal enhancers, which is why they are not often used in label-free optical biosensors. In Table 3, an overview of the optical techniques applied to MSN-based biosensing of cancer biomarkers is given.

Table 3. An overview of optical techniques used for the biosensing of cancer biomarkers with MSNs applied to enable or enhance the signal of detection.

Type	Method	MSN Role	Target Biomarker	Key Performances	Reference
Colorimetric	Enzyme based	AuNC-loaded MSNs for improved signal	HER2	10–1000 cells [1]; 10 cells [2]	[113]
	Non-enzyme based	DMSN-enabled signal development using CPT/DM-FA nanozyme	GSH	5–80 µM [1]; 0.654 µM [2]	[114]
		PQQ-decorated MSNs for sandwich-type signal enhancer	PSA	5–500 pg/mL [1]; 1 pg/mL [2]	[106]
Fluorescence	Inner filter effect	CuNC-loaded MSNs for improved fluorescence signal	ACP	0.5–28 U/L [1]; 0.47 U/L [2]	[115]
		Nanoreactor based on Cu-MOF-MSNs for signal enhancement	GSH	0–0.1 mM [1]; 25 µM [2]	[116]
		Release of Rh6G from MSNs upon ssDNA-AuNP cleaving by target	Flap endonuclease 1	0.05–1.75 U [1]; 0.03 U [2]	[117]
		Hybridization-manipulated signal on Luc/CS/MSNs	let-7a (miRNA)	30 fM–9 pM [1]; 10 fM [2]	[118]
		Aptamer-enabled signal on/off in MSN nanosystem with CS(cur)NPs and AuNPs	MUC-1 (CA 15-3)	-	[119]
	Forster resonance energy transfer (FRET)	Aptamer-enabled signal development using QD@MSNs	PSA and CEA	PSA: 1 fg/mL–0.1 ng/mL [1]; 0.9 fg/mL [2]; CEA: 1 fg/mL–10 pg/mL [1]; 0.7 fg/mL [2]	[120]
	Lateral-flow immunoassay	Sandwich-type signal development using BDMSNs	CA 125 and HE4	CA125: 0.1–1000 U/mL [1]; 5 U/mL [2]; HE4: 1–1000 pM [1]; 5 pM [2]	[121]
Chemiluminescence		Signal amplification by HRP-Ab1@MSNs	CEA	10 pg/mL–20 ng/mL [1]; 3 pg/mL [2]	[122]
Electrochemiluminescence		Signal enhancement by CS-Lu-modified SBMMs	SKBR-3	20–2000 cells/mL [1]; 20 cells/mL [2]	[37]
		DMSN-enabled signal development using CPT/DM-FA nanozyme	GSH	10–250 µM [1]; 0.654 µM [2]	[114]
		Controlled release of Ru(dcbpy)$_3^{2+}$ from PBA-MSNs	MCF-7	3×10^2–10^5 cells [1]; 208 cells [2]	[123]
		Ru(dcbpy)$_3^{2+}$-loaded MSNs with dual-quenching signal development	CA 15-3	5.0×10^{-5}–6.0×10^2 U/mL [1]; 2.4×10^{-6} U/mL [2]	[124]
		Controlled release of luminol-Ab2 from MSN-PEI upon target binding and pH-stimuli response	CYFRA 21-1	1 fg/mL–100 ng/mL [1]; 0.4 fg/mL [2]	[125]
		TPE-TEA-encapsulated MSNs for signal enhancement using DNA strand displacement strategy	MCF-7 cells	10 pg/mL–100 ng/mL [1]	[126]

Table 3. Cont.

Type	Method	MSN Role	Target Biomarker	Key Performances	Reference
Surface plasmon resonance	Plasmonic energy resonance transfer	MSN-enabled Au nanocrescent antenna (MONA)	MCF-7 cancer cells	-	[127]
Other	UV-Vis spectrometry	DMSN-enabled signal development using CPT/DM-FA nanozyme	GSH	2–60 µM [1] 0.654 µM [2]	[114]
	Surface-enhanced Raman spectroscopy	Target-enabled signal development by specific DNA release from MSNs	Methyltransferase	0.1–10 U/mL [1] 0.02 U/mL [2]	[128]

[1] Linear range. [2] LOD. Abbreviations: AuNC—gold nanocluster; DMSN—dendritic mesoporous silica nanoparticle; CPT—camptothecin; FA—folic acid; GSH—glutathione; PQQ—pyrroloquinoline quinone; CuNC—copper nanocluster; ACP—acid phosphatase; MOF—metal–organic framework; Rh6G—rhodamine 6G; Luc—lucigenin; CS—chitosan; cur—curcumin; QDs—quantum dots; BDMSNs—biotin-enriched dendritic mesoporous silica nanoparticles; Lu—luminol; Ru(dcbpy)$_3^{2+}$—Tris(bipyridine)ruthenium(II) chloride; PBA—phenylboronic acid; PEI—polyethylenimine; TPE—tetraphenylethylene; and TEA—triethylamine.

4. MSN-Based Biosensors by Cancer Type

4.1. Lung Cancer

In 2022, lung cancer was the most diagnosed cancer, representing 12.4% of all cancers worldwide [17]. In 2020, the highest incidence subtypes of lung cancer were adenocarcinoma (39%), squamous cell carcinoma (25%), small-cell carcinoma (11%), and large-cell carcinoma (8%) [129].

A core–shell ultrasensitive nanozyme (CPT/DM-FA) was developed for fluorescence, UV–vis, and color brightness triple-mode GSH sensing and specific cancer cell detection [114]. The nanozyme consisted of a dendritic mesoporous silica nanoparticle (DMSN) core serving as a camptothecin carrier and a platform for synthesizing a MnO$_2$ shell. Integration of FRET and oxidase-mimic-mediated ^{1}O$_2$, O$_2^-$ generation facilitated fluorescence, UV-vis, and colorimetric GSH sensing with a linear range from 2 to 250 µM and a limit of detection of 0.654 µM. The surface folic acid modification enabled specific cancer cell detection. The platform exhibited switch-on signal response and high sensitivity, suitable for real serum samples (A549 cells (lung cancer) and PC-12 cells). Challenges include nonspecific response due to similar sulfhydryl groups in cysteine and homocysteine.

A self-on ECL biosensor was developed for the efficient detection of CYFRA 21-1 [125]. The biosensor utilized a pH stimulus response-controlled release strategy, employing polyethylenimine-modified silica (SiO$_2$-PEI) as a carrier, BSA/luminol-Ab2 as the encapsulated substance, and AuNPs as the blocking agent (Figure 1). Glucose served as the inducer for controlled release. The glucose oxidation led to the production of gluconic acid, triggering a decrease in pH, which caused the release of BSA/luminol-Ab2 from SiO$_2$-PEI due to the detachment of AuNPs. The specific binding between CYFRA 21-1 antibody and antigen facilitated ECL signal generation. The biosensor demonstrated detection capabilities within a range of 0.001–100,000 ng/L and a limit of detection of 0.4 fg/mL.

Another electrochemical immunosensor for detecting CYFRA 21-1 was developed by Yola et al. [95]. The sensor utilized a silicon nitride (Si$_3$N$_4$)–molybdenum disulfide (MoS$_2$) composite on multi-walled carbon nanotubes (MWCNTs) as a sensor platform, along with core–shell-type magnetic mesoporous silica nanoparticles@gold nanoparticles (MMSNs@AuNPs) as a signal amplifier. The process involved immobilizing capture antibodies on the sensor platform via stable electrostatic/ionic interactions, followed by specific antibody–antigen interactions with the signal amplifier to form a sandwich-type voltammetric immunosensor. The immunosensor exhibited a linear detection range of 0.01–1.0 pg/mL and a detection limit of 2.00 fg/mL. The sensor demonstrated selectivity and sensitivity in plasma samples, highlighting its potential for early detection of lung cancer. An immunosensor based on a personal glucose meter (PGM) was also designed for the detection of CYFRA 21-1 [85]. Glucose was entrapped into polyethyleneimine-modified mesoporous silica nanoparticles (MSN-PEI) using CYFRA 21-1 antibody-labeled gold nanoparticles (AuNPs-Ab) (Figure 2). In the presence of the CYFRA 21-1 antigen, AuNPs-Ab leaves the surface, caused by recognition and binding processes between the antibody and the antigen. Consequently, glucose molecules were released from the pores of MSNs, which are measured by PGM. The proposed immunosensing system exhibited a linear response to CYFRA 21-1 ranging from 1.3 ng/mL to 160 ng/mL with a detection limit of 0.79 ng/mL.

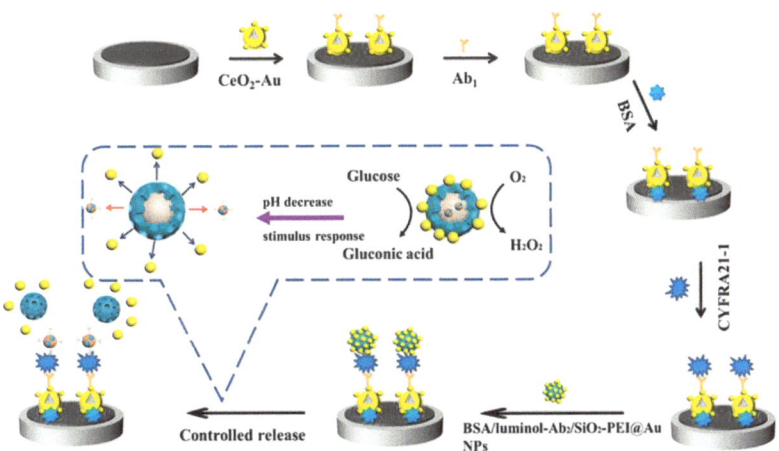

Figure 1. Schematic representation of glucose oxidation-induced pH-responsive MSN-based ECL biosensor for detection of CYFRA 21-1. Reproduced with permission [125].

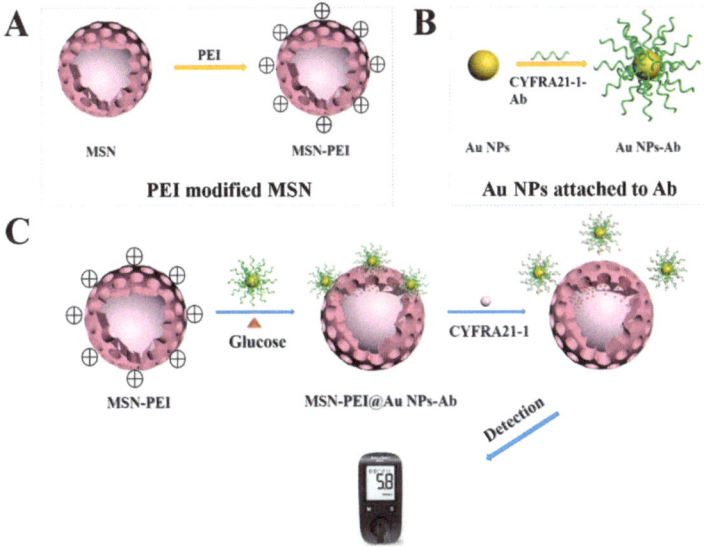

Figure 2. Schematic representation of a controlled-release MSN-based immunosensor for the rapid detection of CYFRA21–1 with the help of a personal glucose meter. Reproduced with permission [85].

4.2. Breast Cancer

In 2022, breast cancer was the fourth most diagnosed cancer, representing 7.3% of all cancers worldwide. Additionally, breast cancer is the most frequently diagnosed cancer and the primary cause of cancer-related death among women [17].

An electrochemically synthesized vertically oriented silica-based mesoporous material (SBMM) modified electrode, combined with a chitosan–luminol (CS-Lu) composite, was utilized for the cytosensing of breast cancer cells [37]. An ECL cyto-immunosensing method was developed for the detection of metastatic breast cancer cells, specifically SKBR-3 cells. The method utilizes a silica-based mesoporous nanostructure synthesized via an environmentally friendly in situ electrosynthesis approach, offering high loading capacity and mechanical strength. Luminol, combined with chitosan, forms a stable lumino composite film on the electrode surface, enhancing stability and sensitivity. Chitosan serves as an adhesive, enhancing stability and sensitivity, while also facilitating the covalent attachment of antibodies for specific cell detection. The protocol demonstrated a lower limit of quantitation of 20 cells/mL and a linear dynamic range of 20 to 2000 cells/mL. Specificity was confirmed against other breast cancer cell lines (MCF-7 and MDA-MB-231), while repeatability was shown with a relative standard deviation of about 1.6% for 500 cells/mL.

Another ECL immunosensor was developed for the specific detection of CA 15-3, a biomarker associated with breast cancer (Figure 3) [124]. The sensor utilized a dual-quenching strategy, incorporating Ru(dcbpy)$_3^{2+}$, PEI, and AuNPs immobilized on DMSNs to enhance ECL efficiency. The high loading amounts of Ru(dcbpy)$_3^{2+}$, conductivity, and localized surface plasmon resonance (LSPR) effect of AuNPs contributed to improved ECL intensity. Specifically, a sandwich structural sensing platform (Ab1-CA 15-3-Ab2) was formed, where CA 15-3 served as the target antigen. Cu$_2$O nanoparticles coated with poly(dopamine) (Cu$_2$O@PDA) were introduced to the sensor through antigen–antibody interaction, leading to significant ECL quenching due to the dual quenchers of Cu$_2$O and PDA. The sensor exhibited sensitivity with a linear detection range from 5.0×10^{-5} to 6.0 ± 10^2 U/mL and a limit of detection of 2.4×10^{-6} U/mL. Moreover, the sensor demonstrated good selectivity and stability for CA 15-3 detection in serum samples, indicating its potential for clinical applications in the diagnosis and monitoring of breast cancer biomarkers.

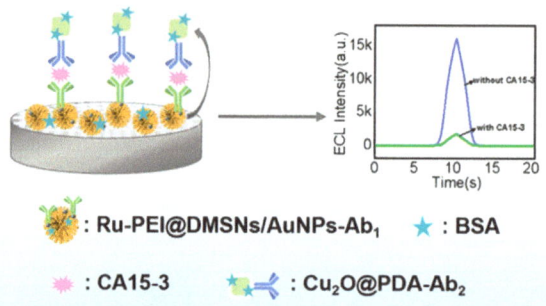

Figure 3. Schematic representation of a dual-quenching ECL immunosensor for the detection of CA15-3 based on dendritic mesoporous silica nanoparticles. Reproduced with permission [124].

Li et al. prepared a peroxidase-mimicking mesoporous silica–gold nanocluster hybrid platform (MSN–AuNC–anti-HER2) modified with recognizable biomolecules for colorimetric detection of HER2-positive (HER2+) breast cancer cells [113]. They immobilized anti-HER2 antibodies onto the surface of MSNs while loading gold nanoclusters inside the pores of MSNs. The prepared MSN–AuNC–anti-HER2 platform was able to catalyze H$_2$O$_2$ reduction and oxidation of the peroxidase substrate, colorimetric agent, 3,3′,5,5′-

tetramethylbenzidine (TMB). It has been suggested that MSN enzyme immobilization and enrichment are crucial for achieving low detection limits. Additionally, it has been demonstrated that the designed system has a high affinity to HER2 receptors.

In another study, mesoporous silica-based ECL was utilized for MCF-7 breast cancer cell detection. MSNs were modified with phenyl-boronic acid, loaded with ECL-active molecules (ruthenium-based dye, Ru(dcbpy)$_3^{2+}$), and capped by polyhydroxy-functionalized AuNPs [123]. In the presence of ascorbic acid, MCF-7 cells endogenously produce a large number of H_2O_2, which subsequently induces the oxidation of arylboronic ester linker causing the release of Ru(phen)$_3^{2+}$ and increasing the ECL signal. The system exhibited a detection limit of 208 cells/mL for MCF-7 breast cancer cells. A more recent study introduced MONA (Mesoporous silica with Optical Au Nanocrescent Antenna), an integrated nanostructure designed for multifunctional cellular targeting, drug delivery, and molecular imaging. MONA combines an asymmetric Au nanocrescent (AuNC) antenna with a mesoporous silica nanosphere [127].

The MSN serves as a molecular carrier with a large pore volume, facilitating efficient drug delivery, while the AuNC functions as a nanosensor and optical switch. Key findings include specific targeting of EpCAM in MCF-7 breast cancer cells, achieved through conjugation of anti-EpCAM onto MONA, rapid apoptosis of MCF-7 cells facilitated by light-driven molecular, doxorubicin (DOX) delivery, utilizing a highly focused photothermal gradient generated by the asymmetric AuNC, and monitoring of apoptotic events, particularly cytochrome c activity in response to DOX releases by measuring plasmonic energy resonance transfer (PRET) between the AuNC and cytochrome c molecules. A novel strategy to enable EL on MSNs is based on the encapsulation of aggregation-induced EL molecules TPE and TEA as a co-reactant, developing an MSN-TPE-TPA self-enhanced EL system [126]. Furthermore, the detection of MCF-7 cells is realized through strategic capture of CD44 transmembrane glycoprotein via novel WC-7 heptapeptide additionally functionalized with double-stranded DNA probes, of which one is modified with ferrocene, an EL quencher, and acts as a signal initiator. In the presence of target cells, a complex peptide-dsDNA binds to CD44 protein, and by the strand displacement strategy, an Fc-carrying DNA probe is released and extracted making space for its hybridization to a capture probe on the MSN-based EL system.

4.3. Prostate Cancer

In 2022, prostate cancer was the fourth most diagnosed cancer, representing 7.3% of all cancers worldwide. Further, prostate cancer was the second most common cancer globally and the fifth leading cause of cancer-related deaths among men [17].

Fluorescent nanoprobes composed of N-acetyl-l-cysteine capped-copper nanoclusters (NAC-CuNCs) were incorporated into three-dimensional mesoporous silica particles (M-SiO2) through the electrostatic assembly for detecting prostate cancer (PCa) biomarker acid phosphatase (ACP) (Figure 4) [115]. This process enhanced the fluorescence emission and quantum yield of the NAC-CuNCs due to the confinement effect of M-SiO$_2$. These nanoprobes were then combined with MnO$_2$ nanosheets, a fluorescence quencher, resulting in a fluorescence quenching effect through the inner filter effect. Subsequently, the addition of ACP triggered the hydrolysis of l-ascorbic acid-2-phosphate (AAP) into ascorbic acid (AA). This AA, in turn, facilitated the reduction of MnO$_2$ nanosheets into Mn^{2+}, thus restoring the fluorescence emission and creating a turn-off/turn-on fluorescent detection platform for ACP. The platform exhibited a detection limit of 0.47 U/L for ACP activity and demonstrated high accuracy in measuring ACP levels in real serum samples.

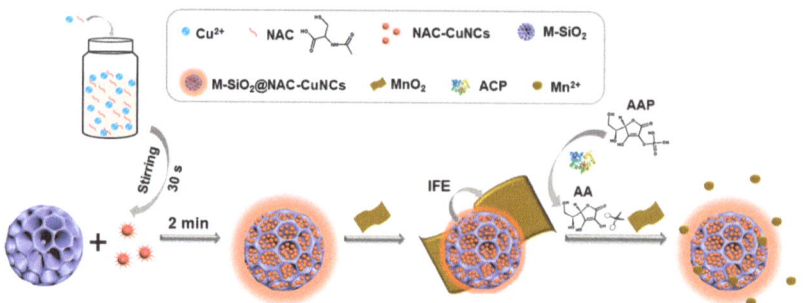

Figure 4. Schematic representation of construction and working principle of a fluorescent nanoprobe based on CuNC-incorporated 3D MSNs. Reproduced with permission [115].

An all-solid-state (ASS) potentiometric sensor for sarcosine, a biomarker for prostate cancer, has been developed using a molecularly imprinted polymer (MIP) polymerized over silica nanoparticles (Si) [87]. This MIP-Si sensor exhibits high selectivity in phosphate-buffered solution (PBS) and simulated body fluid (SBF). It demonstrates a linear response in the concentration range of 10^{-5}–10^{-8} mol/L, with a detection limit of 7.8×10^{-8} mol/L and a response time of approximately 30 s. The sensor remains stable for at least 150 days, showcasing its stability, reproducibility, and sensitivity for PCa detection. This work introduced the miniaturized potentiometric ASS sensor for sarcosine recognition and highlighted its low limit of detection, quick response time, and wide linear range. The MIP synthesized on silica nanoparticles enables the development of a selective sensor for sarcosine with analytical applicability in PCa diagnostic applications. One principle to overcome the use of enzyme-based colorimetric systems and apply the nanomaterial technology is an improved strategy for nanozyme catalytic performance in color reaction, where PQQ-decorated MSNs act as a nanocatalyst in the reduction of Fe(III)-ferrozine into Fe(II)-ferrozine by Tris(2-carboxyethyl)phosphine (TCEP) [106]. PQQ-decorated MSNs functionalized with anti-PSA antibody 2 are used as an enhanced catalyst of a colorimetric signal, while magnetic beads functionalized with anti PSA antibody 1 are used as a capture probe. Using the sandwich-type mechanism, with PSA as a bridge between the capture nanoprobe and nanocatalyst, a colorimetric signal was measured using UV-Vis absorption spectra, and LOD was estimated to be 1 pg/mL. MSNs can play a significant role in fluorescent signal amplification, i.e., the "turn-on" approach, where they are loaded with fluorophores [120]. Particularly, MSNs are functionalized with luminous CdTe quantum dots with two emission wavelengths and adsorbed on the quenching surface of MoS_2 nanosheets via target-specific aptamers; once aptamers bind target molecules, namely PSA and CEA, MSNs are desorbed, and fluorescence is turned-on. This dual-fluorescence mechanism enabled the ultrasensitive detection of two cancer biomarkers, with LOD of 0.7 fg/mL for CEA and 0.9 fg/mL for PSA.

4.4. Cervical Cancer

Cervical cancer ranks as the fourth most common cancer in women, both in terms of new cases and deaths. In 2022, there were approximately 660,000 new diagnoses and 350,000 fatalities globally [17].

MSNs can serve as a redox probe nano-depot, which can be released to amplify the electrochemical signal upon target capture by the bioreceptor. In that sense, MSNs are loaded with MB, capped with chitosan, and additionally functionalized with anti-E6 antibody 2 for the detection of HPV16 E6 oncoprotein [92]. Moreover, a glassy carbon electrode (WE) was modified with innovative dendritic palladium–boron–phosphorus nanospheres (PdBP-NSs) and anti-E6 antibody 1 for specific E6 capture. Owing to the sandwich-type interaction mechanism, the biosensor was able to detect as low as 34.1 fg/mL in a broad dynamic range of 50 fg/mL–4 ng/mL. The use of MSNs reduces the electropolymerization of MB during the reaction and amplifies the signal response.

A sandwich-type electrochemical immunosensor was developed for ultrasensitive detection of squamous cell carcinoma antigen (SCCA), a common biomarker for cervical cancer [96]. Highly branched PtCo nanocrystals (PtCo BNCs) were synthesized via a solvothermal reaction to serve as electrode substrates, enhancing conductivity and providing active sites for antibody (Ab1) loading. Dendritic mesoporous SiO_2@AuPt nanoparticles (DM-SiO_2@AuPt NPs) were prepared through wet chemical methods and used to adsorb thionine (Thi) as a signal label, increasing detection sensitivity. PtCo BNCs facilitated electron transfer and Ab1 loading, while DM-SiO2@AuPt NPs enhanced Thi loading and captured the secondary antibody (Ab2). The combination of PtCo BNCs and DM-SiO2@AuPt NPs amplified electrochemical signals, enabling sensitive SCCA detection. The sensor exhibited a linear range from 0.001 to 120 ng/mL and a detection limit of 0.33 pg/mL with high reproducibility and acceptable recovery in diluted human serum samples.

4.5. Pancreatic Cancer

A controlled release of glucose from MSN pores, which is used as an active component in electrochemical reactions on the modified WE, is achieved to successfully detect CA 19-9 [90]. Glucose-loaded MSNs are capped with ZnS, modified with anti-CA19-9 antibody 2 (ZnS@MSN-Glu-Ab2), and act as a signal amplifier when bound to CA19-9 previously captured by antibody 1 in a reaction well. Only CA19-9-anchored MSNs will undergo uncapping via DTT cleaving of disulfide bonds, which releases glucose. Finally, an electrochemical signal was developed using novel 3D cactus-like nickel–cobalt-layered double hydroxide on copper selenide nanosheet-modified carbon cloth (NiCo-LDH/CuSe/CC) with enhanced electrochemical activity for glucose oxidation. Glucose oxidation was monitored using a DPV and a very low concentration of only 0.0005 U/mL was calculated as a limit of detection. Researchers also introduced a novel approach for the ultrasensitive detection of GPC1, a potential biomarker for pancreatic cancer, through a photoelectrochemical (PEC) immunosensor utilizing gold nanoclusters (AuNCs) [101]. Furthermore, the study extended this technique to develop a multichannel light-addressable PEC sensor capable of simultaneously detecting GPC1, carcinoembryonic antigen (CEA), and glutathione (GSH). This sensor combines AuNC/GO-based PEC immunosensors for GPC1 and CEA detection with carbon dots@mesoporous silica bead (CDs@MSB)-based PEC sensors for GSH detection. The combined sensor demonstrates high sensitivity and specificity, achieving accurate and simultaneous detection of the biomarkers in cell, mouse, and patient models of pancreatic cancer. Compared to commercial kits, the light-addressable sensor offers superior sensitivity, lower detection limits, and faster detection times, with robust anti-interference capabilities in complex biological environments. Overall, this innovative sensor holds promise for advancing the diagnosis of pancreatic cancer, and the authors suggest future expansion to incorporate additional biomarkers for enhanced diagnostic accuracy and sensitivity.

4.6. Ovarian Cancer

Liu et al. investigated an approach utilizing BDMSNs combined with multiplex lateral flow immunoassay (MLFIA) for the simultaneous detection of ovarian cancer biomarkers CA 125 and HE4 [121]. The BDMSNs serve as fluorescent signal reporters and demonstrate robust antibody enrichment properties due to their aggregation-induced emission property and high affinity for the biotin–streptavidin system. The linear ranges for CA125 and HE4 detection were found to be 0.1–1000 U/mL and 1–1000 pM, respectively, with corresponding limits of detection of 5 U/mL and 5 pM. The coefficient of variation for intra-assay and inter-assay were both less than 15%. Furthermore, the developed BDMSN-MLFIA showed no cross-reactivity with common tumor markers (AFP, CA 199, CEA), and the clinical test results demonstrated a correlation coefficient of over 98% when compared with commercial electrochemiluminescence methods. A sandwich-type magneto-immunosensor was developed for the simultaneous detection and quantification of three ovarian cancer biomarkers: HE4, AFP, and CA 125 [100]. The immunosensor employs bioaffinity interactions of target molecules with specific antibodies and uses screen-printed electrodes combined with electroactive nanomaterials, including gold nanoparticles (AuNPs) and CdTe and PbS QDs. These nanomaterials are conjugated with specific antibodies and integrated with mesoporous silica nanoparticles (SiNPs) for enhanced electrochemical signals.

4.7. Other Cancers

Fei et al. constructed and evaluated a GSH-triggered nanoreactor, developed using mesoporous silica nanoparticles (MSNs) coated with a MOF shell formed by coordinating Cu(II) with trimesic acid [116]. The Cu(I) species, generated via GSH-mediated reduction, acts as a catalyst to accelerate azide–alkyne 1,3-cycloaddition (CuAAC) reactions. The nanoreactor demonstrates good biocompatibility and efficacy in GSH sensing, both in cellular environments and in wheat plumules. Specifically, it exhibits high specificity and sensitivity to GSH, with a minimum detection concentration of 0.025 mM in vitro. Additionally, it enables the visualization of GSH distribution within single living cells, unlike traditional electrochemical methods. Moreover, fluorescence signals indicate the influence of Cd^{2+} and Pb^{2+} ions on GSH expression in wheat plumules. The nanoreactor's unique properties suggest promising applications in intracellular sensing of various analytes, disease diagnostics, and agricultural research.

An electrochemical cytosensor was developed to detect HT-29 colorectal cancer stem cells (CSCs) using a nanocomposite of mesoporous silica nanoparticles (MSNs) and platinum nanoparticles (PtNPs) on a GCE [99]. The PtNPs, approximately 100 nm in size, were electrodeposited onto the MSN substrate, providing high-rate porosity and increased surface-to-volume ratio, facilitating efficient binding of biotinylated monoclonal antibodies targeting CD133, a CSC marker. DPV and SWV confirmed reduced charge transfer and electrical current upon interaction with CD133+ cells. The cytosensor demonstrated sensitivity to detect CSCs ranging from 5 to 20 cells/5 µL, outperforming flow cytometry. The integration of MSNs and PtNPs enhanced mass and charge transfer rates, providing active sites for antibody binding.

Another paper introduced a novel dual-signal-amplified sandwich-type electrochemical immunoassay for the detection of CEA [91]. By utilizing dual-labeled mesoporous silica nanospheres (amine-functionalized SBA-15 entrapping Au nanorod followed by covalent conjugation of HRP and antibody (anti-CEA, Ab2)) as signal amplifiers, combined with NiO@Au- and anti-CEA (Ab1)-decorated graphene as a conductive layer, they achieved remarkable sensitivity enhancement. The synthesized dual-labeled mesoporous silica (DLMS) nanospheres demonstrated ultra-low limits of detection (5.25 fg/mL) and a wide linear range (0.1–5 pg/mL) measured by DPV. The developed immunosensor also showed as an appropriate system in terms of selectivity, detecting no significant impact of different interfering proteins. Furthermore, the DLMS-based immunosensor exhibited excellent performance in real-time CEA determination, with significantly improved recoveries (>98%), confirmed by a typical spiking technique on human serum samples and a commercially accessible method (ELISA). This innovative approach holds promise for meeting the clinical demand for ultrasensitive detection of CEA biomarkers, thereby contributing to early cancer diagnosis and disease progression monitoring.

In a separate study, researchers developed a 3D electrochemical sensing interface for sialic acid (SA) utilizing a mesoporous–macroporous structure created through a layer-by-layer assembly method [94]. The interface was constructed on electrode surfaces using polystyrene (PS) microtubes coated with mesoporous silica and loaded with sambucus nigra agglutinin (SNA). The detection was based on the specific recognition of SNA and SA. The interface demonstrated enhanced cellular capture efficiency and specific recognition of SA overexpressed on cancer cell surfaces. By employing a layer-by-layer assembly method, the exposure of active substances was maximized, resulting in better cellular capture performance compared to direct mixing methods. The 3D structure of the PS nanotubes increased the electrode's specific surface area, improving its efficiency in capturing cancer cells. Additionally, the mesoporous structure facilitated the loading of more SNA, enhancing the specific recognition of cancer cells. The developed cytosensor exhibited a linear detection range of $1-1.0 \times 10^7$ cells/m and a detection limit of 4 cells/mL (S/N = 3).

A controlled-release MSN-based nanoprobe was developed for detecting Flap endonuclease 1 (FEN1), a structure-specific nuclease that catalyzes the removal of a 5′ overhanging DNA flap from a specific DNA structure [117]. They entrapped the fluorescence molecule Rh6G using gold nanoparticles linked to specific single-stranded DNA (AuNPs-ssDNA) as a molecular gate. The presence of FEN1 cleaves the ssDNA, resulting in the release of Rh6G and the recovery of fluorescence. They demonstrated a good linear relationship with the logarithm of FEN1 activity ranging from 0.05 to 1.75 U with a detection limit of 0.03 U. Furthermore, it has been suggested that biosensors could distinguish tumor cells from normal cells. Further, a mesoporous silica-based nanotheranostic system targeting MUC-1-positive tumor cells (MCF-7 and HT-29) was developed [119]. It involves encapsulating curcumin into chitosan–triphosphate nanoparticles, which are then loaded into a nanosystem consisting of mesoporous silica, chitosan, and gold, targeted by an aptamer. The nanosystem enables targeted imaging and drug delivery, with the aptamer triggering drug release upon binding to MUC-1 receptors. The system shows selective toxicity towards MUC-1-positive cells and it is proposed for cancer diagnosis, imaging, and therapy. However, to form the highly sensitive biosensor, optimization of the threshold concentration of the aptamer is needed.

Another paper presents a reverse-phase microemulsion synthesizing method for obtaining silica nanoparticles and incorporating chitosan and the fluorescent dye lucigenin during the reaction [118]. Chitosan addition enhances nanoparticle porosity and facilitates lucigenin molecule integration, increasing fluorescence quantum yield compared to lucigenin/silica NPs without chitosan. Target DNA/miRNA was hybridized with biotin-labeled probe DNA fixed onto the surface of the magnetic beads. Target DNA/miRNA detection relied on the distinct fluorescence responses observed between single-stranded DNA (ssDNA) and double-stranded DNA (dsDNA). The composite nanoparticles exhibit discriminative fluorescence intensity based on the charge difference between single-stranded DNA (ssDNA) and double-stranded DNA (dsDNA), enabling direct detection of let-7a in human gastric cancer cell samples without enzymes, labeling, or immobilization. The method demonstrates a detection limit of 10 fM and selectivity, with lucigenin/chitosan/silica composite nanoparticles serving as efficient DNA hybrid indicators. These composite nanoparticles amplify fluorescence signals through mass transfer nanochannels, resulting in enhanced sensitivity for let-7a detection in tumor cells compared to existing methods. Additionally, by modifying the probe DNA on magnetic beads, the composite nanoprobes can detect other biomolecules. Dendritic-large MSNs are synthesized to improve antibody and horseradish peroxidase (HRP) immobilization used for two-step detection of CEA [122]. Namely, HRP is involved in CL intensity enhancement of luminol only in the presence of CEA, which is achieved by magnetic separation of $Fe_3O_4@SiO_2$-Ab2 microspheres conjugated to MSN-HRP/Ab1 through the antibody 2-CEA-antibody 1 bridge.

MSNs are employed to amplify the SERS signal for methyltransferase activity determination [128]. Here, MSN pores are loaded by a loading DNA and capped by a specifically designed dsDNA, which can be opened by a trigger DNA produced upon the presence of the target enzyme and nicking endonuclease. The loading and trigger DNAs are released, where trigger DNA can repeat the uncapping cycle (amplification step), and the loading DNA undergoes further SERS signal development. For that, functionalized magnetic beads (MBs) with capture DNA and functionalized AuNPs with reporter DNA having a SERS probe (rhodamine-based) are used. The loading DNA is hybridized to both capture and reporter DNAs, which is then separated, and the Raman spectra are recorded. The 0.02 U/mL detection limit of the target enzyme is reached using the novel principle of this method.

A biosensor for the determination of L-lactic acid (LA) has also been developed [88]. The biosensor uses a flow injection analysis (FIA) system with a lactate oxidase (LOx)-

based mini-reactor connected to a silver amalgam screen-printed electrode (AgA-SPE) for detection. The mini-reactor contains mesoporous silica (SBA-15) coated with covalently immobilized LOx, enabling a large enzyme loading of approximately 270 µg. This setup ensures high stability, with 93.8% of the initial signal retained after 350 measurements and 96.9% after 7 months. The detection principle is based on the amperometric monitoring of oxygen consumption due to LA oxidation, measured by the four-electron reduction of oxygen at −900 mV vs. Ag pseudo-reference electrode. This method avoids interference from common oxidizing substances like ascorbic and uric acid. The biosensor was tested for LA quantification in saliva, wine, and dairy products, showing high selectivity, stability, and sensitivity, with a limit of detection of 12.0 µmol/L. The design allows for easy replacement of the mini-reactor or reuse of the electrode, making it versatile and practical for clinical diagnostics and food quality control. In another study, an electrochemical aptasensor for detecting lysozyme (Lys) was developed using a nanocomposite of amino-reduced graphene oxide (Amino-rGO), an ionic liquid (1-Butyl-3-methylimidazolium bromide), and amino-mesosilica nanoparticles (Amino-MSNs) [93]. This nanocomposite, integrated into a screen-printed carbon electrode, offers thermal and chemical stability, conductivity, surface-to-volume ratio, cost efficiency, biocompatibility, and bioelectrocatalytic properties. Anti-lysozyme aptamers (anti-Lys aptamers) were covalently coupled to the nanocomposite using glutaraldehyde as a linker, enhancing the electrochemical signal and sensitivity. The aptasensor's performance was characterized by CV, DPV, and EIS. The presence of lysozyme increased charge transfer resistance in EIS and decreased DPV peak currents, providing analytical signals for lysozyme detection. Two calibration curves were established, demonstrating LOD of 2.1 and 4.2 fmol/L.

5. Perspectives and Outlook

Due to the outstanding properties of MSNs, they have substantial benefits in sensing cancer biomarkers (Table 4).

Table 4. The impact of the physicochemical properties of MSNs on their performance as detection materials.

Properties	Benefits	Challenges	Applications Related to Sensing Cancer Biomarkers
High surface area	Surface functionalization with different molecules.	Controlling the amount and distribution of surface functional groups.	High amount of receptors for interaction with analytes or for attachment to sensing surfaces for optical or electrochemical detection with low LOD.
Porosity	Uniform distribution of pores with small diameter (2–3 nm), which can be used to load and entrap cargo molecules.	Optimization of porous structure to enhance the capacity for storing and entrapping molecules.	Loading signaling molecules (analytes) and their controlled release for optical or electrochemical sensing.
High stability	Facile formation of stable covalent linkages in reaction with organosilanes. Stability in testing media.	Achieving enhanced degradation for in vivo applications. Long-term stability in weakly alkaline media can present a challenge to achieving sensors for prolonged operation.	Formation of stable sensing surfaces for possible reusable detection.
Biocompatibility	Due to its biocompatibility, the use of silica is approved for cosmetics use.	Achieving approvement for in vivo diagnostics.	Possible construction of wearable biosensors.
Low costs	Highly scalable synthesis with cheap reactants and does not require high purity of chemicals.	The need for the use of expensive recognition elements in post-synthesis modification for specific and selective sensing.	Possible application for affordable POCT detection.

A novel generation of biosensors employing the use of MSNs is on the rise. Besides the high surface area, which is typical for nanoparticles, MSNs offer several unique attributes that bring substantial benefits for devising biosensors. One of these properties is the possibility of stable (covalent) functionalization of their surface. Hence, different types of functional groups, such as thiol, amine, hydroxyl, halogen, and others are easily grafted on the silica surface in a one-step reaction with different alkoxysilanes, which is typically performed in organic solvents and dry conditions, preferably at elevated temperatures to stimulate the evaporation of as-formed nus products (alcohols). Further modification of the surface is subsequently achieved through different possible covalent coupling reactions, such as click reactions (e.g., thiols with maleimide groups [130] or hydroxyl groups with isocyanates), carbodiimide-catalyzed coupling reactions (between amines and carboxylates [131]), or substitution reactions (e.g., the substitution of halogen with nucleophiles such as amine groups [132]). This feature allows the employment of versatile functionalization strategies for achieving the desired final functionalization on the NP surface. Moreover, each functionalization step can be performed through heterogeneous reactions, and hence the modified NPs can be isolated and washed by simple centrifugation. The same covalent functionalization strategies can be used for attaching the desired NPs to the desired 2D surfaces, thus yielding stable biosensing platforms.

The ordered porosity of MSNs is another unique property that brings substantial incentive to their use in devising biosensors. The pores can be loaded with signaling molecules such as dyes or electroactive species for devising optical or electrochemical biosensors, respectively. More importantly, the release of the loaded cargo molecules can be governed by surface-functionalization and pore-blocking species. Thus, large molecules or nanoparticles have been demonstrated for successful capping of the pore entrances and entrapping cargo molecules [133,134]. Furthermore, the on-desire release of cargo molecules can be achieved by binding the pore-blocking species to the MSNs through stimuli-cleavable linkers. For this purpose, the employed linkers contain functional groups within their structure that can be cleaved upon reaction with specific reagents (such as disulfide groups for cleavage by reduction agents, e.g., glutathione), change in pH (such as hydrazone or acetal linkages for cleavage by acidification), or upon exposure to other incentives such as magnetic field or light irradiation.

The advantage of using the loaded MSNs for triggering the release of signaling molecules lies in the possibility of releasing a substantial amount of the loaded molecules per cleavage event, which could lead to highly sensitive detection. Thus, a substantial amplification factor is expected as one cleavage-triggering agent (analyte) could release an abundance of the pore-loaded signaling molecules. It can be envisioned that such a property would be beneficial for releasing dyes or electroactive species for optical or electrochemical sensing platforms, respectively. The fact that MSNs are not optically active and non-conductive without the loaded signaling molecules is also beneficial for enhancing the signal/noise ratio. Nevertheless, the non-conductive nature of these nanoparticles may limit their applicability in some electrochemical sensors. To address this issue, surface modifications with conductive species (polymer, graphene, or noble metal layers) should be considered. In this case, having the MSNs on the surface of the electrodes would be beneficial for the sensing process by enhancing the surface roughness and hence the sensitivity of the sensor.

The development of biosensors for cancer biomarkers based on MSNs is also promising from the aspect of the known procedures for their affordable large-scale production [135–137]. However, the use of expensive recognition elements such as antibodies or aptamers could increase the cost of the final products. Nevertheless, the condition of heterogenous post-modification could allow the reuse of the non-reacted recognition elements after centrifugation of MSNs, which could decrease the final costs of biosensors.

Finally, even though silica is known for its stability, such as its low degradability in neutral or acidic conditions, its hydrolysis and dissolution in the presence of basic molecules could limit its applicability in such environments. Thus, the long-term stability of MSNs could be an issue for reusable or prolonged detection of cancer biomarkers in the weakly alkaline physiological environment of blood (pH 7.4) or in urine (pH up to 8). Hence, the development of standard protocols is still needed for suitable functionalization and passivation of the MSNs to increase their stability in weakly basic conditions.

6. Conclusions

In general, the unique characteristics of MSNs warrant their vast potential for the construction of efficient optical and electrochemical sensors, which is yet to be realized in full measure through further research. The porosity of MSNs allows loading of the signaling molecules and their possible release triggering in the presence of desired analytes. The formation of stable covalent bonds on the surface of MSNs and between the MSNs and the sensing substrates offers opportunities for the construction of stable biosensing structures. However, the limiting factor in the case of the electrochemical sensors could be their low conductivity, while the low stability of MSNs in alkaline environments could limit their use for prolonged and reusable sensing in weakly alkaline blood and urine samples. Nevertheless, the exceptional capabilities for covalent functionalization of the MSNs surface could enhance their conductivity as well as their stability in alkaline conditions and hence allow the construction of affordable and efficient POCT sensors in the future.

Author Contributions: Conceptualization, N.Ž.K.; methodology, M.M. (Minja Mladenović) and S.J.; validation, M.M. (Mirjana Mundžić) and A.P.; investigation, M.M. (Minja Mladenović), S.J., M.M. (Mirjana Mundžić) and A.P.; resources, N.Ž.K. data curation, M.M. (Minja Mladenović); writing—original draft preparation, M.M. (Minja Mladenović), S.J., M.M. (Mirjana Mundžić), A.P. and N.Ž.K.; writing—review and editing, I.B. and N.Ž.K. supervision, I.B. and N.Ž.K.; project administration, N.Ž.K.; funding acquisition, N.Ž.K. All authors have read and agreed to the published version of the manuscript.

Funding: This work was funded by The European Union's Horizon 2020 research and innovation program under grant agreement 952259 (NANOFACTS) and from the Ministry of Education, Science, and Technological Development of the Republic of Serbia (Grant No. 451-03-66/2024-03/ 200358). I.B. acknowledged the ANTARES project that has received funding from the European Union's Horizon 2020 research and innovation program under Grant Agreement SGA-CSA No. 739570 under FPA No. 664387 (doi.org/10.3030/739570).

Institutional Review Board Statement: Not applicable.

Informed Consent Statement: Not applicable.

Data Availability Statement: Not applicable.

Conflicts of Interest: The authors declare no conflicts of interest.

References

1. Wu, S.-H.; Mou, C.-Y.; Lin, H.-P. Synthesis of Mesoporous Silica Nanoparticles. *Chem. Soc. Rev.* **2013**, *42*, 3862–3875. [CrossRef]
2. Narayan, R.; Nayak, U.Y.; Raichur, A.M.; Garg, S. Mesoporous Silica Nanoparticles: A Comprehensive Review on Synthesis and Recent Advances. *Pharmaceutics* **2018**, *10*, 118. [CrossRef] [PubMed]
3. Nooney, R.I.; Thirunavukkarasu, D.; Chen, Y.; Josephs, R.; Ostafin, A.E. Synthesis of Nanoscale Mesoporous Silica Spheres with Controlled Particle Size. *Chem. Mater.* **2002**, *14*, 4721–4728. [CrossRef]
4. Kankala, R.K.; Han, Y.; Na, J.; Lee, C.; Sun, Z.; Wang, S.; Kimura, T.; Ok, Y.S.; Yamauchi, Y.; Chen, A.; et al. Nanoarchitectured Structure and Surface Biofunctionality of Mesoporous Silica Nanoparticles. *Adv. Mater.* **2020**, *32*, 1907035. [CrossRef] [PubMed]
5. Sancenón, F.; Pascual, L.; Oroval, M.; Aznar, E.; Martínez-Máñez, R. Gated Silica Mesoporous Materials in Sensing Applications. *ChemistryOpen* **2015**, *4*, 418–437. [CrossRef] [PubMed]
6. Manzano, M.; Vallet-Regí, M. Mesoporous Silica Nanoparticles in Nanomedicine Applications. *J. Mater. Sci. Mater. Med.* **2018**, *29*, 65. [CrossRef] [PubMed]
7. Rastogi, A.; Tripathi, D.K.; Yadav, S.; Chauhan, D.K.; Živčák, M.; Ghorbanpour, M.; El-Sheery, N.I.; Brestic, M. Application of Silicon Nanoparticles in Agriculture. *3 Biotech* **2019**, *9*, 90. [CrossRef]
8. Zamboulis, A.; Moitra, N.; Moreau, J.J.E.; Cattoën, X.; Wong Chi Man, M. Hybrid Materials: Versatile Matrices for Supporting Homogeneous Catalysts. *J. Mater. Chem.* **2010**, *20*, 9322. [CrossRef]
9. Aguilar, Z.P. Nanobiosensors. In *Nanomaterials for Medical Applications*; Elsevier: Amsterdam, The Netherlands, 2013; pp. 127–179. ISBN 978-0-12-385089-8.

10. Bhalla, N.; Jolly, P.; Formisano, N.; Estrela, P. Introduction to Biosensors. *Essays Biochem.* **2016**, *60*, 1–8. [CrossRef]
11. Naresh, V.; Lee, N. A Review on Biosensors and Recent Development of Nanostructured Materials-Enabled Biosensors. *Sensors* **2021**, *21*, 1109. [CrossRef]
12. Tothill, I.E. Biosensors for Cancer Markers Diagnosis. *Semin. Cell Dev. Biol.* **2009**, *20*, 55–62. [CrossRef] [PubMed]
13. Singhal, J.; Verma, S.; Kumar, S.; Mehrotra, D. Recent Advances in Nano-Bio-Sensing Fabrication Technology for the Detection of Oral Cancer. *Mol. Biotechnol.* **2021**, *63*, 339–362. [CrossRef] [PubMed]
14. Jafari, S.; Derakhshankhah, H.; Alaei, L.; Fattahi, A.; Varnamkhasti, B.S.; Saboury, A.A. Mesoporous Silica Nanoparticles for Therapeutic/Diagnostic Applications. *Biomed. Pharmacother.* **2019**, *109*, 1100–1111. [CrossRef] [PubMed]
15. Kholafazad Kordasht, H.; Pazhuhi, M.; Pashazadeh-Panahi, P.; Hasanzadeh, M.; Shadjou, N. Multifunctional Aptasensors Based on Mesoporous Silica Nanoparticles as an Efficient Platform for Bioanalytical Applications: Recent Advances. *TrAC Trends Anal. Chem.* **2020**, *124*, 115778. [CrossRef]
16. Qasim Almajidi, Y.; Althomali, R.H.; Gandla, K.; Uinarni, H.; Sharma, N.; Hussien, B.M.; Alhassan, M.S.; Mireya Romero-Parra, R.; Singh Bisht, Y. Multifunctional Immunosensors Based on Mesoporous Silica Nanomaterials as Efficient Sensing Platforms in Biomedical and Food Safety Analysis: A Review of Current Status and Emerging Applications. *Microchem. J.* **2023**, *191*, 108901. [CrossRef]
17. Bray, F.; Laversanne, M.; Sung, H.; Ferlay, J.; Siegel, R.L.; Soerjomataram, I.; Jemal, A. Global Cancer Statistics 2022: GLOBOCAN Estimates of Incidence and Mortality Worldwide for 36 Cancers in 185 Countries. *CA Cancer J. Clin.* **2024**, *74*, 229–263. [CrossRef]
18. Tabassum, D.P.; Polyak, K. Tumorigenesis: It Takes a Village. *Nat. Rev. Cancer* **2015**, *15*, 473–483. [CrossRef] [PubMed]
19. Wodarz, A.; Näthke, I. Cell Polarity in Development and Cancer. *Nat. Cell Biol.* **2007**, *9*, 1016–1024. [CrossRef]
20. Soper, S.A.; Brown, K.; Ellington, A.; Frazier, B.; Garcia-Manero, G.; Gau, V.; Gutman, S.I.; Hayes, D.F.; Korte, B.; Landers, J.L.; et al. Point-of-Care Biosensor Systems for Cancer Diagnostics/Prognostics. *Biosens. Bioelectron.* **2006**, *21*, 1932–1942. [CrossRef]
21. Henry, N.L.; Hayes, D.F. Cancer Biomarkers. *Mol. Oncol.* **2012**, *6*, 140–146. [CrossRef]
22. Barhoum, A.; Altintas, Z.; Devi, K.S.S.; Forster, R.J. Electrochemiluminescence Biosensors for Detection of Cancer Biomarkers in Biofluids: Principles, Opportunities, and Challenges. *Nano Today* **2023**, *50*, 101874. [CrossRef]
23. Wagner, P.D.; Verma, M.; Srivastava, S. Challenges for Biomarkers in Cancer Detection. *Ann. N. Y. Acad. Sci.* **2004**, *1022*, 9–16. [CrossRef] [PubMed]
24. Jayanthi, V.S.P.K.S.A.; Das, A.B.; Saxena, U. Recent Advances in Biosensor Development for the Detection of Cancer Biomarkers. *Biosens. Bioelectron.* **2017**, *91*, 15–23. [CrossRef] [PubMed]
25. Wu, L.; Qu, X. Cancer Biomarker Detection: Recent Achievements and Challenges. *Chem. Soc. Rev.* **2015**, *44*, 2963–2997. [CrossRef] [PubMed]
26. Gan, S.D.; Patel, K.R. Enzyme Immunoassay and Enzyme-Linked Immunosorbent Assay. *J. Investig. Dermatol.* **2013**, *133*, e12. [CrossRef] [PubMed]
27. Kubista, M.; Andrade, J.M.; Bengtsson, M.; Forootan, A.; Jonák, J.; Lind, K.; Sindelka, R.; Sjöback, R.; Sjögreen, B.; Strömbom, L.; et al. The Real-Time Polymerase Chain Reaction. *Mol. Asp. Med.* **2006**, *27*, 95–125. [CrossRef] [PubMed]
28. Land, K.J.; Boeras, D.I.; Chen, X.-S.; Ramsay, A.R.; Peeling, R.W. REASSURED Diagnostics to Inform Disease Control Strategies, Strengthen Health Systems and Improve Patient Outcomes. *Nat. Microbiol.* **2018**, *4*, 46–54. [CrossRef] [PubMed]
29. Khan, H.; Shah, M.R.; Barek, J.; Malik, M.I. Cancer Biomarkers and Their Biosensors: A Comprehensive Review. *TrAC Trends Anal. Chem.* **2023**, *158*, 116813. [CrossRef]
30. Füzéry, A.K.; Levin, J.; Chan, M.M.; Chan, D.W. Translation of Proteomic Biomarkers into FDA Approved Cancer Diagnostics: Issues and Challenges. *Clin. Proteom.* **2013**, *10*, 13. [CrossRef]
31. Wu, J.; Fu, Z.; Yan, F.; Ju, H. Biomedical and Clinical Applications of Immunoassays and Immunosensors for Tumor Markers. *TrAC Trends Anal. Chem.* **2007**, *26*, 679–688. [CrossRef]
32. Sanchez-Carbayo, M. Recent Advances in Bladder Cancer Diagnostics. *Clin. Biochem.* **2004**, *37*, 562–571. [CrossRef] [PubMed]
33. Gann, P.H. A Prospective Evaluation of Plasma Prostate-Specific Antigen for Detection of Prostatic Cancer. *JAMA* **1995**, *273*, 289. [CrossRef] [PubMed]
34. Gold, P.; Freedman, S.O. Demonstration of Tumor-Specific Antigens in Human Colonic Carcinomata by Immunological Tolerance and Absorption Techniques. *J. Exp. Med.* **1965**, *121*, 439–462. [CrossRef] [PubMed]
35. Mordente, A.; Meucci, E.; Martorana, G.E.; Silvestrini, A. Cancer Biomarkers Discovery and Validation: State of the Art, Problems and Future Perspectives. In *Advances in Cancer Biomarkers*; Scatena, R., Ed.; Advances in Experimental Medicine and Biology; Springer: Dordrecht, The Netherlands, 2015; Volume 867, pp. 9–26. ISBN 978-94-017-7214-3.
36. Ross, J.S.; Fletcher, J.A.; Bloom, K.J.; Linette, G.P.; Stec, J.; Symmans, W.F.; Pusztai, L.; Hortobagyi, G.N. Targeted Therapy in Breast Cancer. *Mol. Cell. Proteom.* **2004**, *3*, 379–398. [CrossRef]
37. Nasrollahpour, H.; Mahdipour, M.; Isildak, I.; Rashidi, M.-R.; Naseri, A.; Khalilzadeh, B. A Highly Sensitive Electrochemiluminescence Cytosensor for Detection of SKBR-3 Cells as Metastatic Breast Cancer Cell Line: A Constructive Phase in Early and Precise Diagnosis. *Biosens. Bioelectron.* **2021**, *178*, 113023. [CrossRef]
38. Zheng, T.; Pierre-Pierre, N.; Yan, X.; Huo, Q.; Almodovar, A.J.O.; Valerio, F.; Rivera-Ramirez, I.; Griffith, E.; Decker, D.D.; Chen, S.; et al. Gold Nanoparticle-Enabled Blood Test for Early Stage Cancer Detection and Risk Assessment. *ACS Appl. Mater. Interfaces* **2015**, *7*, 6819–6827. [CrossRef]
39. Epstein, J.I. PSA and PAP as Immunohistochemical Markers IN Prostate Cancer. *Urol. Clin. N. Am.* **1993**, *20*, 757–770. [CrossRef]

40. Li, X.; Tan, J.; Yu, J.; Feng, J.; Pan, A.; Zheng, S.; Wu, J. Use of a Porous Silicon–Gold Plasmonic Nanostructure to Enhance Serum Peptide Signals in MALDI-TOF Analysis. *Anal. Chim. Acta* **2014**, *849*, 27–35. [CrossRef] [PubMed]
41. Schuerle, S.; Dudani, J.S.; Christiansen, M.G.; Anikeeva, P.; Bhatia, S.N. Magnetically Actuated Protease Sensors for in Vivo Tumor Profiling. *Nano Lett.* **2016**, *16*, 6303–6310. [CrossRef]
42. Rao, H.; Wu, H.; Huang, Q.; Yu, Z.; Zhang, Q.; Zhong, Z. Clinical Diagnostic Value for Colorectal Cancer Based on Serum CEA, CA24-2 and CA19-9. *Clin. Lab.* **2021**, *67*, 4. [CrossRef]
43. Tokunaga, R.; Sakamoto, Y.; Nakagawa, S.; Yoshida, N.; Baba, H. The Utility of Tumor Marker Combination, Including Serum P53 Antibody, in Colorectal Cancer Treatment. *Surg. Today* **2017**, *47*, 636–642. [CrossRef] [PubMed]
44. Lu, N.; Gao, A.; Dai, P.; Mao, H.; Zuo, X.; Fan, C.; Wang, Y.; Li, T. Ultrasensitive Detection of Dual Cancer Biomarkers with Integrated CMOS-Compatible Nanowire Arrays. *Anal. Chem.* **2015**, *87*, 11203–11208. [CrossRef] [PubMed]
45. Yang, Y.; Chang, S.; Wang, N.; Song, P.; Wei, H.; Liu, J. Clinical Utility of Six Serum Tumor Markers for the Diagnosis of Lung Cancer. *iLABMED* **2023**, *1*, 132–141. [CrossRef]
46. Qiao, Z.; Perestrelo, R.; Reyes-Gallardo, E.M.; Lucena, R.; Cárdenas, S.; Rodrigues, J.; Câmara, J.S. Octadecyl Functionalized Core–Shell Magnetic Silica Nanoparticle as a Powerful Nanocomposite Sorbent to Extract Urinary Volatile Organic Metabolites. *J. Chromatogr. A* **2015**, *1393*, 18–25. [CrossRef] [PubMed]
47. Wang, Y.W.; Doerksen, J.D.; Kang, S.; Walsh, D.; Yang, Q.; Hong, D.; Liu, J.T.C. Multiplexed Molecular Imaging of Fresh Tissue Surfaces Enabled by Convection-Enhanced Topical Staining with SERS-Coded Nanoparticles. *Small* **2016**, *12*, 5612–5621. [CrossRef] [PubMed]
48. Zhang, X.; Chen, B.; He, M.; Zhang, Y.; Peng, L.; Hu, B. Boronic Acid Recognition Based-Gold Nanoparticle-Labeling Strategy for the Assay of Sialic Acid Expression on Cancer Cell Surface by Inductively Coupled Plasma Mass Spectrometry. *Analyst* **2016**, *141*, 1286–1293. [CrossRef] [PubMed]
49. Zhao, Y.-J.; Ju, Q.; Li, G.-C. Tumor Markers for Hepatocellular Carcinoma. *Mol. Clin. Oncol.* **2013**, *1*, 593–598. [CrossRef] [PubMed]
50. Rao, S.; Smith, D.A.; Guler, E.; Kikano, E.G.; Rajdev, M.A.; Yoest, J.M.; Ramaiya, N.H.; Tirumani, S.H. Past, Present, and Future of Serum Tumor Markers in Management of Ovarian Cancer: A Guide for the Radiologist. *RadioGraphics* **2021**, *41*, 1839–1856. [CrossRef] [PubMed]
51. Khong, H.T.; Rosenberg, S.A. Pre-Existing Immunity to Tyrosinase-Related Protein (TRP)-2, a New TRP-2 Isoform, and the NY-ESO-1 Melanoma Antigen in a Patient with a Dramatic Response to Immunotherapy. *J. Immunol.* **2002**, *168*, 951–956. [CrossRef]
52. Manne, U.; Srivastava, R.-G.; Srivastava, S. Keynote Review: Recent Advances in Biomarkers for Cancer Diagnosis and Treatment. *Drug Discov. Today* **2005**, *10*, 965–976. [CrossRef]
53. Balendiran, G.K.; Dabur, R.; Fraser, D. The Role of Glutathione in Cancer. *Cell Biochem. Funct.* **2004**, *22*, 343–352. [CrossRef]
54. Morris, P.E.; Bernard, G.R. Significance of Glutathione in Lung Disease and Implications for Therapy. *Am. J. Med. Sci.* **1994**, *307*, 119–127. [CrossRef]
55. Kennedy, L.; Sandhu, J.K.; Harper, M. E.; Cuperlovic-Culf, M. Role of Glutathione in Cancer: From Mechanisms to Therapies. *Biomolecules* **2020**, *10*, 1429. [CrossRef] [PubMed]
56. Gamcsik, M.P.; Kasibhatla, M.S.; Teeter, S.D.; Colvin, O.M. Glutathione Levels in Human Tumors. *Biomarkers* **2012**, *17*, 671–691. [CrossRef] [PubMed]
57. Moll, R. Epithelial Tumor Markers: Cytokeratins and Tissue Polypeptide Antigen (TPA). In *Morphological Tumor Markers*; Seifer, G., Ed.; Current Topics in Pathology; Springer: Berlin/Heidelberg, Germany, 1987; Volume 77, pp. 71–101. ISBN 978-3-642-71358-3.
58. Sundström, B.E.; Stigbrand, T.I. Cytokeratins and Tissue Polypeptide Antigen. *Int. J. Biol. Markers* **1994**, *9*, 102–108. [CrossRef]
59. Wieskopf, B.; Demangeat, C.; Purohit, A.; Stenger, R.; Gries, P.; Kreisman, H.; Quoix, E. Cyfra 21-1 as a Biologic Marker of Non-Small Cell Lung Cancer. *Chest* **1995**, *108*, 163–169. [CrossRef]
60. Duffy, M.J.; Shering, S.; Sherry, F.; McDermott, E.; O'Higgins, N. CA 15–3: A Prognostic Marker in Breast Cancer. *Int. J. Biol. Markers* **2000**, *15*, 330–333. [CrossRef]
61. Mitri, Z.; Constantine, T.; O'Regan, R. The HER2 Receptor in Breast Cancer: Pathophysiology, Clinical Use, and New Advances in Therapy. *Chemother. Res. Pract.* **2012**, *2012*, 743193. [CrossRef] [PubMed]
62. Maetzel, D.; Denzel, S.; Mack, B.; Canis, M.; Went, P.; Benk, M.; Kieu, C.; Papior, P.; Baeuerle, P.A.; Munz, M.; et al. Nuclear Signalling by Tumour-Associated Antigen EpCAM. *Nat. Cell Biol.* **2009**, *11*, 162–171. [CrossRef]
63. Went, P.T.H.; Lugli, A.; Meier, S.; Bundi, M.; Mirlacher, M.; Sauter, G.; Dirnhofer, S. Frequent EpCam Protein Expression in Human Carcinomas. *Hum. Pathol.* **2004**, *35*, 122–128. [CrossRef]
64. Gutman, A.B.; Gutman, E.B. An "Acid" Phosphatase Occurring in The Serum Of Patients with Metastasizing Carcinoma of the Prostate Gland. *J. Clin. Investig.* **1938**, *17*, 473–478. [CrossRef]
65. Henneberry, M.O.; Engel, G.; Grayhack, J.T. Acid Phosphatase. *Urol. Clin. N. Am.* **1979**, *6*, 629–641. [CrossRef]
66. Singh, J.; Pasi, D.K.; Bala, M.; Kumar, A.; Singh, A.; Jakhar, R.; Sharma, A.; Saini, A. Evaluation of prostate-specific antigen and total serum acid phosphatase in prostatic carcinoma. *Natl. J. Physiol. Pharm. Pharmacol.* **2023**, *13*, 1065–1071.
67. Lilja, H. A Kallikrein-like Serine Protease in Prostatic Fluid Cleaves the Predominant Seminal Vesicle Protein. *J. Clin. Investig.* **1985**, *76*, 1899–1903. [CrossRef]
68. Gjertson, C.K.; Albertsen, P.C. Use and Assessment of PSA in Prostate Cancer. *Med. Clin. N. Am.* **2011**, *95*, 191–200. [CrossRef]
69. Burd, E.M. Human Papillomavirus and Cervical Cancer. *Clin. Microbiol. Rev.* **2003**, *16*, 1–17. [CrossRef]

70. Chang, Y.-F.; Yan, G.-J.; Liu, G.-C.; Hong, Y.; Chen, H.-L.; Jiang, S.; Zhong, Y.; Xiyang, Y.-B.; Hu, T. HPV16 E6 Promotes the Progression of HPV Infection-Associated Cervical Cancer by Upregulating Glucose-6-Phosphate Dehydrogenase Expression. *Front. Oncol.* **2021**, *11*, 718781. [CrossRef] [PubMed]
71. Ghittoni, R.; Accardi, R.; Hasan, U.; Gheit, T.; Sylla, B.; Tommasino, M. The Biological Properties of E6 and E7 Oncoproteins from Human Papillomaviruses. *Virus Genes* **2010**, *40*, 1–13. [CrossRef] [PubMed]
72. Kato, H.; Torigoe, T. Radioimmunoassay for Tumor Antigen of Human Cervical Squamous Cell Carcinoma. *Cancer* **1977**, *40*, 1621–1628. [CrossRef]
73. Ohara, K. Assessment of Cervical Cancer Radioresponse by Serum Squamous Cell Carcinoma Antigen and Magnetic Resonance Imaging. *Obstet. Gynecol.* **2002**, *100*, 781–787. [CrossRef]
74. Hasan, S.; Jacob, R.; Manne, U.; Paluri, R. Advances in Pancreatic Cancer Biomarkers. *Oncol. Rev.* **2019**, *13*, 69–76. [CrossRef]
75. Duan, L.; Hu, X.; Feng, D.; Lei, S.; Hu, G. GPC-1 May Serve as a Predictor of Perineural Invasion and a Prognosticator of Survival in Pancreatic Cancer. *Asian J. Surg.* **2013**, *36*, 7–12. [CrossRef]
76. Atallah, G.A.; Abd Aziz, N.H.; Teik, C.K.; Shafiee, M.N.; Kampan, N.C. New Predictive Biomarkers for Ovarian Cancer. *Diagnostics* **2021**, *11*, 465. [CrossRef] [PubMed]
77. Cesewski, E.; Johnson, B.N. Electrochemical Biosensors for Pathogen Detection. *Biosens. Bioelectron.* **2020**, *159*, 112214. [CrossRef]
78. Tanwar, A.; Gandhi, H.A.; Kushwaha, D.; Bhattacharya, J. A Review on Microelectrode Array Fabrication Techniques and Their Applications. *Mater. Today Chem.* **2022**, *26*, 101153. [CrossRef]
79. Grieshaber, D.; MacKenzie, R.; Vörös, J.; Reimhult, E. Electrochemical Biosensors—Sensor Principles and Architectures. *Sensors* **2008**, *8*, 1400–1458. [CrossRef]
80. Chaubey, A.; Malhotra, B.D. Mediated Biosensors. *Biosens. Bioelectron.* **2002**, *17*, 441–456. [CrossRef]
81. Cho, I.H.; Kim, D.H.; Park, S. Electrochemical biosensors: Perspective on functional nanomaterials for on-site analysis. *Biomater. Res.* **2020**, *24*, 6. [CrossRef] [PubMed]
82. Dong, T.; Matos Pires, N.M.; Yang, Z.; Jiang, Z. Advances in Electrochemical Biosensors Based on Nanomaterials for Protein Biomarker Detection in Saliva. *Adv. Sci.* **2023**, *10*, 2205429. [CrossRef]
83. Huang, X.; Zhu, Y.; Kianfar, E. Nano Biosensors: Properties, Applications and Electrochemical Techniques. *J. Mater. Res. Technol.* **2021**, *12*, 1649–1672. [CrossRef]
84. Deng, K.; Zhang, Y.; Tong, X. Sensitive Electrochemical Detection of microRNA-21 Based on Propylamine-Functionalized Mesoporous Silica with Glucometer Readout. *Anal. Bioanal. Chem.* **2018**, *410*, 1863–1871. [CrossRef] [PubMed]
85. Lv, F.; Wang, M.; Ma, H.; Hu, Y.; Wei, Q.; Wu, D. Sensitive Detection of CYFRA21-1 by a Controlled Release Sensor Based on Personal Glucose Meter. *Sens. Actuators B Chem.* **2022**, *371*, 132543. [CrossRef]
86. Gai, P.; Gu, C.; Hou, T.; Li, F. Integration of Biofuel Cell-Based Self-Powered Biosensing and Homogeneous Electrochemical Strategy for Ultrasensitive and Easy-To-Use Bioassays of MicroRNA. *ACS Appl. Mater. Interfaces* **2018**, *10*, 9325–9331. [CrossRef] [PubMed]
87. Fernández-Puig, S.; Lazo-Fraga, A.R.; Korgel, B.A.; Oza, G.; Dutt, A.; Vallejo-Becerra, V.; Valdés-González, A.C.; Chávez-Ramírez, A.U. Molecularly Imprinted Polymer-Silica Nanocomposite Based Potentiometric Sensor for Early Prostate Cancer Detection. *Mater. Lett.* **2022**, *309*, 131324. [CrossRef]
88. Tvorynska, S.; Barek, J.; Josypcuk, B. High-Performance Amperometric Biosensor for Flow Injection Analysis Consisting of a Replaceable Lactate Oxidase-Based Mini-Reactor and a Silver Amalgam Screen-Printed Electrode. *Electrochim. Acta* **2023**, *445*, 142033. [CrossRef]
89. Wang, H.; Zhang, Y.; Yu, H.; Wu, D.; Ma, H.; Li, H.; Du, B.; Wei, Q. Label-Free Electrochemical Immunosensor for Prostate-Specific Antigen Based on Silver Hybridized Mesoporous Silica Nanoparticles. *Anal. Biochem.* **2013**, *434*, 123–127. [CrossRef] [PubMed]
90. Li, B.; Li, Y.; Li, C.; Yang, J.; Liu, D.; Wang, H.; Xu, R.; Zhang, Y.; Wei, Q. An Ultrasensitive Split-Type Electrochemical Immunosensor Based on Controlled-Release Strategy for Detection of CA19-9. *Biosens. Bioelectron.* **2023**, *227*, 115180. [CrossRef]
91. Krishnan, S.; He, X.; Zhao, F.; Zhang, Y.; Liu, S.; Xing, R. Dual Labeled Mesoporous Silica Nanospheres Based Electrochemical Immunosensor for Ultrasensitive Detection of Carcinoembryonic Antigen. *Anal. Chim. Acta* **2020**, *1133*, 119–127. [CrossRef] [PubMed]
92. Wen, T.; Xia, C.; Yu, Q.; Yu, Y.; Li, S.; Zhou, C.; Sun, K.; Yue, S. A Dual-Signal Electrochemical Immunosensor for the Detection of HPV16 E6 Oncoprotein Based on PdBP Dendritic Ternary Nanospheres and MBSi-Chi Nanocomposites. *Analyst* **2022**, *147*, 2272–2279. [CrossRef]
93. Jamei, H.R.; Rezaei, B.; Ensafi, A.A. An Ultrasensitive Electrochemical Anti-Lysozyme Aptasensor with Biorecognition Surface Based on Aptamer/Amino-rGO/Ionic Liquid/Amino-Mesosilica Nanoparticles. *Colloids Surf. B Biointerfaces* **2019**, *181*, 16–24. [CrossRef]
94. Chen, H.; Wang, Z.; Liu, Z.; Niu, Q.; Wang, X.; Miao, Z.; Zhang, H.; Wei, J.; Wan, M.; Mao, C. Construction of 3D Electrochemical Cytosensor by Layer-by-Layer Assembly for Ultra-Sensitive Detection of Cancer Cells. *Sens. Actuators B Chem.* **2021**, *329*, 128995. [CrossRef]
95. Yola, M.L.; Atar, N.; Özcan, N. A Novel Electrochemical Lung Cancer Biomarker Cytokeratin 19 Fragment Antigen 21-1 Immunosensor Based on Si_3N_4/MoS_2 Incorporated MWCNTs and Core–Shell Type Magnetic Nanoparticles. *Nanoscale* **2021**, *13*, 4660–4669. [CrossRef] [PubMed]

96. Zhang, J.-X.; Lv, C.-L.; Tang, C.; Jiang, L.-Y.; Wang, A.-J.; Feng, J.-J. Ultrasensitive Sandwich-Typed Electrochemical Immunoassay for Detection of Squamous Cell Carcinoma Antigen Based on Highly Branched PtCo Nanocrystals and Dendritic Mesoporous SiO2@AuPt Nanoparticles. *Microchim. Acta* **2022**, *189*, 416. [CrossRef] [PubMed]
97. Zhang, D.; Li, W.; Ma, Z. Improved Sandwich-Format Electrochemical Immunosensor Based on "Smart" SiO_2@polydopamine Nanocarrier. *Biosens. Bioelectron.* **2018**, *109*, 171–176. [CrossRef] [PubMed]
98. Cheng, H.; Li, W.; Duan, S.; Peng, J.; Liu, J.; Ma, W.; Wang, H.; He, X.; Wang, K. Mesoporous Silica Containers and Programmed Catalytic Hairpin Assembly/Hybridization Chain Reaction Based Electrochemical Sensing Platform for MicroRNA Ultrasensitive Detection with Low Background. *Anal. Chem.* **2019**, *91*, 10672–10678. [CrossRef] [PubMed]
99. Sadi, S.; Khalilzadeh, B.; Mahdipour, M.; Sokouti Nasimi, F.; Isildak, I.; Davaran, S.; Rashidi, M.-R.; Bani, F. Early Stage Evaluation of Cancer Stem Cells Using Platinum Nanoparticles/CD133+ Enhanced Nanobiocomposite. *Cancer Nano* **2023**, *14*, 55. [CrossRef]
100. Kovarova, A.; Kastrati, G.; Pekarkova, J.; Metelka, R.; Drbohlavova, J.; Bilkova, Z.; Selesovska, R.; Korecka, L. Biosensor with Electrochemically Active Nanocomposites for Signal Amplification and Simultaneous Detection of Three Ovarian Cancer Biomarkers. *Electrochim. Acta* **2023**, *469*, 143213. [CrossRef]
101. Xiao, G.; Ge, H.; Yang, Q.; Zhang, Z.; Cheng, L.; Cao, S.; Ji, J.; Zhang, J.; Yue, Z. Light-Addressable Photoelectrochemical Sensors for Multichannel Detections of GPC1, CEA and GSH and Its Applications in Early Diagnosis of Pancreatic Cancer. *Sens. Actuators B Chem.* **2022**, *372*, 132663. [CrossRef]
102. Rasheed, S.; Kanwal, T.; Ahmad, N.; Fatima, B.; Najam-ul-Haq, M.; Hussain, D. Advances and Challenges in Portable Optical Biosensors for Onsite Detection and Point-of-Care Diagnostics. *TrAC Trends Anal. Chem.* **2024**, *173*, 117640. [CrossRef]
103. Zhao, X.; Dai, X.; Zhao, S.; Cui, X.; Gong, T.; Song, Z.; Meng, H.; Zhang, X.; Yu, B. Aptamer-Based Fluorescent Sensors for the Detection of Cancer Biomarkers. *Spectrochim. Acta Part A Mol. Biomol. Spectrosc.* **2021**, *247*, 119038. [CrossRef]
104. Wang, H.; Wu, T.; Li, M.; Tao, Y. Recent Advances in Nanomaterials for Colorimetric Cancer Detection. *J. Mater. Chem. B* **2021**, *9*, 921–938. [CrossRef] [PubMed]
105. Khansili, N.; Rattu, G.; Krishna, P.M. Label-Free Optical Biosensors for Food and Biological Sensor Applications. *Sens. Actuators B Chem.* **2018**, *265*, 35–49. [CrossRef]
106. Xia, N.; Deng, D.; Mu, X.; Liu, A.; Xie, J.; Zhou, D.; Yang, P.; Xing, Y.; Liu, L. Colorimetric Immunoassays Based on Pyrroloquinoline Quinone-Catalyzed Generation of Fe(II)-Ferrozine with Tris(2-Carboxyethyl)Phosphine as the Reducing Reagent. *Sens. Actuators B Chem.* **2020**, *306*, 127571. [CrossRef]
107. Son, M.H.; Park, S.W.; Sagong, H.Y.; Jung, Y.K. Recent Advances in Electrochemical and Optical Biosensors for Cancer Biomarker Detection. *BioChip J.* **2023**, *17*, 44–67. [CrossRef]
108. Chen, S.; Yu, Y.-L.; Wang, J.-H. Inner Filter Effect-Based Fluorescent Sensing Systems: A Review. *Anal. Chim. Acta* **2018**, *999*, 13–26. [CrossRef] [PubMed]
109. Qi, H.; Zhang, C. Electrogenerated Chemiluminescence Biosensing. *Anal. Chem.* **2020**, *92*, 524–534. [CrossRef] [PubMed]
110. Kaur, B.; Kumar, S.; Kaushik, B.K. Recent Advancements in Optical Biosensors for Cancer Detection. *Biosens. Bioelectron.* **2022**, *197*, 113805. [CrossRef]
111. Piliarik, M.; Vaisocherová, H.; Homola, J. Surface Plasmon Resonance Biosensing. In *Biosensors and Biodetection*; Rasooly, A., Herold, K.E., Eds.; Methods in Molecular Biology; Humana Press: Totowa, NJ, USA, 2009; Volume 503, pp. 65–88. ISBN 978-1-60327-566-8.
112. Lin, C.; Li, Y.; Peng, Y.; Zhao, S.; Xu, M.; Zhang, L.; Huang, Z.; Shi, J.; Yang, Y. Recent Development of Surface-Enhanced Raman Scattering for Biosensing. *J. Nanobiotechnol.* **2023**, *21*, 149. [CrossRef] [PubMed]
113. Li, M.; Lao, Y.-H.; Mintz, R.L.; Chen, Z.; Shao, D.; Hu, H.; Wang, H.-X.; Tao, Y.; Leong, K.W. A Multifunctional Mesoporous Silica–Gold Nanocluster Hybrid Platform for Selective Breast Cancer Cell Detection Using a Catalytic Amplification-Based Colorimetric Assay. *Nanoscale* **2019**, *11*, 2631–2636. [CrossRef]
114. Zhang, Y.; Meng, W.; Li, X.; Wang, D.; Shuang, S.; Dong, C. Dendritic Mesoporous Silica Nanoparticle-Tuned High-Affinity MnO_2 Nanozyme for Multisignal GSH Sensing and Target Cancer Cell Detection. *ACS Sustain. Chem. Eng.* **2022**, *10*, 5911–5921. [CrossRef]
115. Chen, S.; Li, Z.; Xue, R.; Huang, Z.; Jia, Q. Confining Copper Nanoclusters in Three Dimensional Mesoporous Silica Particles: Fabrication of an Enhanced Emission Platform for "Turn off-on" Detection of Acid Phosphatase Activity. *Anal. Chim. Acta* **2022**, *1192*, 339387. [CrossRef] [PubMed]
116. Zeng, F.; Pan, Y.; Luan, X.; Gao, Y.; Yang, J.; Wang, Y.; Song, Y. Copper Metal-Organic Framework Incorporated Mesoporous Silica as a Bioorthogonal Biosensor for Detection of Glutathione. *Sens. Actuators B Chem.* **2021**, *345*, 130382. [CrossRef]
117. Bagheri Hashkavayi, A.; Cha, B.S.; Hwang, S.H.; Kim, J.; Park, K.S. Highly Sensitive Electrochemical Detection of Circulating EpCAM-Positive Tumor Cells Using a Dual Signal Amplification Strategy. *Sens. Actuators B Chem.* **2021**, *343*, 130087. [CrossRef]
118. Chang, Z.; Feng, J.; Zheng, X. A Highly Sensitive Fluorescence Sensor Based on Lucigenin/Chitosan/SiO_2 Composite Nanoparticles for microRNA Detection Using Magnetic Separation. *Luminescence* **2020**, *35*, 835–844. [CrossRef]
119. Esmaeili, Y.; Khavani, M.; Bigham, A.; Sanati, A.; Bidram, E.; Shariati, L.; Zarrabi, A.; Jolfaie, N.A.; Rafienia, M. Mesoporous Silica@chitosan@gold Nanoparticles as "on/off" Optical Biosensor and pH-Sensitive Theranostic Platform against Cancer. *Int. J. Biol. Macromol.* **2022**, *202*, 241–255. [CrossRef] [PubMed]
120. Sun, Y.; Fan, J.; Cui, L.; Ke, W.; Zheng, F.; Zhao, Y. Fluorometric Nanoprobes for Simultaneous Aptamer-Based Detection of Carcinoembryonic Antigen and Prostate Specific Antigen. *Microchim. Acta* **2019**, *186*, 152. [CrossRef] [PubMed]

121. Liu, H.; Cao, J.; Ding, S.-N. Simultaneous Detection of Two Ovarian Cancer Biomarkers in Human Serums with Biotin-Enriched Dendritic Mesoporous Silica Nanoparticles-Labeled Multiplex Lateral Flow Immunoassay. *Sens. Actuators B Chem.* **2022**, *371*, 132597. [CrossRef]
122. Liu, S.; Li, J.; Zou, Y.; Jiang, Y.; Wu, L.; Deng, Y. Construction of Magnetic Core–Large Mesoporous Satellite Immunosensor for Long-Lasting Chemiluminescence and Highly Sensitive Tumor Marker Determination. *Small* **2023**, *19*, 2304631. [CrossRef]
123. Chen, Y.; Chen, Z.; Fang, L.; Weng, A.; Luo, F.; Guo, L.; Qiu, B.; Lin, Z. Electrochemiluminescence Sensor for Cancer Cell Detection Based on H2O2-Triggered Stimulus Response System. *J. Anal. Test.* **2020**, *4*, 128–135. [CrossRef]
124. Wang, W.; Feng, D.; Wang, Y.; Kan, X. Ruthenium Poly(Ethylenimine)/Gold Nanoparticles Immobilized on Dendritic Mesoporous Silica Nanoparticles for a CA15-3 Electrochemiluminescence Immunosensor via Cu$_2$O@PDA Dual Quenching. *ACS Appl. Nano Mater.* **2023**, *6*, 19271–19278. [CrossRef]
125. Huang, Z.; Zhao, L.; Li, Y.; Wang, H.; Ma, H.; Wei, Q.; Wu, D. Glucose Oxidation Induced pH Stimuli Response Controlled Release Electrochemiluminescence Biosensor for Ultrasensitive Detection of CYFRA 21-1. *Talanta* **2024**, *266*, 124955. [CrossRef] [PubMed]
126. Jia, Y.; Du, Y.; Ru, Z.; Fan, D.; Yang, L.; Ren, X.; Wei, Q. Aggregation-Induced Electrochemiluminescence Frame of Silica-Confined Tetraphenylethylene Derivative Matrixes for CD44 Detection via Peptide Recognition. *Anal. Chem.* **2023**, *95*, 6725–6731. [CrossRef] [PubMed]
127. Park, Y.; Yoon, H.J.; Lee, S.E.; Lee, L.P. Multifunctional Cellular Targeting, Molecular Delivery, and Imaging by Integrated Mesoporous-Silica with Optical Nanocrescent Antenna: MONA. *ACS Nano* **2022**, *16*, 2013–2023. [CrossRef] [PubMed]
128. Wang, X.; Cui, M.; Zhou, H.; Zhang, S. DNA-Hybrid-Gated Functional Mesoporous Silica for Sensitive DNA Methyltransferase SERS Detection. *Chem. Commun.* **2015**, *51*, 13983–13985. [CrossRef] [PubMed]
129. Zhang, Y.; Vaccarella, S.; Morgan, E.; Li, M.; Etxeberria, J.; Chokunonga, E.; Manraj, S.S.; Kamate, B.; Omonisi, A.; Bray, F. Global Variations in Lung Cancer Incidence by Histological Subtype in 2020: A Population-Based Study. *Lancet Oncol.* **2023**, *24*, 1206–1218. [CrossRef]
130. Mundžić, M.; Lazović, J.; Mladenović, M.; Pavlović, A.; Ultimo, A.; Gobbo, O.L.; Ruiz-Hernandez, E.; Santos-Martinez, M.J.; Knežević, N.Ž. MRI-Based Sensing of pH-Responsive Content Release from Mesoporous Silica Nanoparticles. *J. Sol-Gel Sci. Technol.* **2024**. [CrossRef]
131. Mladenović, M.; Morgan, I.; Ilić, N.; Saoud, M.; Pergal, M.V.; Kaluđerović, G.N.; Knežević, N.Ž. pH-Responsive Release of Ruthenium Metallotherapeutics from Mesoporous Silica-Based Nanocarriers. *Pharmaceutics* **2021**, *13*, 460. [CrossRef] [PubMed]
132. Miyajima, T.; Abry, S.; Zhou, W.; Albela, B.; Bonneviot, L.; Oumi, Y.; Sano, T.; Yoshitake, H. Estimation of Spacing between 3-Bromopropyl Functions Grafted on Mesoporous Silica Surfaces by a Substitution Reaction Using Diamine Probe Molecules. *J. Mater. Chem.* **2007**, *17*, 3901. [CrossRef]
133. Yi, S.; Zheng, J.; Lv, P.; Zhang, D.; Zheng, X.; Zhang, Y.; Liao, R. Controlled Drug Release from Cyclodextrin-Gated Mesoporous Silica Nanoparticles Based on Switchable Host–Guest Interactions. *Bioconjugate Chem.* **2018**, *29*, 2884–2891. [CrossRef]
134. Cheng, C.-A.; Deng, T.; Lin, F.-C.; Cai, Y.; Zink, J.I. Supramolecular Nanomachines as Stimuli-Responsive Gatekeepers on Mesoporous Silica Nanoparticles for Antibiotic and Cancer Drug Delivery. *Theranostics* **2019**, *9*, 3341–3364. [CrossRef]
135. Kim, C.; Yoon, S.; Lee, J.H. Facile large-scale synthesis of mesoporous silica nanoparticles at room temperature in a monophasic system with fine size control. *Microporous Mesoporous Mater.* **2019**, *288*, 109595. [CrossRef]
136. Zhang, K.; Xu, L.-L.; Jiang, J.-G.; Calin, N.; Lam, K.-F.; Zhang, S.-J.; Wu, H.-H.; Wu, G.-D.; Albela, B.; Bonneviot, L.; et al. Facile Large-Scale Synthesis of Monodisperse Mesoporous Silica Nanospheres with Tunable Pore Structure. *J. Am. Chem. Soc.* **2013**, *135*, 2427–2430. [CrossRef] [PubMed]
137. Jundale, R.B.; Sonawane, J.R.; Palghadmal, A.V.; Jaiswal, H.K.; Deore, H.S.; Kulkarni, A.A. Scaling-up continuous production of mesoporous silica particles at kg scale: Design & operational strategies. *React. Chem. Eng.* **2024**, *9*, 1914–1923. [CrossRef]

Disclaimer/Publisher's Note: The statements, opinions and data contained in all publications are solely those of the individual author(s) and contributor(s) and not of MDPI and/or the editor(s). MDPI and/or the editor(s) disclaim responsibility for any injury to people or property resulting from any ideas, methods, instructions or products referred to in the content.

Article

Investigation of the Impact of Hydrogen Bonding Degree in Long Single-Stranded DNA (ssDNA) Generated with Dual Rolling Circle Amplification (RCA) on the Preparation and Performance of DNA Hydrogels

Xinyu Wang [1], Huiyuan Wang [1], Hongmin Zhang [1], Tianxi Yang [2], Bin Zhao [3,*] and Juan Yan [1,*]

[1] Laboratory of Quality and Safety Risk Assessment for Aquatic Products on Storage and Preservation (Shanghai), Ministry of Agriculture, Shanghai Engineering Research Center of Aquatic-Product Process & Preservation, College of Food Science and Technology, Shanghai Ocean University, Shanghai 201306, China; m210300896@st.shou.edu.cn (X.W.); d220300089@st.shou.edu.cn (H.W.); hmzhang@shou.edu.cn (H.Z.)

[2] Faculty of Land and Food Systems, The University of British Columbia, Vancouver, BC V6T 1Z4, Canada; tianxi.yang@ubc.ca

[3] Department of Biochemistry and Molecular Biology, The University of British Columbia, Vancouver, BC V6T 1Z4, Canada

* Correspondence: bin.zhao@ubc.ca (B.Z.); j-yan@shou.edu.cn (J.Y.)

Citation: Wang, X.; Wang, H.; Zhang, H.; Yang, T.; Zhao, B.; Yan, J. Investigation of the Impact of Hydrogen Bonding Degree in Long Single-Stranded DNA (ssDNA) Generated with Dual Rolling Circle Amplification (RCA) on the Preparation and Performance of DNA Hydrogels. *Biosensors* **2023**, *13*, 755. https://doi.org/10.3390/bios13070755

Received: 4 July 2023
Revised: 20 July 2023
Accepted: 21 July 2023
Published: 23 July 2023

Copyright: © 2023 by the authors. Licensee MDPI, Basel, Switzerland. This article is an open access article distributed under the terms and conditions of the Creative Commons Attribution (CC BY) license (https://creativecommons.org/licenses/by/4.0/).

Abstract: DNA hydrogels have gained significant attention in recent years as one of the most promising functional polymer materials. To broaden their applications, it is critical to develop efficient methods for the preparation of bulk-scale DNA hydrogels with adjustable mechanical properties. Herein, we introduce a straightforward and efficient molecular design approach to producing physically pure DNA hydrogel and controlling its mechanical properties by adjusting the degree of hydrogen bonding in ultralong single-stranded DNA (ssDNA) precursors, which were generated using a dual rolling circle amplification (RCA)-based strategy. The effect of hydrogen bonding degree on the performance of DNA hydrogels was thoroughly investigated by analyzing the preparation process, morphology, rheology, microstructure, and entrapment efficiency of the hydrogels for Au nanoparticles (AuNPs)–BSA. Our results demonstrate that DNA hydrogels can be formed at 25 °C with simple vortex mixing in less than 10 s. The experimental results also indicate that a higher degree of hydrogen bonding in the precursor DNA resulted in stronger internal interaction forces, a more complex internal network of the hydrogel, a denser hydrogel, improved mechanical properties, and enhanced entrapment efficiency. This study intuitively demonstrates the effect of hydrogen bonding on the preparation and properties of DNA hydrogels. The method and results presented in this study are of great significance for improving the synthesis efficiency and economy of DNA hydrogels, enhancing and adjusting the overall quality and performance of the hydrogel, and expanding the application field of DNA hydrogels.

Keywords: three-dimensional network; nucleic acid material; physical cross-linking; pure DNA hydrogel; nucleic acid signal amplification technique; biosensing

1. Introduction

Deoxyribonucleic acid (DNA) hydrogels are a promising class of macroscopic three-dimensional (3D) materials that offer a unique combination of high hydrophilicity and mechanical properties in polymer hydrogels [1] along with the remarkable biological functions of DNA such as structural designability, biocompatibility, selection specificity, molecular recognition ability, and responsiveness to environmental factors [2,3]. Because of these advantages, DNA hydrogels have wide-ranging applications in diverse fields such as food safety [4,5], medical diagnostics [6–8], environmental analysis [9,10], and controllable drug delivery and release [11,12]. Despite the availability of different chemical covalent linkages [13,14] and non-covalent physical cross-linking methods for preparing DNA hydrogels [15,16], the substantial cost of bulk-scale fabrication of pure DNA hydrogels and their low mechanical properties significantly limit their potential applications [1,17].

Rolling circle amplification (RCA) is a widely used isothermal nucleic acid amplification technique [18,19] that utilizes DNA polymerase to repeatedly copy a circular DNA template [20], resulting in the production of a long single-stranded DNA (ssDNA) molecule with a periodic sequence with high efficiency under mild reaction conditions [21,22]. RCA technology has attracted significant attention in the development of multifunctional assemblies for disease diagnosis and treatment, food safety testing, environmental quality monitoring, and other applications [23]. Zhao et al. provided a comprehensive overview of the fundamental engineering principles employed in the design of RCA technologies. They also discussed the latest advancements in RCA-based diagnostics and bioanalytical tools. Additionally, the authors summarized the utilization of RCA for the construction of multivalent molecular scaffolds and nanostructures, highlighting their applications in biology, diagnostics, and therapeutics [24]. In 2012, in a groundbreaking study, Luo and his colleagues developed a mechanical metamaterial made from a pure DNA hydrogel using RCA technology and multi-primed chain amplification [25]. The DNA hydrogel exhibited unusual mechanical properties, with a solid-like nature in water and a liquid-like nature after the removal of water. Since then, RCA technology has been extensively explored for the self-assembly of ultralong ssDNA to construct hydrogels with the desired functions and performance for use in various biosensing [26–28] and medical scenarios [29,30]. For instance, Yang et al. developed a DNA hydrogel using the double RCA assembly strategy for the separation of bone marrow stromal cells [31]. Long ssDNA generated using RCA with aptamer sequences ensured the specific anchoring of cells, while the physically cross-linked network exhibited a moderate storage modulus (G') of about 12 Pa, which could minimize mechanical damage to cells. These studies have shown that RCA technology is a simple, rapid, and cost-effective synthetic method that can be used for the bulk production of DNA hydrogels, and the functionalization of hydrogels can be achieved by programming the circular template sequences and integrating different DNA bio-functional modules (e.g., aptamer, G-quadruplex, i-motif structure) [32–35]. However, regulatory analysis of the mechanical properties of pure DNA hydrogels made with RCA technology has not been thoroughly explored.

In fact, the integration of various nanomaterials, including gold/silver nanoparticles (Au/Ag NPs), carbon materials, magnetic nanoparticles, and clays, into DNA hydrogels [36,37] has become a commonly used approach to regulating the properties of DNA hydrogels [38]. This is attributed to the fact that nanomaterials offer a broader spectrum of opportunities and potential for the utilization of DNA hydrogels, as they enhance mechanical properties, improve stability, regulate morphology and structure, and provide responsiveness. The mechanical properties of the hydrogel are largely determined by the concentration and quantity ratio of the components used in the hydrogel preparation. While this strategy is highly promising, it is important to note that the introduction of nanomaterials may negatively impact the biocompatibility and biodegradability of DNA hydrogels, and the presence of selected nanomaterials may interfere with hydrogel matrix cross-linking. In addition, there have been reports of toughening hydrogels in response to stimulation using heat, light, pH, and salt [39], but the precise control of these stimuli is also a challenge [40].

It is widely recognized that a DNA strand comprises a phosphate–deoxyribose backbone and one of four bases: adenine (A), guanine (G), cytosine (C), or thymine (T). Moreover, single-stranded DNA (ssDNA) can form stable double-stranded DNA (dsDNA) structures through hydrogen bonds between A and T and between C and G, which is known as "Watson–Crick base pairing" [41]. Hydrogen bonding, as a dynamic and weak non-covalent bond, is a significant form of physical cross-linking in addition to chain entanglement, hydrophobic interaction, and other interactions. It plays a pivotal role in the interaction between DNA molecules and is the primary driving force for DNA strand hybridization [18]. Because of its dynamic nature and degree of bonding, hydrogen bonding provides great flexibility in regulating molecular structure and properties. Hence, it is imperative to investigate the relationship between the degree of hydrogen bonding, the preparation of DNA hydrogels, and the resulting mechanical properties of DNA hydrogels. However, there is a dearth of relevant reports on this topic, highlighting the need for further research in this area.

In this study, we have experimentally demonstrated the influence of hydrogen bonding on the preparation and performance of DNA hydrogels. This study involves the utilization of ssDNA generated using dual-RCA technology as precursors for the preparation of DNA hydrogels. The degree of hydrogen bonding between the ssDNA precursors was regulated by designing circular template sequences. The entrapment efficiency for AuNPs varied in DNA hydrogels prepared with different degrees of hydrogen bonding, which reflects the differing mechanical properties of the prepared DNA hydrogels. Our findings not only present a novel approach to developing DNA hydrogels with distinct properties but also establish a foundation for their potential applications in diverse fields.

2. Materials and Methods

2.1. Materials

All synthetic DNA sequences were tabulated as shown in Table 1. DNA sequences were synthesized and HPLC purified by Shenggong Biotechnology Co., Ltd. (Shanghai, China). T4 DNA ligase (1000 U/µL), phi29 DNA polymerase (5 KU), deoxynucleotides (dNTP, 10 mM) were from Sangon Biotechnology Co., Ltd. (Shanghai, China). Tetrachloroauric acid hydrate ($HAuCl_4 \cdot 3H_2O$) was purchased from J&K Scientific (Shanghai, China). Bovine serum albumin (BSA) was purchased from Strem Chemicals, Inc., Massachusetts USA. In addition, 50× TAE buffer (2 M Tris-HCl, 100 mM EDTA, pH 8.4), 4S GelRed and 4S GelGreen nucleic acid staining agents (10,000× aqueous solution), sodium citrate ($C_6H_5Na_3O_7 \cdot 2H_2O$), and other reagents and chemicals were purchased from Shenggong Biotechnology (Shanghai) Co., Ltd. and were at least analytical-reagent grade. All aqueous solutions were prepared using Milli-Q water (18.2 M$\Omega \cdot$cm).

Table 1. DNA sequences used in this work.

DNA	Sequences (5′~3′)
Primer 1	CACAGCTGAGGATAGGACAT
Primer 2	GGACATGCAAGCAGAGCACA
Primer 3	ATGTCCTATCCTCAGCTGTG
PL-DNA-1	Phosphorylated-CTCAGCTGTGATTCATAC<u>GTACCAAC</u>GCACACAGAATTTTTTTATGTCCTATC
PL-DNA-2	Phosphorylated-TTGCATGTCCAGTTCTTTGTCCTGAGTTTTACTGTGCCTGCTGCTGTGCTCTGC
PL-DNA-3	Phosphorylated-CTCAGCTGTGATTCATAC<u>GTTGGTAC</u>GCACACAGAATTTTTTTATGTCCTATC
PL-DNA-4	Phosphorylated-GATAGGACATAAAAAAAATTCTGTGTGCGTTGGTACGTATGAATCACAGCTGAG

The underlined bases are complementary; PL-DNA: phosphorylated linear DNA.

2.2. Apparatus

Constant temperature oscillation metal bath (HCM100-Pro) and handheld Centrifuge (D1008) were provided by Dalong Xingchuang Experimental Instrument Co., Ltd. (Beijing, China). Vortex mixer (Mixer 4K, Shenggong Biotechnology Co., Ltd., Shanghai, China); fluorescence microscope (CKX-41, Olympus, Tokyo, Japan); S-3400-emission scanning electron microscope (Hitachi, Tokyo, Japan); Haake Mars rheometer (Thermo Fisher, Karlsruhe, Germany); UV–visible spectrophotometer (UV-2540, Shimazu Co., Kyoto, Japan); gel imaging and analysis system (Bio-Rad Co., Ltd., Hercules, American).

2.3. Preparation of the Circular Templates

A volume of 5 µL of 100 µM phosphorylated linear DNA was mixed with 10 µL of 100 µM primer and annealed at 90 °C for 5 min. The solution was then slowly cooled to room temperature. Next, 2 µL of T4 ligase (1000 U/µL) and 3 µL of 10× T4 ligase buffer were added to the reaction mixture, which was mixed well and incubated overnight at 25 °C. The T4 ligase was subsequently heat-inactivated at 65 °C for 10 min, after which the prepared circular DNA template solution was stored at 4 °C.

2.4. Preparation of the ssDNA Precursors Using Dual Rolling Circle Amplification (RCA)

Amplification was carried out in a 60 µL reaction system containing 4 µL of circular DNA template solution, 1 µL of phi29 polymerase (5 KU), 5 µL of phi29 polymerase buffer (10×), 5 µL of dNTPs (10 mM), and 45 µL of TE buffer (1×). The reaction mixture was thoroughly mixed and then incubated at 37 °C for 3 h, resulting in the generation of ssDNA products. In parallel, another RCA reaction system was performed using a circular template with a different DNA sequence, which also produced an ssDNA product. These two kinds of ssDNA products, having different DNA composition sequences, were used as precursors for the subsequent preparation of DNA hydrogel.

2.5. Preparation of DNA Hydrogels with ssDNA Precursors

Two types of ssDNA precursors with different sequences (60 µL) were mixed together thoroughly for 10 s at 25 °C, resulting in the formation of a visible DNA hydrogel.

2.6. Preparation of BSA-Coated Au Nanoparticles (AuNPs–BSA)

Colloidal gold nanoparticles (AuNPs) with an average diameter of 30 nm were prepared using the conventional Frens method [42]. Specifically, 50 mL of Milli-Q water and 50 µL of 10% HAuCl4 solution were added to a 250 mL round-bottom flask equipped with a reflux device and heated to boiling. Then, 1% trisodium citrate (0.7 mL) was rapidly added with vigorous stirring for 30 min [43]. During this process, the colorless solution gradually changed to a purplish-red color. The heating was stopped, and the solution was stirred overnight at room temperature. Finally, the prepared AuNPs solution was filtered through a filter membrane with a pore size of 0.22 µm and stored at 4 °C.

To prepare the AuNPs–BSA compound, 1 mL of 30 nm AuNPs and 200 µL of 30% BSA (diluted with Milli-Q water) were mixed well and incubated overnight at room temperature. The mixture was then centrifuged at 4 °C, 11,000 r/min for 20 min. The supernatant was discarded, and the precipitate was resuspended in 1 mL of 1× PBS solution. This centrifugation step was repeated three times to remove any free BSA. Finally, the precipitate was resuspended in 100 µL of 1× PBS, and the obtained AuNPs–BSA compound was stored in the dark at 4 °C.

2.7. Characterization Methods

Rheological tests: Rheological experiments were conducted on DNA hydrogel samples using the time scan mode. Storage modulus (G′) and loss modulus (G″) values were measured at a fixed frequency (1 Hz) and fixed strain (1%) at 25 °C. Time-scan rheological tests were also performed using a 20 mm parallel-plate geometry (gap size 0.01 mm) at a fixed frequency (1 Hz) and strain (1%) at 25 °C for 3 min.

Scanning electron microscope (SEM) imaging: Different DNA hydrogel samples were flashed frozen in liquid nitrogen and further freeze-dried for at least 48 h. The microstructures of these samples were then sputtered with gold and studied under a scanning electron microscope at a voltage of 5 kV.

UV–visible spectroscopy characterization: The UV–visible absorption spectral measurements were conducted using a UV-2540 spectrometer. The UV absorbance of 20 µL of supernatant AuNPs–BSA samples from three distinct DNA hydrogels was measured at 520 nm.

3. Results

3.1. Mechanism of the Degree of Hydrogen Bonding Based on Dual RCA Strategy for Regulating the Mechanical Properties of DNA Hydrogels

The schematic illustration of the investigation of the effect of hydrogen bonding degree on the performance of DNA hydrogels prepared with dual RCA technology is shown in Scheme 1. The method comprises three major steps: preparation of four types of circular DNA templates (CT-1, CT-2, CT-3, and CT-4), followed by four independent but simultaneous RCA reactions in the presence of phi29 DNA polymerase that generate four ultralong ssDNA chains with repeated periodic sequences complementary to the circular DNA template (ssDNA-1, ssDNA-2, ssDNA-3, and ssDNA-4). By programming the circular template sequences, ssDNA-1 is expected to construct varied degrees of hydrogen bonding with the other three chains (ssDNA-2, ssDNA-3, and ssDNA-4). Specifically, ssDNA-1 is completely non-complementary to ssDNA-2, partially complementary to ssDNA-3, and fully complementary to ssDNA-4. In the final step, those three groups of DNA strands (ssDNA-1 + ssDNA-2, ssDNA-1 + ssDNA-3, and ssDNA-1 + ssDNA-4) with different degrees of hydrogen bonding were used as precursors and mixed in one test tube to prepare three groups of DNA hydrogels (DNA hydrogel-1, DNA hydrogel-2, and DNA hydrogel-3) through self-assembly. The performance of three sets of DNA hydrogels, prepared using ultralong ssDNA chains with different degrees of hydrogen bonding in accordance with the dual RCA strategy, was compared and analyzed.

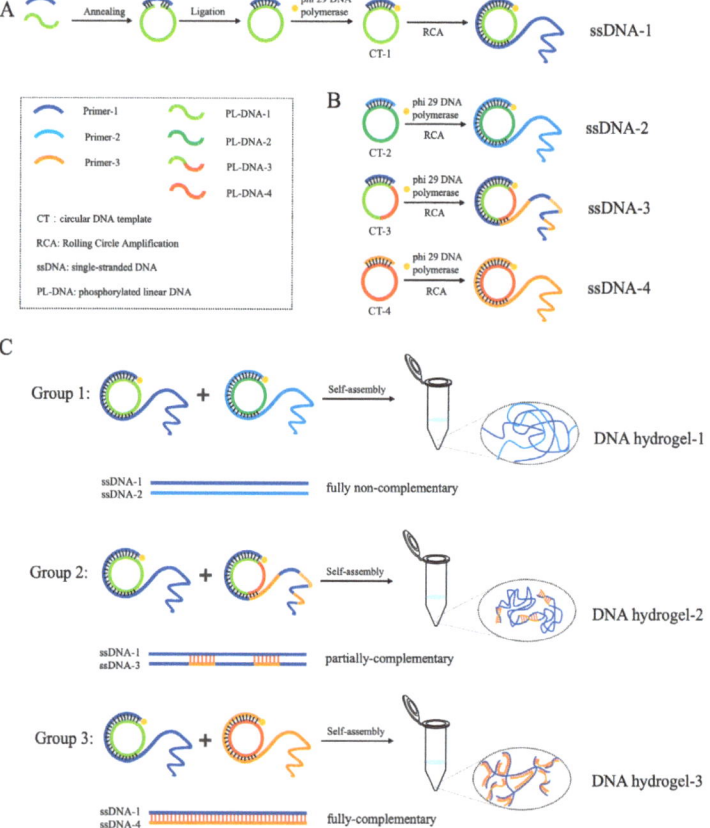

Scheme 1. Schematic illustration of DNA hydrogels prepared using ultralong single-stranded DNA (ssDNA) with varying hydrogen bonding degrees generated with dual rolling circle amplification (RCA).

(**A**): Preparation of circular DNA template (CT-1) and generation of ssDNA chain (ssDNA-1) by RCA reaction. (**B**): Generation of three ssDNA chains (ssDNA-2, ssDNA-3, and ssDNA-4) by RCA reaction based on three types of circular DNA templates (CT-2, CT-3, and CT-4). (**C**): Group 1 shows DNA hydrogel-1 prepared after the self-assembly of two fully non-complementary DNA strands, ssDNA-1 and ssDNA-2; group 2 shows DNA hydrogel-2 prepared after the self-assembly of two partially-complementary DNA strands, ssDNA-1 and ssDNA-3; group 3 shows DNA hydrogel-3 prepared after the self-assembly of two fully-complementary DNA strands, ssDNA-1 and ssDNA-4.

3.2. Preparation and Characterization of RCA Products

The successful formation of DNA hydrogels relies on the efficacy of the RCA reaction. To validate the feasibility of the study, agarose gel electrophoresis was utilized to characterize the RCA products, as shown in Figure 1A. The circular DNA template (CT) was composed of a phosphorylated linear single-stranded DNA (PL-DNA) and another single-stranded DNA serving as a linker. The linker DNA had complementary ends to the linear DNA, which brought the 5′ end and 3′ end of the linear DNA closer together, thereby forming a phosphodiester bond under the action of T4 ligase and obtaining a ligated circular DNA [44]. In addition, the linear DNA sequence was used as the RCA reaction template, and the linker DNA was used as the RCA primer. The primer was extended along the CT with the aid of phi29 DNA polymerase to produce tandem repeated sequences complementary to the CT. The formation of ligated CT-1 was verified by the gel results, where CT-1 (lane 3) migrated slower than linear DNA-1 (lane 2) and primer-1 (lane 1). Furthermore, the RCA products were trapped in the loading well (yellow arrow, lane 4) and unable to migrate downward when exposed to an electric field, indicating the large molecular weight of the products and thus the success of the RCA reaction. The ligation of other CTs and their mediated RCA were also verified to be successful (Figure S1).

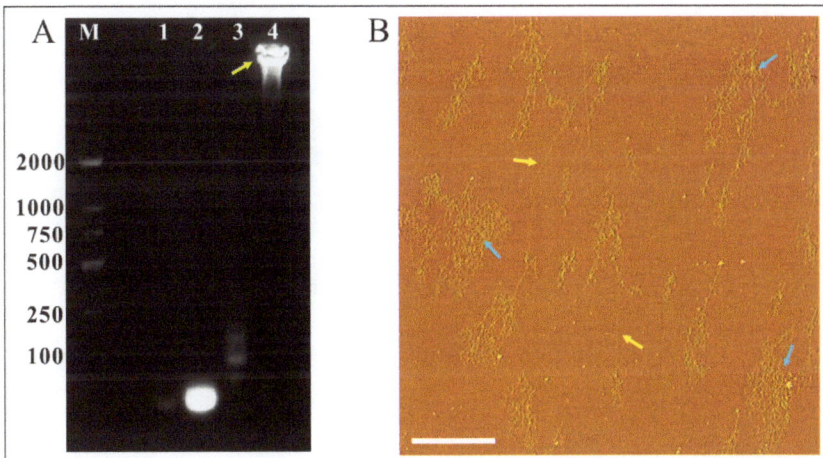

Figure 1. (**A**): Agarose gel electrophoresis results of the RCA product. M: DNA marker; 1: primer-1; 2: phosphorylated linear DNA-1 (PL-DNA-1); 3: circular DNA template-1 (CT-1); 4: RCA products (ssDNA-1). (**B**): AFM phase image of RCA products. Yellow arrows indicate single-stranded DNA and blue arrows indicate nanostructures by DNA random coiling. The scale bar represents 1 µm.

The Atomic Force Microscope (AFM) was used to further characterize the RCA reaction and provide a visual representation of the structural details of the ssDNA chains. As depicted in Figure 1B, the DNA strands were observed to intertwine and form disorderly, concentrated, coiled complexes (blue arrows), instead of being fully stretched and in a single-stranded linear state (yellow arrows). This was mainly attributed to the flexibility of ssDNA, which is influenced by non-covalent forces such as electrostatic interactions, hydrophobic interaction, and intra-stranded base pairing, leading to the formation of non-specific secondary structures. The successful RCA reaction provided the precursors for DNA hydrogels and served as the foundation for their preparation.

3.3. Preparation and Morphological Characterization of DNA Hydrogels

To visualize the different states of DNA molecules in solution before and after the formation of DNA hydrogel, imaging results under white light and ultraviolet (UV) modes are presented separately. Figure 2A shows two chains of ssDNA-1 and ssDNA-2, which were generated with RCA and were completely non-complementary in terms of sequence composition, mixed to produce DNA hydrogel-1 (group 1). Before mixing, the precursor DNA solution in the two centrifuge tubes appeared slightly cloudy, but no obvious precipitate was found. However, after mixing, a clear white flocculent aggregation rapidly appeared in the upper layer of the solution in the tube. Upon exposure to UV, red and green filaments were observed in the two ssDNA tubes (ssDNA-1 stained with 4sGelRed and ssDNA-2 stained with 4sGelGreen) (group 1, Figure 2B), indicating that RCA produced large amounts of ssDNA products that were not visible to the naked eye under white light. Similarly, DNA hydrogel-2 and DNA hydrogel-3 were formed, observed, and compared under white light and 365 nm UV light exposure (group 2 and group 3).

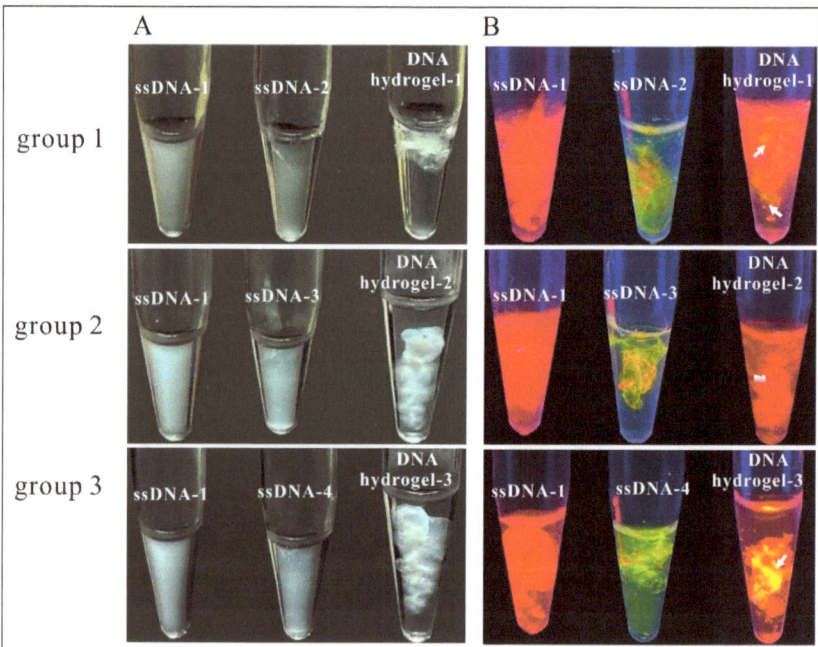

Figure 2. Images of DNA hydrogels prepared using assembly of ssDNAs with different degrees of hydrogen bonding under natural light (**A**) and UV light (**B**). ssDNA-1 was stained with 4sGelRed; ssDNA-2, ssDNA-3, and ssDNA-4 were stained with 4sGelGreen. Group 1 shows DNA hydrogel-1 prepared after the mixture of two completely non-complementary DNA strands, ssDNA-1 and ssDNA-2; group 2 shows DNA hydrogel-2 prepared after the mixture of two partially complementary DNA strands, ssDNA-1 and ssDNA-3; group 3 shows DNA hydrogel-3 prepared after the mixture of two fully complementary DNA strands, ssDNA-1 and ssDNA-4.

As the degree of hydrogen bonding between ssDNA-1 and its counterpart DNA increased, the resulting DNA hydrogels gradually sank to the bottom of the tube, indicating a more compact structure. This was further confirmed through fluorescence imaging. DNA hydrogel-1 appeared as a fluffy, sponge-like structure with the largest volume, while DNA hydrogel-2 resembled a twisted wool ball, and DNA hydrogel-3 displayed firm lamellar structures (right column, Figure 2B). Additionally, a comparison of fluorescence color changes after gel preparation revealed that DNA hydrogel-1 maintained a certain degree of independence between the red and green DNA strands (white arrows, group 1, Figure 2B), while DNA hydrogel-2 exhibited a color change to orange (white arrow, group 2, Figure 2B), and DNA hydrogel-3 was dominated by a yellow product (white arrow, group 3, Figure 2B). In accordance with these results, we tentatively hypothesized that non-complementary long ssDNA molecules would form a loosely structured DNA hydrogel because of the physical entanglement of ssDNA chains, while partially complementary ssDNA would form a DNA hydrogel through base pairing and physical intertwining. Fully complementary ssDNA, on the other hand, would primarily form a relatively dense hydrogel through DNA hybridization based on base pairing.

3.4. Characterization of Mechanical Properties and Microstructure of DNA Hydrogels

To investigate this hypothesis, the rheological characteristics of the DNA hydrogels were further examined. As illustrated in Figure 3A–C, it is evident that for the hydrogels produced from entirely non-complementary DNA chains (Figure 3A), partially complementary DNA chains (Figure 3B), and fully complementary DNA chains (Figure 3C), the storage modulus (G') was largely higher than the loss modulus (G''), indicating the solid nature of these three types of DNA hydrogels. Furthermore, the G' values of these hydrogels were compared, which can reflect the mechanical properties of the hydrogel. As shown in Figure 3D, the G' value for the DNA hydrogel based on fully complementary ssDNA chains was the highest, followed by partially and non-complementary ones. Additionally, it was observed that the G' values of gels based on fully and partially complementary ssDNA were significantly greater than that based on non-complementary ssDNA, which is consistent with the above morphology results and suggests that the mechanical properties of DNA hydrogel can be adjusted by introducing complementary base pairs into ssDNA chains. Interestingly, the G' value of the hydrogel based on fully complementary DNA strands was not substantially larger than that of the hydrogel based on partially complementary DNA chains. This may be attributed to the hydrogel preparation method, in which two types of ssDNA were mixed with vortexing and incubated at room temperature for only 10 s. It is likely that this process was insufficient in enabling the complete hybridization of all complementary sequences in the mixture, leaving some ssDNA unpaired.

The mechanical properties of hydrogels are closely correlated with their internal microstructure. Therefore, scanning electron microscopy (SEM) was employed next to observe the microstructure of the three types of prepared DNA hydrogels. Although all three exhibited a characteristic nanoflower microstructure (Figure 4A–C) comprised of RCA-generated ssDNA and had similar diameters (~500 nm) [8], the isolated nanoflower microstructures tended to connect with each other (yellow arrows, Figure 4B,C) with an increased degree of hydrogen bonding, gradually forming a porous sheet structure (red arrows, Figure 4C). This structure was similar to the microstructure of a DNA hydrogel produced with the hybridization chain reaction (HCR) technique [45,46], which can yield double helices [47]. The results intuitively demonstrate that as the degree of hydrogen bonding increased, the DNA cross-linking points became denser, and the internal microstructure of the gel became more complex.

Despite the complete complementarity of the ssDNA precursors in Figure 4C, the microstructure of the resulting nanoflowers was still observed. The incomplete hybridization of ssDNA by hydrogen bonding may account for this phenomenon. Other forces, such as hydrophilic–hydrophobic interaction, electrostatic interaction, and π–π stacking [15,48], contribute to the entanglement of ssDNA within and between strands (Scheme 2) [49]. From the preparation and characterization data of the aforementioned three gels, several conclusions can be inferred. For DNA hydrogel-1, the absence of hydrogen bond interactions between the two long DNA strands necessitates complete reliance on physical entanglement, including intra-chain and inter-chain physical entanglement, to form the gel. DNA hydrogel-2, on the other hand, is formed through a combination of hydrogen bond interactions and physical entanglement due to the presence of partially complementary bases between the two strands. Although the precursor DNA chains of DNA hydrogel-3 consist of completely complementary long ssDNA, it may not fully form dsDNA and retains some level of physical entanglement primarily influenced by hydrogen bond interactions. Non-covalent hydrogen bonds serve as network enhancers in DNA hydrogels, enhancing inter-chain interactions and promoting the mechanical strength and stability of the gel. Physical entanglement enables the gel to undergo sliding rather than fracturing when subjected to external forces, thereby contributing to the gel's favorable tensile properties. Inter-chain entanglement may be more conducive to hydrogel stability and extensibility compared to intra-chain entanglement. In the absence of inter-chain entanglement, a distinct DNA hydrogel may not form or only a very loosely structured DNA hydrogel may be observed. This finding demonstrates the ability to finely adjust the mechanical properties of pure DNA hydrogels through precise control of factors such as the degree of base complementary pairing, hybridization temperature, hybridization time, oscillation time, and the method of regulating hydrogen bonding and physical entanglement. This level of control is important for facilitating specialized applications and plays a crucial role in the design of various functional DNA hydrogels.

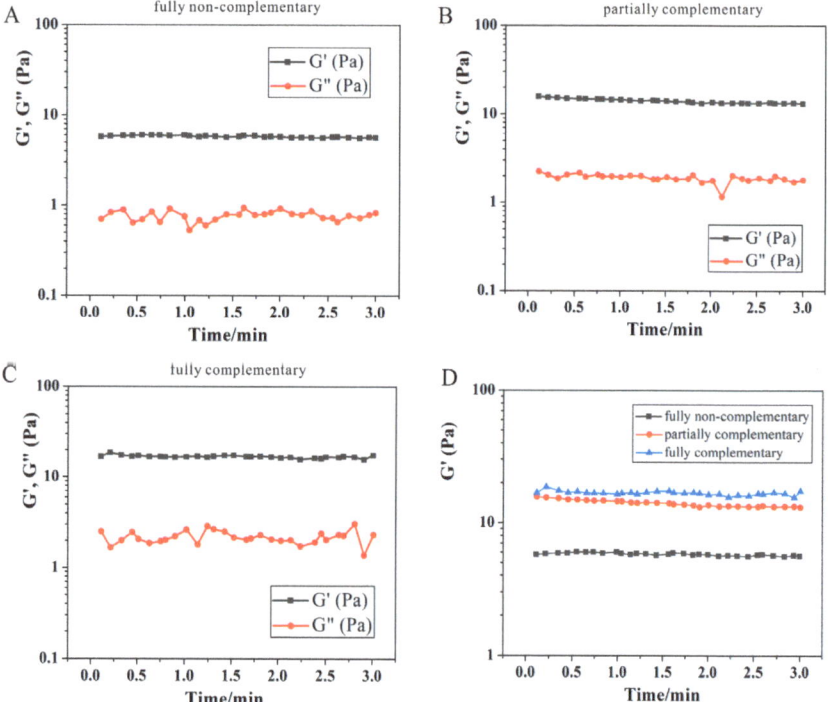

Figure 3. Time-scan rheological performance of DNA hydrogels prepared using assembly of ssDNAs with different degrees of hydrogen bonding. (**A**): DNA hydrogel-1 prepared with two completely non-complementary DNA strands (ssDNA-1 and ssDNA-2). (**B**): DNA hydrogel-2 prepared with two partially complementary DNA strands (ssDNA-1 and ssDNA-3). (**C**): DNA hydrogel-3 prepared with two fully complementary DNA strands (ssDNA-1 and ssDNA-4). (**D**): Comparison of the energy storage modulus (G′) of DNA hydrogels with different degrees of hydrogen bonding.

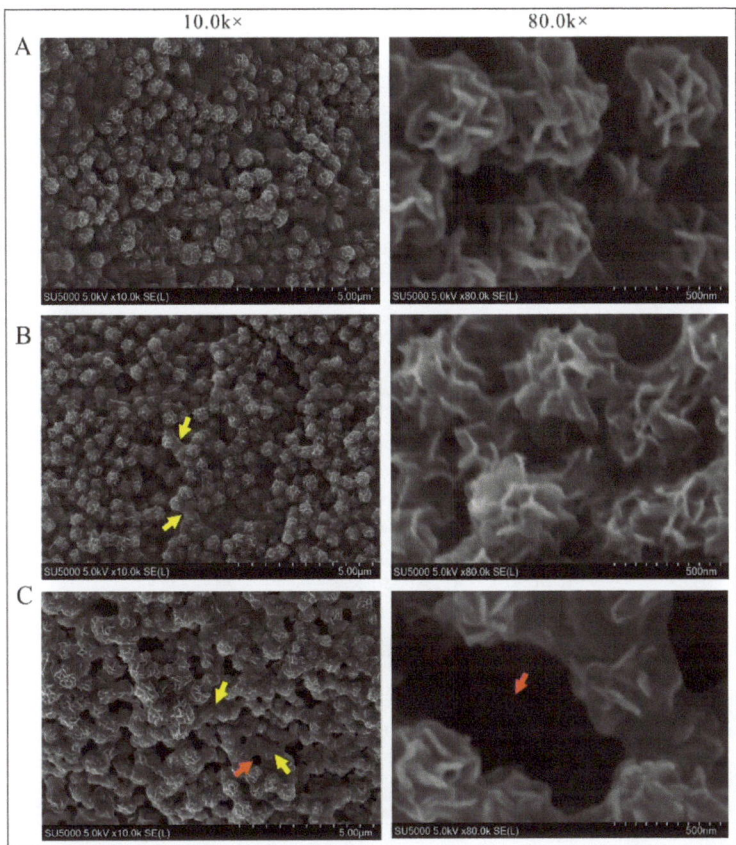

Figure 4. SEM characterizations of DNA hydrogels prepared using assembly of ssDNAs with different degrees of hydrogen bonding. (**A**): DNA hydrogel-1 prepared with two completely non-complementary DNA strands. (**B**): DNA hydrogel-2 prepared with two partially complementary ssDNA chains. (**C**): DNA hydrogel-3 prepared with two fully complementary DNA strands. Yellow arrows: locations where isolated nanoflower microstructures tended to connect with each other; red arrows: porous sheet structures.

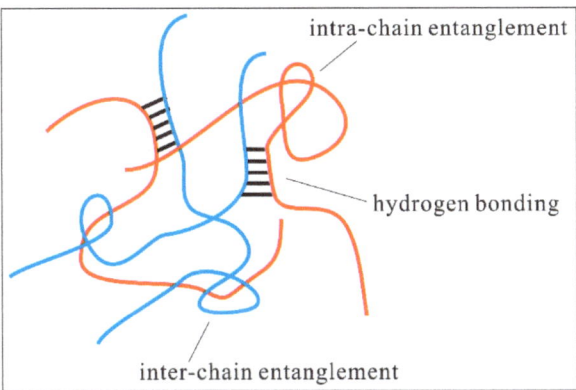

Scheme 2. Schematic diagram of DNA hydrogels based on DNA chain entanglement (including intra- and inter-chain entanglement) and hydrogen bonding of long single-stranded DNAs.

3.5. Entrapment Efficiency Test of DNA Hydrogels

Because of their unique porous structure that can carry a large number of water-soluble compounds as well as the physical entanglement of DNA strands, DNA hydrogels have become a promising platform for encapsulating diverse particles and biomolecules in biosensing systems [50] and sustainable drug-delivery applications [51]. To further compare the entrapment efficiency of the hydrogels, we analyzed the ultraviolet absorption spectra of the varying amounts of gold nanoparticles (AuNPs) [52] present in the supernatant following gel entrapment. To prevent aggregation of the bare AuNPs in the salt-containing experimental system, we coated them with bovine serum albumin (BSA) before incorporating them into the hydrogel network. The AuNPs–BSA were then trapped inside the cross-linked network during hydrogel formation.

The entrapment efficiency of the DNA hydrogel based on full complementary ssDNA chains was initially assessed by examining the entrapment of varying amounts of AuNPs–BSA (0, 5, 10, 15, 20, 25, 30 μL) using seven hydrogels prepared under identical conditions. Figure 5A shows the results, whereby the supernatants of the tubes were transparent when the amount of AuNPs–BSA was less than 25 μL, indicating that most of the AuNPs were wrapped in the hydrogel. However, when the amount exceeded 25 μL, the supernatant of the tube turned red (yellow arrow, Figure 5A), indicating that a significant amount of free AuNPs–BSA remained. Therefore, 25 μL of AuNPs–BSA (dashed frame, Figure 5A) was chosen as the optimized indicator to test the entrapment ability of the different DNA hydrogels.

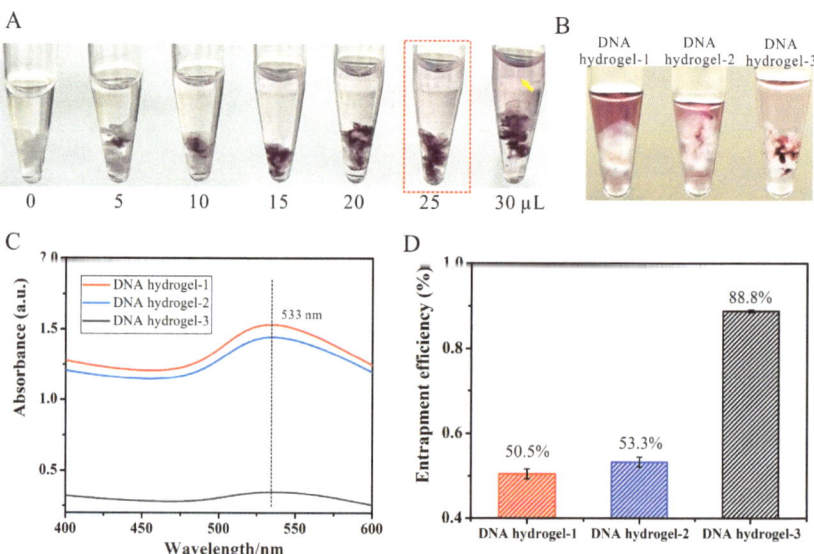

Figure 5. (**A**): Images of a series of different amounts of AuNPs–BSA (0, 5, 10, 15, 20, 25, 30 μL) trapped by DNA hydrogels prepared with full complementary ssDNA chains under identical conditions. (**B**): Images of the three kinds of DNA hydrogels loaded with 25 μL AuNPs–BSA. (**C**): UV–visible absorption spectra of three tubes of DNA hydrogels supernatant (AuNPs–BSA) in Figure B. The maximum absorption wavelength of the 30 nm AuNPs prepared in this study is 533 nm. (**D**): Corresponding entrapment efficiency of the three kinds of DNA hydrogels.

Three groups of DNA precursor solutions were supplemented with 25 μL of AuNPs–BSA and briefly vortexed to prepare hydrogels. As shown in Figure 5B, distinct hydrogels were formed in each of the three tubes and were observed to wrap a certain amount of AuNPs–BSA, as evidenced by the red color of the gels. UV absorption spectra were acquired from the supernatant of each tube, and the results are presented in Figure 5C. The three supernatants exhibited distinct UV–vis values at a wavelength of 533 nm, where

30-nm AuNPs have the maximum absorption. Of the three DNA hydrogels tested, DNA hydrogel-1 (based on completely non-complementary DNA strands) exhibited the highest absorbance, followed by DNA hydrogel-2 (based on partially complementary DNA strands), while DNA hydrogel-3 based on fully complementary DNA chains showed the lowest absorbance. Then, the entrapment efficiency of the DNA hydrogels for AuNPs–BSA was calculated using the following formula:

$$E = (A_{total} - A_{hydrogel})/A_{total};$$

where E represents the entrapment efficiency, A_{total} is the UV absorption value of 25 µL AuNPs–BSA which is shown in Figure S2, and $A_{hydrogel}$ is the UV absorption value of the supernatant after entrapment of AuNPs–BSA with different hydrogels. After performing calculations, we determined that DNA hydrogel-3 exhibited the highest entrapment efficiency of 88.8%, followed by DNA hydrogel-2 at 53.3% and DNA hydrogel-1 at the lowest efficiency of 50.5% (Figure 5D). This result is consistent with the mechanical properties of the hydrogels, which suggests that in the presence of robust hydrogen bonding, the physical cross-linking force between the ssDNA precursor chains in DNA hydrogels is amplified, leading to the formation of a more condensed gel structure that is also capable of effectively incorporating more signaling molecules for biological detection or more drug molecules for efficient drug-delivery applications.

4. Conclusions

In summary, the dual rolling circle amplification (RCA) technique offers a promising strategy for producing physical pure DNA hydrogels with diverse mechanical properties. By utilizing ultralong ssDNA prepared with dual RCA, we synthesized three types of DNA hydrogels with varying degrees of hydrogen bonding comprising completely non-complementary, partially complementary, and fully complementary ssDNA chains. Our results demonstrate a close association between the performance of DNA hydrogels and the degree of hydrogen bonding, with an increase in hydrogen bonding degree leading to a corresponding enhancement of the gel's mechanical strength and entrapment efficiency. Furthermore, we observed that large amounts of unrelated ssDNA precursors can also form hydrogel via intra- and inter-chain entanglement, although the mechanical properties of such gels are limited. Interestingly, even when using fully complementary DNA chains, the long single DNA chains tend to be partially wound rather than fully hybridized through base complementary hybridization. Compared to previous works, the advantages of using the dual RCA method for the preparation and regulation of DNA hydrogels are mainly as follows: (1). High efficiency. The long single-stranded DNA produced with RCA amplification can generate a large amount of DNA hydrogel by simply shaking at room temperature for 10 s. (2). Simple operation. The preparation process does not require any large instruments or complex operating steps. (3). Exclusion of other materials. By adjusting the sequence design of circular DNA, the mechanical properties of DNA hydrogels can be achieved, ensuring the excellent biocompatibility of the DNA hydrogel. At the same time, it should be noted that the mechanical properties of DNA hydrogels prepared with the current methods are still relatively low, and these hydrogels cannot be universally applied to demanding specialized fields. Additionally, the regulatory mechanisms still need further exploration, such as the proportion of hydrogen bonding interactions and physical entanglements, as well as the balance between intra-chain and inter-chain physical entanglements. The control of these proportions and their relationship with the performance of DNA hydrogels require further investigation. Additionally, the ratio of complementary sequences and the number of A–T and C–G base pairs may also affect the mechanical properties of the DNA hydrogel, which will be explored in our future studies.

Supplementary Materials: The following supporting information can be downloaded at: https://www.mdpi.com/article/10.3390/bios13070755/s1, Figure S1: Agarose gel electrophoresis results of the RCA product; Figure S2: UV–visible absorption spectra of AuNPs–BSA.

Author Contributions: X.W., conceptualization, investigation, and writing—original draft. H.W., investigation and methodology. H.Z., investigation and data curation. T.Y., methodology. B.Z., conceptualization, data curation. J.Y., conceptualization, writing—review and editing, validation, and supervision. All authors have read and agreed to the published version of the manuscript.

Funding: This research was funded by the Natural Science Foundation of Shanghai Municipal (No. 20ZR1424100), the National Natural Science Foundation of China (No. 21775102), and the Key Laboratory of Bioanalysis and Metrology for State Market Regulation (No. KLSMR2022-01).

Institutional Review Board Statement: Not applicable.

Informed Consent Statement: Not applicable.

Data Availability Statement: Not applicable.

Conflicts of Interest: The authors declare that they have no known competing financial interests or personal relationship that could have appeared to influence the work reported in this paper.

References

1. Li, F.; Tang, J.; Geng, J.; Luo, D.; Yang, D. Polymeric DNA hydrogel: Design, synthesis and applications. *Prog. Polym. Sci.* **2019**, *98*, 101163. [CrossRef]
2. Xu, P.F.; Noh, H.; Lee, J.H.; Domaille, D.W.; Nakatsuka, M.A.; Goodwin, A.P.; Cha, J.N. Imparting the unique properties of DNA into complex material architectures and functions. *Mater. Today* **2013**, *16*, 290–296. [CrossRef] [PubMed]
3. Alemdaroglu, F.E.; Herrmann, A. DNA meets synthetic polymers—Highly versatile hybrid materials. *Org. Biomol Chem* **2007**, *5*, 1311–1320. [CrossRef]
4. Cheng, W.; Wu, X.; Zhang, Y.; Wu, D.; Meng, L.; Chen, Y.; Tang, X. Recent applications of hydrogels in food safety sensing: Role of hydrogels. *Trends Food Sci. Technol.* **2022**, *129*, 244–257. [CrossRef]
5. Yang, Z.; Chen, L.; McClements, D.J.; Qiu, C.; Li, C.; Zhang, Z.; Miao, M.; Tian, Y.; Zhu, K.; Jin, Z. Stimulus-responsive hydrogels in food science: A review. *Food Hydrocoll.* **2022**, *124*, 107218. [CrossRef]
6. Li, J.; Mo, L.; Lu, C.H.; Fu, T.; Yang, H.H.; Tan, W. Functional nucleic acid-based hydrogels for bioanalytical and biomedical applications. *Chem. Soc. Rev.* **2016**, *45*, 1410–1431. [CrossRef] [PubMed]
7. Kahn, J.S.; Hu, Y.; Willner, I. Stimuli-Responsive DNA-Based Hydrogels: From Basic Principles to Applications. *Accounts Chem. Res.* **2017**, *50*, 680–690. [CrossRef]
8. Sheng, J.; Pi, Y.; Zhao, S.; Wang, B.; Chen, M.; Chang, K. Novel DNA nanoflower biosensing technologies towards next-generation molecular diagnostics. *Trends Biotechnol.* **2023**, *41*, 653–668. [CrossRef]
9. Wang, Y.; Zhu, Y.; Hu, Y.; Zeng, G.; Zhang, Y.; Zhang, C.; Feng, C. How to Construct DNA Hydrogels for Environmental Applications: Advanced Water Treatment and Environmental Analysis. *Small* **2018**, *14*, 1703305. [CrossRef]
10. Nnachi, R.C.; Sui, N.; Ke, B.; Luo, Z.; Bhalla, N.; He, D.; Yang, Z. Biosensors for rapid detection of bacterial pathogens in water, food and environment. *Environ. Int.* **2022**, *166*, 107357. [CrossRef]
11. Sun, Z.; Song, C.; Wang, C.; Hu, Y.; Wu, J. Hydrogel-Based Controlled Drug Delivery for Cancer Treatment: A Review. *Mol. Pharm.* **2020**, *17*, 373–391. [CrossRef]
12. Quazi, M.Z.; Park, N. DNA Hydrogel-Based Nanocomplexes with Cancer-Targeted Delivery and Light-Triggered Peptide Drug Release for Cancer-Specific Therapeutics. *Biomacromolecules* **2023**, *24*, 2127–2137. [CrossRef] [PubMed]
13. Um, S.H.; Lee, J.B.; Park, N.; Kwon, S.Y.; Umbach, C.C.; Luo, D. Enzyme-catalysed assembly of DNA hydrogel. *Nat. Mater.* **2006**, *5*, 797–801. [CrossRef]
14. Geng, J.; Yao, C.; Kou, X.; Tang, J.; Luo, D.; Yang, D. A Fluorescent Biofunctional DNA Hydrogel Prepared by Enzymatic Polymerization. *Adv. Healthc. Mater.* **2018**, *7*, 1700998. [CrossRef] [PubMed]
15. Hu, W.; Wang, Z.; Xiao, Y.; Zhang, S.; Wang, J. Advances in crosslinking strategies of biomedical hydrogels. *Biomater. Sci.* **2019**, *7*, 843–855. [CrossRef] [PubMed]
16. Pardo, Y.A.; Yancey, K.G.; Rosenwasser, D.S.; Bassen, D.M.; Butcher, J.T.; Sabin, J.E.; Ma, M.; Hamada, S.; Luo, D. Interfacing DNA hydrogels with ceramics for biofunctional architectural materials. *Mater. Today* **2022**, *53*, 98–105. [CrossRef]
17. Iqbal, S.; Ahmed, F.; Xiong, H. Responsive-DNA hydrogel based intelligent materials: Preparation and applications. *Chem. Eng. J.* **2021**, *420*, 130384. [CrossRef]
18. Xia, X.; Yang, H.; Cao, J.; Zhang, J.; He, Q.; Deng, R. Isothermal nucleic acid amplification for food safety analysis. *TrAC Trends Anal. Chem.* **2022**, *153*, 116641. [CrossRef]
19. Cao, X.; Chen, C.; Zhu, Q. Biosensors based on functional nucleic acids and isothermal amplification techniques. *Talanta* **2023**, *253*, 123977. [CrossRef]

20. Beyer, S.; Nickels, P.; Simmel, F.C. Periodic DNA nanotemplates synthesized by rolling circle amplification. *Nano Lett.* **2005**, *5*, 719–722. [CrossRef]
21. Li, C.; Wang, Y.; Li, P.F.; Fu, Q. Construction of rolling circle amplification products-based pure nucleic acid nanostructures for biomedical applications. *Acta Biomater.* **2023**, *160*, 1–13. [CrossRef] [PubMed]
22. Tang, J.; Liang, A.; Yao, C.; Yang, D. Assembly of Rolling Circle Amplification-Produced Ultralong Single-Stranded DNA to Construct Biofunctional DNA Materials. *Chemistry* **2023**, *29*, e202202673. [CrossRef] [PubMed]
23. Zhang, K.; Lv, S.; Lin, Z.; Li, M.; Tang, D. Bio-bar-code-based photoelectrochemical immunoassay for sensitive detection of prostate specific antigen using rolling circle amplification and enzymatic biocatalytic precipitation. *Biosens. Bioelectron.* **2018**, *101*, 159–166. [CrossRef] [PubMed]
24. Ali, M.M.; Li, F.; Zhang, Z.; Zhang, K.; Kang, D.K.; Ankrum, J.A.; Le, X.C.; Zhao, W. Rolling circle amplification: A versatile tool for chemical biology, materials science and medicine. *Chem. Soc. Rev.* **2014**, *43*, 3324–3341. [CrossRef] [PubMed]
25. Lee, J.B.; Peng, S.; Yang, D.; Roh, Y.H.; Funabashi, H.; Park, N.; Rice, E.J.; Chen, L.; Long, R.; Wu, M.; et al. A mechanical metamaterial made from a DNA hydrogel. *Nat. Nanotechnol.* **2012**, *7*, 816–820. [CrossRef] [PubMed]
26. Mao, X.; Chen, G.; Wang, Z.; Zhang, Y.; Zhu, X.; Li, G. Surface-immobilized and self-shaped DNA hydrogels and their application in biosensing. *Chem. Sci.* **2018**, *9*, 811–818. [CrossRef]
27. Zeng, R.; Huang, Z.; Wang, Y.; Tang, D. Enzyme-encapsulated DNA hydrogel for highly efficient electrochemical sensing glucose. *ChemElectroChem* **2020**, *7*, 1537–1541. [CrossRef]
28. Chen, Q.; Tian, R.; Liu, G.; Wen, Y.; Bian, X.; Luan, D.; Wang, H.; Lai, K.; Yan, J. Fishing unfunctionalized SERS tags with DNA hydrogel network generated by ligation-rolling circle amplification for simple and ultrasensitive detection of kanamycin. *Biosens. Bioelectron.* **2022**, *207*, 114187. [CrossRef]
29. Xu, W.; Huang, Y.; Zhao, H.; Li, P.; Liu, G.; Li, J.; Zhu, C.; Tian, L. DNA Hydrogel with Tunable pH-Responsive Properties Produced by Rolling Circle Amplification. *Chemistry* **2017**, *23*, 18276–18281. [CrossRef]
30. Yao, C.; Zhang, R.; Tang, J.; Yang, D. Rolling circle amplification (RCA)-based DNA hydrogel. *Nat. Protoc.* **2021**, *16*, 5460–5483. [CrossRef]
31. Yao, C.; Tang, H.; Wu, W.; Tang, J.; Guo, W.; Luo, D.; Yang, D. Double Rolling Circle Amplification Generates Physically Cross-Linked DNA Network for Stem Cell Fishing. *J. Am. Chem. Soc.* **2020**, *142*, 3422–3429. [CrossRef]
32. Wang, C.; Zhang, J. Recent Advances in Stimuli-Responsive DNA-Based Hydrogels. *ACS Appl. Bio Mater.* **2022**, *5*, 1934–1953. [CrossRef]
33. Zhang, J.; Huang, H.; Song, G.; Huang, K.; Luo, Y.; Liu, Q.; He, X.; Cheng, N. Intelligent biosensing strategies for rapid detection in food safety: A review. *Biosens. Bioelectron.* **2022**, *202*, 114003. [CrossRef] [PubMed]
34. Lin, X.; Wang, Z.; Jia, X.; Chen, R.; Qin, Y.; Bian, Y.; Sheng, W.; Li, S.; Gao, Z. Stimulus-responsive hydrogels: A potent tool for biosensing in food safety. *Trends Food Sci. Technol.* **2023**, *131*, 91–103. [CrossRef]
35. Wang, H.; Wang, X.; Lai, K.; Yan, J. Stimulus-Responsive DNA Hydrogel Biosensors for Food Safety Detection. *Biosensors* **2023**, *13*, 320. [CrossRef] [PubMed]
36. Yao, S.; Chang, Y.; Zhai, Z.; Sugiyama, H.; Endo, M.; Zhu, W.; Xu, Y.; Yang, Y.; Qian, X. DNA-Based Daisy Chain Rotaxane Nanocomposite Hydrogels as Dual-Programmable Dynamic Scaffolds for Stem Cell Adhesion. *ACS Appl. Mater. Interfaces* **2022**, *14*, 20739–20748. [CrossRef]
37. Hu, Y.; Fan, C. Nanocomposite DNA hydrogels emerging as programmable and bioinstructive materials systems. *Chem* **2022**, *8*, 1554–1566. [CrossRef]
38. Wahid, F.; Zhao, X.-J.; Jia, S.-R.; Bai, H.; Zhong, C. Nanocomposite hydrogels as multifunctional systems for biomedical applications: Current state and perspectives. *Compos. Part B Eng.* **2020**, *200*, 108208. [CrossRef]
39. Lin, X.; Wang, X.; Zeng, L.; Wu, Z.L.; Guo, H.; Hourdet, D. Stimuli-Responsive Toughening of Hydrogels. *Chem. Mat.* **2021**, *33*, 7633–7656. [CrossRef]
40. Zhao, X.; Chen, X.; Yuk, H.; Lin, S.; Liu, X.; Parada, G. Soft Materials by Design: Unconventional Polymer Networks Give Extreme Properties. *Chem. Rev.* **2021**, *121*, 4309–4372. [CrossRef]
41. Jones, M.R.; Seeman, N.C.; Mirkin, C.A. Programmable materials and the nature of the DNA bond. *Science* **2015**, *347*, 1260901. [CrossRef] [PubMed]
42. Frens, G. Controlled Nucleation for the Regulation of the Particle Size in Monodisperse Gold Suspensions. *Nat. Phys.* **1973**, *241*, 20. [CrossRef]
43. Wang, W.; Tan, L.; Wu, J.; Li, T.; Xie, H.; Wu, D.; Gan, N. A universal signal-on electrochemical assay for rapid on-site quantitation of vibrio parahaemolyticus using aptamer modified magnetic metal-organic framework and phenylboronic acid-ferrocene co-immobilized nanolabel. *Anal. Chim. Acta* **2020**, *1133*, 128–136. [CrossRef] [PubMed]
44. Nilsson, M.; Malmgren, H.; Samiotaki, M.; Kwiatkowski, M.; Chowdhary, B.P.; Landegren, U. Padlock probes: Circularizing oligonucleotides for localized DNA detection. *Science* **1994**, *265*, 2085–2088. [CrossRef]
45. Wang, J.; Chao, J.; Liu, H.; Su, S.; Wang, L.; Huang, W.; Willner, I.; Fan, C. Clamped Hybridization Chain Reactions for the Self-Assembly of Patterned DNA Hydrogels. *Angew. Chem.-Int. Edit.* **2017**, *56*, 2171–2175. [CrossRef]
46. Ye, D.; Li, M.; Zhai, T.; Song, P.; Song, L.; Wang, H.; Mao, X.; Wang, F.; Zhang, X.; Ge, Z.; et al. Encapsulation and release of living tumor cells using hydrogels with the hybridization chain reaction. *Nat. Protoc.* **2020**, *15*, 2163–2185. [CrossRef]

47. Kahn, J.S.; Trifonov, A.; Cecconello, A.; Guo, W.; Fan, C.; Willner, I. Integration of Switchable DNA-Based Hydrogels with Surfaces by the Hybridization Chain Reaction. *Nano Lett.* **2015**, *15*, 7773–7778. [CrossRef]
48. Pan, W.; Wen, H.; Niu, L.; Su, C.; Liu, C.; Zhao, J.; Mao, C.; Liang, D. Effects of chain flexibility on the properties of DNA hydrogels. *Soft Matter* **2016**, *12*, 5537–5541. [CrossRef]
49. Bosnjak, N.; Silberstein, M.N. Pathways to tough yet soft materials. *Science* **2021**, *374*, 150–151. [CrossRef]
50. Chen, M.; Wang, Y.; Zhang, J.; Peng, Y.; Li, S.; Han, D.; Ren, S.; Qin, K.; Li, S.; Gao, Z. Stimuli-responsive DNA-based hydrogels for biosensing applications. *J. Nanobiotechnol.* **2022**, *20*, 40. [CrossRef]
51. Mo, F.; Jiang, K.; Zhao, D.; Wang, Y.; Song, J.; Tan, W. DNA hydrogel-based gene editing and drug delivery systems. *Adv. Drug Deliv. Rev.* **2021**, *168*, 79–98. [CrossRef] [PubMed]
52. Hua, Z.; Yu, T.; Liu, D.; Xianyu, Y. Recent advances in gold nanoparticles-based biosensors for food safety detection. *Biosens. Bioelectron.* **2021**, *179*, 113076. [CrossRef] [PubMed]

Disclaimer/Publisher's Note: The statements, opinions and data contained in all publications are solely those of the individual author(s) and contributor(s) and not of MDPI and/or the editor(s). MDPI and/or the editor(s) disclaim responsibility for any injury to people or property resulting from any ideas, methods, instructions or products referred to in the content.

Article

High-Performance Au@Ag Nanorods Substrate for SERS Detection of Malachite Green in Aquatic Products

Xiaoxiao Zhou [1,2,†], Shouhui Chen [3,4,*,†], Yi Pan [3], Yuanfeng Wang [5], Naifeng Xu [5], Yanwen Xue [3,4], Xinlin Wei [3,*] and Ying Lu [1,2,6,*]

1. College of Food Science and Technology, Shanghai Ocean University, Shanghai 201306, China; 13453949440@163.com
2. Laboratory of Quality & Safety Risk Assessment for Aquatic Products on Storage and Preservation, Ministry of Agriculture, Shanghai 201306, China
3. Department of Food Science & Technology, School of Agriculture and Biology, Shanghai Jiao Tong University, Shanghai 200240, China; panyi1018@sjtu.edu.cn (Y.P.); xueyanwen@sjtu.edu.cn (Y.X.)
4. Food Safety Engineering and Technology Research Centre (Shanghai), Shanghai 200240, China
5. Institute of Food Engineering, College of Life Science, Shanghai Normal University, 100 Guilin Road, Xuhui District, Shanghai 200234, China; yfwang@shnu.edu.cn (Y.W.); xnf1106@shnu.edu.cn (N.X.)
6. Marine Biomedical Science and Technology Innovation Platform of Lingang New Area, Shanghai 201306, China
* Correspondence: chenshouhui1982@sjtu.edu.cn (S.C.); weixinlin@sjtu.edu.cn (X.W.); y-lu@shou.edu.cn (Y.L.)
† These authors contributed equally to this work.

Abstract: In order to improve the detection performance of surface-enhanced Raman scattering (SERS), a low-cost Au@Ag nanorods (Au@Ag NRs) substrate with a good SERS enhancement effect was developed and applied to the detection of malachite green (MG) in aquaculture water and crayfish. By comparing the SERS signal enhancement effect of five kinds of Au@Ag NRs substrates with different silver layer thickness on 4-mercaptobenzoic acid (4-MBA) solution, it was found that the substrate prepared with 100 μL AgNO$_3$ had the smallest aspect ratio (3.27) and the thickest Ag layer (4.1 nm). However, it showed a good signal enhancement effect, and achieved a detection of 4-MBA as low as 1×10^{-11} M, which was 8.7 times higher than that of the AuNRs substrate. In addition, the Au@Ag NRs substrate developed in this study was used for SRES detection of MG in crayfish; its detection limit was 1.58×10^{-9} M. The developed Au@Ag NRs sensor had the advantages of stable SERS signal, uniform size and low cost, which provided a new tool for SERS signal enhancement and highly sensitive SERS detection method development.

Keywords: Au@Ag nanorods substrate; surface-enhanced Raman scattering (SERS); malachite green

1. Introduction

Surface-enhanced Raman Scattering (SERS) has been widely used in the detection and analysis of hazardous substances due to its non-destructive, short-time-consuming, high sensitivity and unique molecular vibration fingerprint [1]. Both the electromagnetic (EM) and chemical (CM) enhancement mechanisms can explain the Raman enhancement effect. It has been found that the electromagnetic enhancement principle of SERS is closely related to surface plasma excitation and the electromagnetic field intensity near the surface, while the chemical enhancement is caused by the charge transfer between molecules and metals [2]. According to the enhancement mechanism, the substrate used for SERS detection had a great impact on the signal strength, and the size, shape and material composition of the substrate would affect the enhancement effect of SERS signal [3,4]. Therefore, a stable, highly bioactive and uniform SERS substrate is crucial for development of sensitive and novel SERS technology.

In the field of SERS detection, metal colloid substrates, flexible substrates and solid substrates have been the most widely used substrates in the last few years [5]. Metal colloid

substrates are prepared by reducing the noble metal (Au, Ag, etc.). By controlling the type and concentration of metal salts and reducing agents, the shape and size of various metal colloid substrates can be tailored. The main method for the synthesis of the metal colloid substrates is chemical synthesis, which is simple to operate, easy to synthesize and less time consuming compared with irradiation reduction and laser ablation. Flexible substrates are developed by the deposition of nanoparticles to the flexible base such as paper base, polymer films and adhesive tape [6,7]. Li et al. combined gold nanotriangles and polydimethylsiloxane film to prepare a flexible substrate for rapid CAP detection in food samples [8].

In general, a solid substrate base includes glass slides, silicon wafers, alumina, polyvinyl chloride (PVC) and so on [9,10]. Yu-San Chien et al. utilized alumina beads as a solid base to develop a SERS sensor and demonstrated its SERS sensitivity of 4-aminophenyl disulfide (4-APDS) [11]. However, the use of solid-based and flexible substrates for SERS preparation is associated with tediousness, high cost and it is time-consuming. In contrast, colloidal metal nanoparticles require a simple material synthesis process that is easy to operate and low-cost. Silver nanoparticles (Ag NPs) and gold nanoparticles (Au NPs) are widely utilized due to their surface plasmon resonances in both the visible light and near-infrared regions, rendering them as the most commonly employed metals [12,13]. Muhammad et al. fabricated a SiO_2@Au composite substrate that enables rapid detection of fipronil residues on egg surfaces, with a lowest detection limit of 0.1 ppm [14]. Chen et al. realized a SERS sensor by combining semiconductors and Ag nanoparticles (NPs) to detect histamine [15].

In addition, the factors such as the size, shape and surface morphology of nanoparticles would have a great influence on the SERS property. Qi et al. [16] utilized gold nanostars to prepare a uniform SERS substrate with PC membrane, which exhibited excellent sensitivity for detecting R6G at concentrations as low as 1×10^{-10} M and achieved an enhancement factor of approximately 3.70×10^5. Javad et al. [17] developed gold-aryl nanocubes for SERS analysis, achieving detection limits of 10^{-11} M R6G. Soumya et al. [18] fabricated AgNPr/ZnO NRs substrates and found that the highest SERS signal was obtained from substrates with a 1500 nm nanorod length, highlighting the importance of substrate shape and composition in improving SERS performance.

On the other hand, bimetallic nanoparticles offer the combined benefits of two or more individual nanoparticles. Silver/gold bimetallic nanoparticles not only exhibit the optical enhancement properties of Ag, but also possess the chemical stability of Au, thereby effectively enhancing substrate strengthening ability. Consequently, developing SERS-active substrates based on Ag/Au binary metal nanostructures has become a prominent research topic in SERS. Liu et al. synthesized Au@Ag nanoparticles with varying thicknesses of Ag shell and investigated the correlation between the SERS signal of thiram and the thickness of the silver shell [19]. Chen et al. have developed Au@Ag nanorods (NRs) substrate for detecting TBZ in apple juice and peach juice, achieving LODs of 0.032 ppm and 0.034 ppm, respectively [20]. Au@Ag nanorods possess broader applicability in optical sensing fields due to their ability to combine the advantages of silver's optical enhancement with their unique shape, compared to Au@Ag nanospheres.

The objective of this study was to fabricate a highly stable and homogeneous Au@Ag NRs sensor with strong SERS signal. Scheme 1 illustrates the fabrication process of the Au@Ag NRs substrate and its SERS detection mechanism for MG. Initially, Au@Ag NRs substrates were chemically synthesized based on the seed growth method by controlling Ag deposition over Au seeds. Subsequently, we investigated the effect of silver layer thickness on the SERS enhancement of Au@Ag NRs using 4-MBA. The fabricated Au@Ag NRs sensor exhibited significant improvement in SERS signal. Finally, detection of MG in aquatic products was performed to discuss the application feasibility of the fabricated Au@Ag NRs substrate.

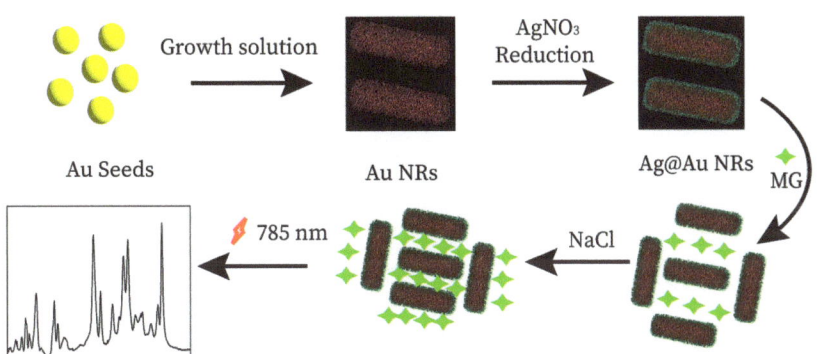

Scheme 1. Schematic diagram for the preparation of Au@Ag nanorods substrate and its SERS detection of malachite green.

2. Materials and Methods

2.1. Reagent

Cetyltrimethylammonium bromide (CTAB), cetyltrimethylammonium chloride (CTAC), sodium oleate (NaOL), chloroauric acid (HAuCl$_4$·4H$_2$O), silver nitrate (AgNO$_3$) and hydrochloric acid (HCl) were obtained from Sinopharm Chemical Reagent Co., Ltd. (Shanghai, China). L-Ascorbic acid (L-AA), 4-mercaptobenzoic acid (4-MBA) and sodium borohydride were obtained from Qiaoyi Biotechnology Co., Ltd. (Shanghai, China). Malachite green, 2,3-dichloro-5,6-dicyano-1,4-benzoquinone and other chemicals were purchased from Aladdin Industrial Co., Ltd. (Shanghai, China). Ultrapure water was used in all experiments from Shanghai Ding-shuo's water purification system (Shanghai Ding-shuo Electronic Technology Co., Ltd., Shanghai, China), which was deionized and ultra-purified to 18.2 MΩ.

2.2. Synthesis of Au@Ag NRs Substrate

The Au@Ag NRs substrate was fabricated on the basis of the method of Pu H et al. with a slight modification [21]. Firstly, AuNRs were synthesized by the seed-mediated approach. Briefly, 0.01 M NaBH$_4$ was rapidly added to the solution by mixing 9.75 mL 0.1 M CTAB solution with 25 µL 0.01 M HAuCl$_4$. The solution, after incubation for 2 h at 37 °C, was used as the Au seed. Next, 30 mL solution containing CTAB (0.768 g) and NaOL (0.547 g) was mixed with 1.35 mL 0.01 M AgNO$_3$, stirred for 5 min and incubated at 37 °C for 15 min. Then 30 mL 1 mM HAuCl$_4$ was added to the above solution and stirred at 700 rpm for 90 min. After the solution turned colorless, 1 M HCl was added and the speed was adjusted to 400 rpm for 15 min. Then, 1 mM ascorbic acid solution was added while stirring vigorously for 30 s. Finally, 100 µL of Au seed solution was added to the solution and kept incubated for 15 h at 37 °C, to grow AuNRs. The synthesized AuNRs were centrifuged twice at 8000 rpm (15 min) and resuspended with 30 mL ultra-water for the next use.

Next, Ag was modified on the surface of AuNRs. Different volumes of 0.01 M AgNO$_3$ (50, 100, 200, 300, 400 µL) were mixed with 100 µL 0.08 M CTAC and incubated for 10 min at 37 °C. Then, 2.5 mL of AuNRs were added and incubated at 37 °C for 5 min. Subsequently, 100 µL 0.1 M ascorbic acid was added to the solution while shaking vigorously. After the solution reacted for 4 h at 37 °C, the resultant solution was purified through centrifugation twice at 7000 rpm (15 min) and 4600 rpm (15 min), and resuspended with 3 mL ultra-water for the next use.

2.3. Characterization of Au@Ag NRs Substrate

The UV spectra of AuNRs and Au@Ag NRs were acquired using a UV-Vis spectrophotometer (Beijing Puxi General Instrument Co., Ltd., Beijing, China). Transmission electron microscopy (TEM) was utilized to obtain TEM images of the aforementioned nanoparticles with Talos L120C G2 from FEI NanoPorts Co., Hillsboro, OR, America being employed for this purpose. The HAADF-STEM images were acquired by utilizing the Talos F200X (FEI NanoPorts Co., Hillsboro, OR, USA), a high-angle annular dark-field scanning transmission electron microscope.

2.4. SERS Detection Based on Au@Ag NRs Substrate

SERS spectra were acquired using a portable Raman spectrometer, the BWS465-785s from BWTEK in Shanghai, China. The SERS enhancement of the synthetic Au@Ag NRs substrate was evaluated using 4-mercaptobenzoic acid (4-MBA) standard. Briefly, 10 µL 10^{-3} M 4-MBA was mixed with 1000 µL synthetic AuNRs and Au@Ag NRs. After incubation for 10 min, 10 µL mixture was dropped on the glass slide for SERS detection. Then, different concentrations of 4-MBA (10^{-7} M, 10^{-8} M, 10^{-9} M, 10^{-10} M, 10^{-11} M) were detected to analyze the SERS detection sensitivity. The parameter settings of the instrument were excitation wavelength (785 nm), power (50%), integration time (10,000 ms) and accumulation (3 times).

Detection of MG based on Au@Ag NRs substrate were performed as following. Firstly, 100 µL of different concentrations of MG solution (5×10^{-9} M~2×10^{-7} M) was mixed with 100 µL Au@Ag NRs substrate and 30 µL of 0.1 M NaCl, respectively. Subsequently, 10 µL of the mixed solution was dropped on the glass slide for detection after incubation for 10 min. The parameters of SERS detection are the same as the 4-MBA standard test, except for the power (100%).

2.5. Detection of Real Samples Based on Au@Ag NRs Substrate

The real sample was prepared following the method of Xu T et al. with slight modification [22]. Initially, 2 g of sample homogenate was mixed with 500 µL NH_2OH-HCl solution. Subsequently, 10 mL acetonitrile, 1g anhydrous magnesium and 4 g alumina were vigorously added and mixed. The supernatant was dried with nitrogen at 50 °C after centrifugation (4000 rpm, 5 min) and the residues were dissolved in 2,3-dichloro-5,6-dicyano-1,4-benzoquinone solution. Subsequently, alumina was added and mixed well and the mixture was centrifuged (12,000 rpm, 10 min). Finally, the supernatant was filtered through a 0.22 µm filter. Then, the real samples were prepared by mixing the MG standard solution with the above solution.

2.6. Statistical Analysis

Detection results were presented as mean value and standard deviation ($n = 3$), and Raman spectra were baseline corrected and smoothed by the BWTEK program. The limit of detection (LOD) for MG was determined by the following formula:

$$LOD = 3S_b + Y_b$$

S_b, above, is the standard deviation of the SERS intensity at 1617 cm^{-1} and Y_b is the mean intensity of the blanks at 1617 cm^{-1}.

3. Results and Discussion

3.1. Characterization of Au@Ag NRs Substrate

In this study, Au@Ag NRs substrate were fabricated by three steps, which were the synthesis of Au seeds, Au NRs and Au@Ag NRs. After the successful preparation of each step, the optical properties of each product change [23]. The colors of Au seeds, Au NRs and Au@Ag NRs changed from brown, to light red, then to green (Figure 1A), which was consistent with the report of Pu et al. [21]. It has been shown that Au@Ag NRs substrates have been successfully fabricated. In addition, a clear wavelength shift of the AuNRs and Au@Ag NRs substrates was observed in the UV spectrophotometry (Figure 1B). When the AuNRs were covered with Ag, the longitudinal plasmon resonance had a blue shift from 832 nm to 687 nm. When the longitudinal plasmon resonance wavelengths was near

690 nm, the Au@Ag NRs substrate was green [24]. Hence, the blue shifts of the longitudinal plasmon resonance wavelengths was consistent with the color change of AuNRs and Au@Ag NRs. It was found that when Ag was modified to the surface of AuNRs, Ag had stronger plasma absorption characteristics and a shorter plasma resonance wavelength, leading to the blue shift [25]. The wavelengths of the transverse plasmon resonance also showed a blue shift from 512 nm (AuNRs) to 462 nm (Au@Ag NRs) (Figure 1B). These results proved that Au@Ag NRs was successfully synthesized.

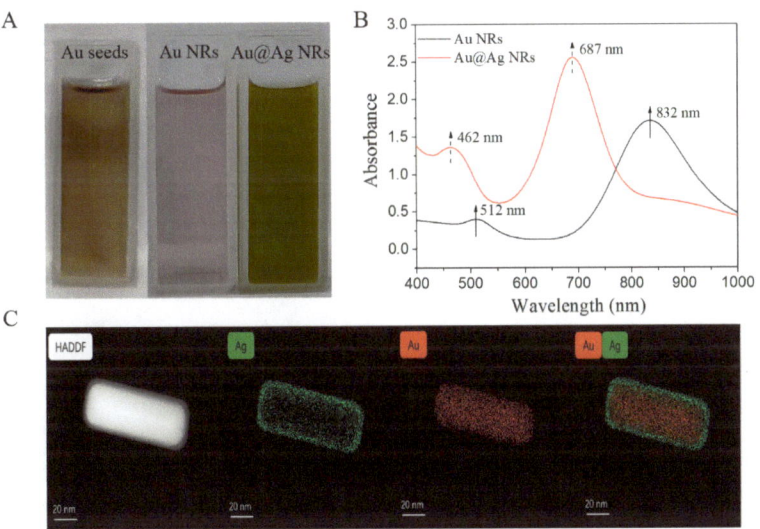

Figure 1. (**A**) Images of Au seed solution, Au NRs solution and Au@Ag NRs solution; (**B**) The UV spectra of AuNRs and Au@Ag NRs substrates; (**C**) A HADDF-STEM image of Au@Ag NRs (left); elemental mapping of Ag, Au and both elements of Au@Ag NRs substrate.

In addition, HAADF-STEM and EDS were employed to achieve atomic-scale realization of the construction of Au@Ag NRs substrate. As depicted in Figure 1C, a distinct core-shell structure (grey shell and light core) was observed by HADDF. This is attributed to the difference in atomic number of Au and Ag [26]. Furthermore, in EDS analysis, different colors represent different elements, with red representing Au and green representing Ag. EDS elements also demonstrated the successful synthesis of Au@Ag NRs substrate.

3.2. SERS Enhancement Evaluation of Au@Ag NRs Substrates with Different Silver Thickness

The SERS enhancement capability could be affected by the thickness of the Ag shell in Au@Ag NRs substrates [20]. Therefore, Au@Ag NRs substrates were fabricated by adding varying AgNO$_3$ amounts and evaluated for their ability to enhance SERS signals. As shown in Figure 2A, 4-MBA standard solution was detected by AuNRs and Au@Ag NRs substrate. With 10^{-6} M 4-MBA standard solution, AuNRs and Au@Ag NRs substrates were detected individually, with which no peaks appeared. This indicated that the presence of AuNRs and Au@Ag NRs substrates does not affect SERS detection. The addition of 4-MBA to Au NRs substrates could enhance the peaks at 1078 cm^{-1} and 1581 cm^{-1}, which belong to the circular breathing pattern and C=N stretching, respectively [27]. While the addition of 4-MBA to Au@Ag NRs substrate, the SERS intensity of peaks at 1078 cm^{-1} and 1581 cm^{-1} was significantly enhanced. The SERS intensity of 4-MBA at 1581 cm^{-1} indicated an increase in SERS intensity by a factor of 8.7 when using Au@Ag NRs substrate instead of AuNRs substrate. The Au@Ag NRs substrate exhibits a higher enhancement effect on 4-MBA, which can be attributed to the stronger bonding between Ag and S in 4–MBA [28].

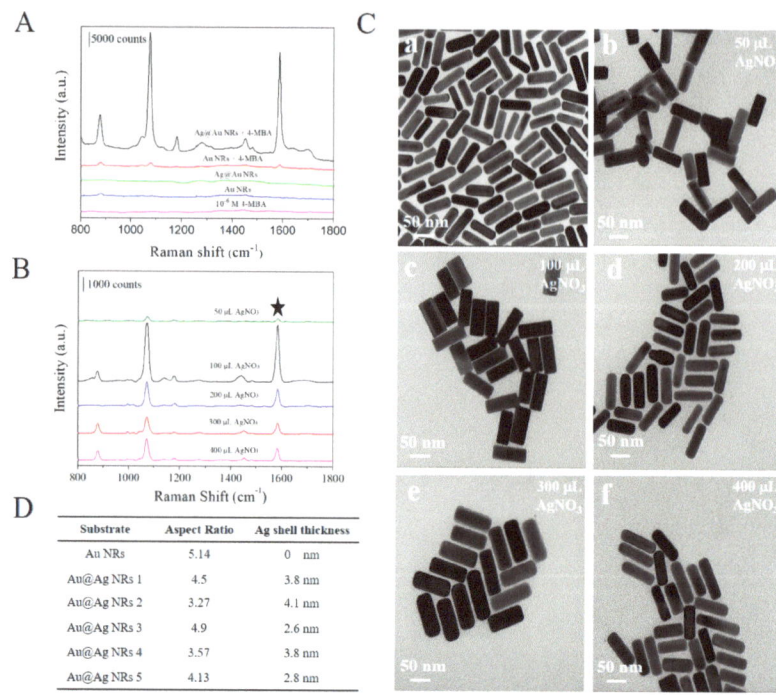

Figure 2. (**A**) SERS intensity of 4−MBA using AuNRs and Au@Ag NRs substrates. (**B**) SERS spectra of 4−MBA based on Au@Ag NRs substrate fabricated with varying amounts of AgNO$_3$. The black star refers to utilize the SERS intensity at 1581 cm^{-1} to assess the amounts of AgNO$_3$. (**C**) TEM images of Au NRs (**a**) and Au@Ag NRs fabricated with varying amounts of AgNO$_3$; (**b**) 50 µL, (**c**) 100 µL, (**d**) 200 µL, (**e**) 300 µL, (**f**) 400 µL. (**D**) The aspect ratio and thickness of the Ag shell of Au@Ag NRs substrates fabricated with varying amounts of AgNO$_3$.

Furthermore, in Figure 2B, the SERS intensity of Au@Ag NRs with varying silver thickness was compared. It was observed that the highest SERS intensity for 4−MBA detection was achieved by Au@Ag NRs substrates fabricated using 100 µL AgNO$_3$. The volume of AgNO$_3$ used during the preparation process can affect the aspect ratio, Ag shell thickness and morphology of Au@Ag NRs substrates [29,30]. As shown in the TEM results (Figure 2C), homogeneous and equal morphology was obtained when preparing Au@Ag NRs using 100 µL AgNO$_3$. As shown in Figure 2D, Au@Ag NRs prepared using 100 µL AgNO$_3$ had the smallest aspect ratio (3.27) and the thickest Ag layer (4.1 nm). Therefore, Au@Ag NRs prepared using 100 µL AgNO$_3$ were selected for subsequent experiments.

3.3. SERS Detection Sensitivity of 4-MBA Based on Au@Ag NRs Substrate

Figure 3A showed the SERS spectra using the Au@Ag NRs substrate with different concentrations of 4-MBA. It was observed that the SERS intensity was directly proportional to the logarithm concentration of 4-MBA and the limit of detection (LOD) reached as low as 10^{-11} M. This is significantly lower than the LOD (10^{-7} M) reported by Waiwijit et al. for the detection of 4-MBA on a SERS substrate based on cotton cloth [31]. The results demonstrated the excellent detection sensitivity of the Au@Ag NRs SERS detection platform for 4-MBA. As the strongest SERS intensity was at 1581 cm^{-1}, a plot was generated in Figure 3B to depict the relationship between SERS intensity of 4-MBA at 1581 cm^{-1} and the logarithm of its concentration. The fitting regression equation y = 146.99 logx + 2736.79 was obtained with a linearity correlation coefficient (R^2) of 0.991.

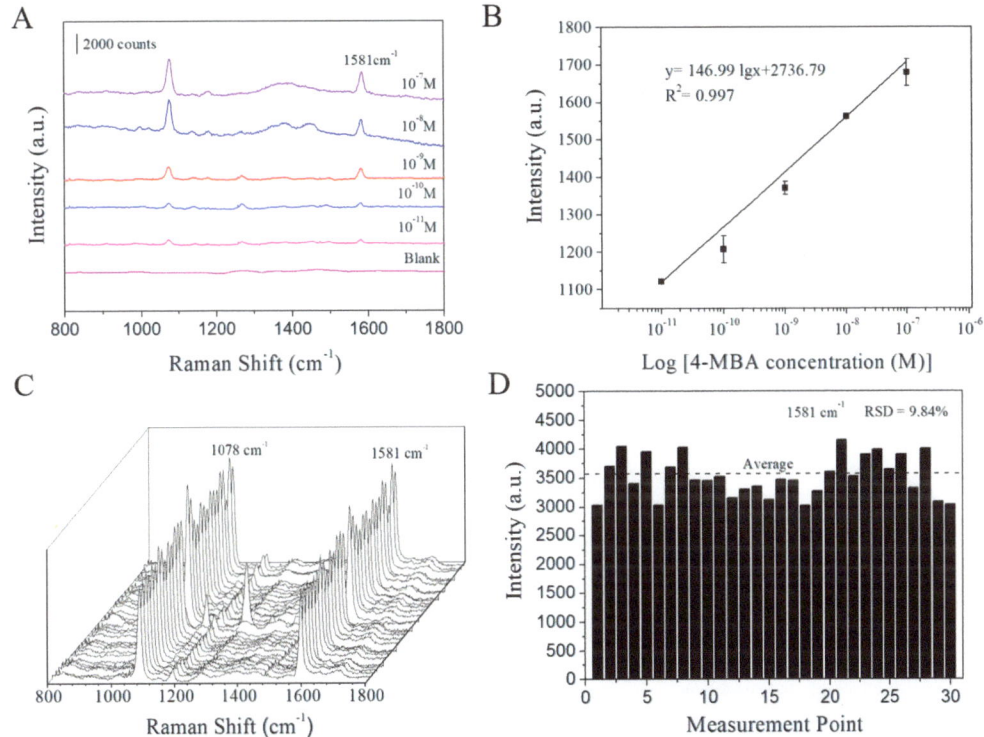

Figure 3. (**A**) SERS spectra detected by Au@Ag NRs substrates for varying concentrations of 4-MBA; (**B**) Linear correlation between SERS intensity at 1581 cm^{-1} and log concentration of 4-MBA; (**C**) SERS spectra of 30 detection points using different Au@Ag NRs substrates; (**D**) SERS intensity distribution of 30 points at 1581 cm^{-1}.

The stability of SERS signals on Au@Ag NRs substrates was assessed using 4−MBA as a probe molecule. Three distinct batches of Au@Ag NRs were selected, and ten points from each batch were randomly tested. As depicted in Figure 3C, two significant Raman peaks were observed at 1078 cm^{-1} and 1581 cm^{-1}, respectively. These peaks showed little change in SERS intensity. Furthermore, the SERS intensities at 1581 cm^{-1} are also shown in Figure 3D. The relative standard deviation (RSD) of the 30 SERS intensities at 1581 cm^{-1} was determined as 9.84%. It is generally accepted that the relative standard deviation of SERS substrates between batches or between different test points in the same batch should not exceed 20% [32]. Therefore, the Au@Ag NRs substrate developed in this study demonstrated excellent stability of SERS signals.

3.4. Detection Performance Evaluation of Real Samples Based on Au@Ag NRs Substrate
3.4.1. Optimization of SERS Detection for Malachite Green

The optimal enhancement of SERS detection is achieved when the analyte's vibration mode is perpendicular to the substrate, which relies on both the quantity of substrate and analyte added. Therefore, the addition amount of Au@Ag NRs and malachite green (MG) was crucial in SERS detection [33]. In addition, it was found that the aggregation or assembly of Au@Ag NRs would be initiated by the addition of salt, due to the electrostatic neutralization [34]. Therefore, in order to detect the MG in foods based on Au@Ag NRs substrate, for the addition amount of 10^{-6} M MG, Au@Ag NRs substrate and 0.1 M NaCl, the reaction time of the mixture of the above substance were optimized. The highest SERS intensity at 1167 cm^{-1} and 1617 cm^{-1} was found in the 100 μL of Au@Ag NRs substrate

(Figure 4A), 100 μL of MG (Figure 4B) and 30 μL of NaCl (Figure 4C), respectively. The optimal reaction time was observed at 5 min (Figure 4D). Therefore, the optimal condition for SERS based on Au@Ag NRs to detect MG was to use 100 μL Au@Ag NRs substrate, 100 μL 10^{-6} M MG and 30 μL 0.1 M NaCl, and the signal was collected after incubation of the mixture for 5 min.

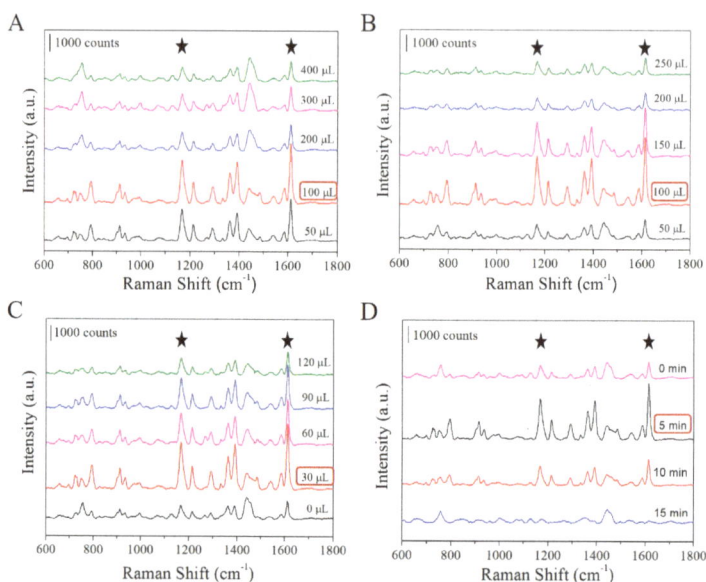

Figure 4. SERS spectra for detection of malachite green using different addition amounts of (**A**) Au@Ag NRs, (**B**) MG and (**C**) 0.1 M NaCl, and (**D**) different incubating time. The black stars denote the SERS intensity of two selected peaks during the optimization process, and the red boxes denote the optimal conditions in detection.

3.4.2. Detection Sensitivity and Accuracy

Under the optimal detection conditions, we investigated the SERS detection sensitivity and accuracy of MG using the developed Au@Ag NRs substrate. In surface-enhanced Raman spectroscopy, the intensity of the characteristic peak corresponding to the analyte in SERS is directly proportional to its concentration. The peak at 1167 cm^{-1} corresponds to the bending vibration of the C-H bond in the benzene ring and the peak at 1617 cm^{-1} corresponds to the stretching vibration of the C-C bond within the benzene ring [35]. During MG determination, we selected the linear relationship between the Raman intensity at 1617 cm^{-1} and the paired value of the MG concentration.

A gradual decline in SERS intensity was observed with decreasing MG concentration in Figure 5A. The fitting regression equation was y = 19,863.15 logx + 176,665.03, the linearity correlation of R^2 was 0.989 and the limit of detection (LOD) for MG was determined to be 1.58 × 10^{-9} M (Figure 5B). Liu et al. [36] prepared Au@SiO$_2$ nanoparticles to detect MG in fish, realizing an LOD of 1.5 × 10^{-9} M. Zhang et al. [33] constructed a nanosphere SERS sensor for the detection of MG in fish with an LOD of 1.37 × 10^{-9} M. Yue X et al. [37] developed a ratio-metric fluorescent sensor combined with a smart phone to achieve MG detection of fish with an LOD of 4.35 × 10^{-6} M. Therefore, the SERS detection sensitivity of MG, based on the developed Au@Ag NRs substrate, was at the same level as the reported methods. However, the developed Au@Ag NRs substrate had the advantages of stable SERS signal, uniform size and low cost, which can be used for the detection of MG in aquatic products.

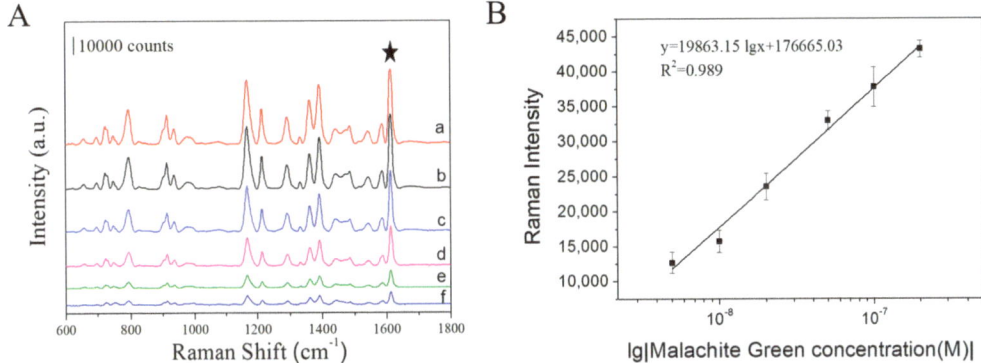

Figure 5. SERS spectra of malachite green at different concentrations using Au@Ag NRs substrate (**A**), and quantitative standard curve of MG based on the SERS intensity at 1617 cm^{-1} (**B**). a: 2×10^{-7} M; b: 10^{-8} M; c: 5×10^{-8} M; d: 2×10^{-8} M; e: 10^{-8} M; f: 5×10^{-9} M. The black star refers to utilize the SERS intensity at 1617 cm^{-1} to assess the concentration of MG.

To evaluate the accuracy of SERS detection of MG by Au@Ag NRs substrate, the spiked MG samples in aquaculture water and crayfish were detected, respectively. The SERS spectra are shown in Figure 6A (aquaculture water) and Figure 6B (crayfish). As shown in Figure 6C, the recovery rate of MG in aquatic water was 89.6~106.0%, and its RSD was 4.74~6.82%. While the recovery rate of MG in crayfish muscle was 93.5~107.0%, with the RSD ranging from 5.19~6.01%. These results indicated the developed Au@Ag NRs sensor had good accuracy for MG detection.

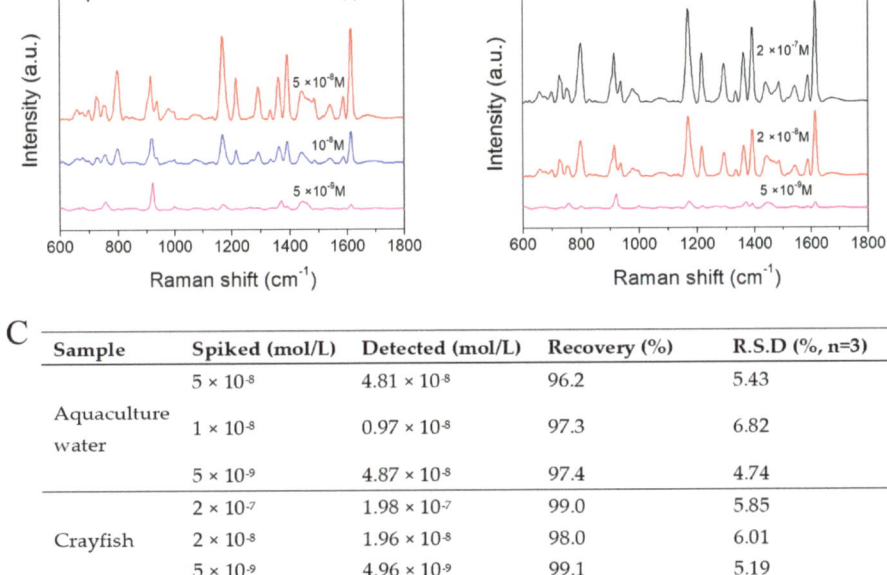

Sample	Spiked (mol/L)	Detected (mol/L)	Recovery (%)	R.S.D (%, n=3)
Aquaculture water	5×10^{-8}	4.81×10^{-8}	96.2	5.43
	1×10^{-8}	0.97×10^{-8}	97.3	6.82
	5×10^{-9}	4.87×10^{-8}	97.4	4.74
Crayfish	2×10^{-7}	1.98×10^{-7}	99.0	5.85
	2×10^{-8}	1.96×10^{-8}	98.0	6.01
	5×10^{-9}	4.96×10^{-9}	99.1	5.19

Figure 6. SERS spectra of malachite green in aquaculture water (**A**) and in crayfish (**B**) at different concentrations based on Au@Ag NRs substrate. (**C**) Recovery results of SERS detection for MG in aquaculture water and crayfish. The black star refers to utilize the SERS intensity at 1617 cm^{-1} to assess the concentration of MG.

4. Conclusions

In this study, Au@Ag NRs substrate was developed with simple preparation and a good SERS enhancement effect. Its SERS signal was 8.7 times higher than that of the AuNRs substrate. The sensitivity for 4-MBA of SERS detection based on Au@Ag NRs substrate was low to 10^{-11} M. In addition, the fabricated Au@Ag NRs substrate had the advantages of stable SERS signal, uniform size and low cost, and could be used for the detection of MG in aquaculture water and crayfish; the limit of detection of MG was calculated as 1.58×10^{-9} M. This study provided a new tool for SERS signal enhancement and a highly sensitive SERS detection method development.

Author Contributions: Conceptualization, X.Z., S.C., Y.P., Y.W., N.X. and Y.X.; methodology, Y.P., Y.W., N.X. and Y.X.; validation, Y.P., Y.W., N.X. and Y.X.; formal analysis, X.Z. and S.C.; investigation, X.Z., S.C. and Y.P.; data curation, X.Z. and S.C.; writing—original draft preparation, X.Z. and S.C.; writing—review and editing, S.C., X.W. and Y.L.; supervision, X.Z., S.C., X.W. and Y.L.; project administration, S.C., X.W. and Y.L.; funding acquisition, S.C., X.W. and Y.L. All authors have read and agreed to the published version of the manuscript.

Funding: This research was funded by the National Key R&D Program of China (No. 2019YFC1606001) and by the Shanghai Agriculture Applied Technology Development Program, China (2019-02-08-00-10-F01143).

Institutional Review Board Statement: Not applicable.

Informed Consent Statement: Not applicable.

Data Availability Statement: Not applicable.

Conflicts of Interest: The authors declare no conflict of interest.

References

1. Zhang, W.; Ma, J.; Sun, D.-W. Raman spectroscopic techniques for detecting structure and quality of frozen foods: Principles and applications. *Crit. Rev. Food Sci. Nutr.* **2021**, *61*, 2623–2639. [CrossRef] [PubMed]
2. Zhu, J.; Jiang, X.; Rong, Y.; Wei, W.; Wu, S.; Jiao, T.; Chen, Q. Label-free detection of trace level zearalenone in corn oil by surface-enhanced Raman spectroscopy (SERS) coupled with deep learning models. *Food Chem.* **2023**, *414*, 135705. [CrossRef] [PubMed]
3. Wang, Y.; Wang, M.; Shen, L.; Sun, X.; Shi, G.; Ma, W.; Yan, X. High-performance flexible surface-enhanced Raman scattering substrates fabricated by depositing Ag nanoislands on the dragonfly wing. *Appl. Surf. Sci.* **2018**, *436*, 391–397. [CrossRef]
4. Willets, K.A.; Van Duyne, R.P. Localized surface plasmon resonance spectroscopy and sensing. *Annu. Rev. Phys. Chem.* **2007**, *58*, 267–297. [CrossRef]
5. Zheng, J.; He, L. Surface-enhanced Raman spectroscopy for the chemical analysis of food. *Compr. Rev. Food Sci. Food Saf.* **2014**, *13*, 317–328. [PubMed]
6. Creedon, N.C.; Lovera, P.; Furey, A.; O'Riordan, A. Transparent polymer-based SERS substrates templated by a soda can. *Sens. Actuators B Chem.* **2018**, *259*, 64–74. [CrossRef]
7. Yang, L.; Zhen, S.J.; Li, Y.F.; Huang, C.Z. Silver nanoparticles deposited on graphene oxide for ultrasensitive surface-enhanced Raman scattering immunoassay of cancer biomarker. *Nanoscale* **2018**, *10*, 11942–11947. [CrossRef]
8. Li, H.; Geng, W.; Zheng, Z.; Haruna, S.A.; Chen, Q. Flexible SERS sensor using AuNTs-assembled PDMS film coupled chemometric algorithms for rapid detection of chloramphenicol in food. *Food Chem.* **2023**, *418*, 135998. [CrossRef]
9. Bai, S.; Du, Y.; Wang, C.; Wu, J.; Sugioka, K. Reusable surface-enhanced Raman spectroscopy substrates made of silicon nanowire array coated with silver nanoparticles fabricated by metal-assisted chemical etching and photonic reduction. *Nanomaterials* **2019**, *9*, 1531. [CrossRef] [PubMed]
10. Gill, H.S.; Thota, S.; Li, L.; Ren, H.; Mosurkal, R.; Kumar, J. Reusable SERS active substrates for ultrasensitive molecular detection. *Sens. Actuators B Chem.* **2015**, *220*, 794–798. [CrossRef]
11. Chien, Y.-S.; Chang, C.-W.; Huang, C.-C. Differential surface partitioning for an ultrasensitive solid-state SERS sensor and its application to food colorant analysis. *Food Chem.* **2022**, *383*, 132415. [CrossRef] [PubMed]
12. Nie, S.; Emory, S.R. Probing single molecules and single nanoparticles by surface-enhanced Raman scattering. *Science* **1997**, *275*, 1102–1106. [CrossRef] [PubMed]
13. Wu, L.; Pu, H.; Huang, L.; Sun, D.-W. Plasmonic nanoparticles on metal-organic framework: A versatile SERS platform for adsorptive detection of new coccine and orange II dyes in food. *Food Chem.* **2020**, *328*, 127105. [CrossRef]
14. Muhammad, M.; Yao, G.; Zhong, J.; Chao, K.; Aziz, M.H.; Huang, Q. A facile and label-free SERS approach for inspection of fipronil in chicken eggs using SiO_2@ Au core/shell nanoparticles. *Talanta* **2020**, *207*, 120324. [CrossRef] [PubMed]

15. Chen, C.; Wang, X.; Waterhouse, G.I.; Qiao, X.; Xu, Z. A surface-imprinted surface-enhanced Raman scattering sensor for histamine detection based on dual semiconductors and Ag nanoparticles. *Food Chem.* **2022**, *369*, 130971. [CrossRef] [PubMed]
16. Qi, X.; Wang, X.; Dong, Y.; Xie, J.; Gui, X.; Bai, J.; Duan, J.; Liu, J.; Yao, H. Fast synthesis of gold nanostar SERS substrates based on ion-track etched membrane by one-step redox reaction. *Spectrochim. Acta Part A Mol. Biomol. Spectrosc.* **2022**, *272*, 120955. [CrossRef]
17. Parambath, J.B.; Kim, G.; Han, C.; Mohamed, A.A. SERS performance of cubic-shaped gold nanoparticles for environmental monitoring. *Res. Chem. Intermed.* **2022**, *49*, 1259–1271. [CrossRef]
18. Columbus, S.; Hamdi, A.; Ramachandran, K.; Daoudi, K.; Dogheche, E.H.; Kaidi, M. Rapid and ultralow level SERS detection of ethylparaben using silver nanoprisms functionalized sea urchin-like Zinc oxide nanorod arrays for food safety analysis. *Sens. Actuators A Phys.* **2022**, *347*, 113962. [CrossRef]
19. Liu, B.; Han, G.; Zhang, Z.; Liu, R.; Jiang, C.; Wang, S.; Han, M.-Y. Shell thickness-dependent Raman enhancement for rapid identification and detection of pesticide residues at fruit peels. *Anal. Chem.* **2012**, *84*, 255–261. [CrossRef]
20. Chen, Z.; Sun, Y.; Shi, J.; Zhang, W.; Zhang, X.; Huang, X.; Zou, X.; Li, Z.; Wei, R. Facile synthesis of Au@Ag core–shell nanorod with bimetallic synergistic effect for SERS detection of thiabendazole in fruit juice. *Food Chem.* **2022**, *370*, 131276. [CrossRef]
21. Pu, H.; Huang, Z.; Xu, F.; Sun, D.-W. Two-dimensional self-assembled Au-Ag core-shell nanorods nanoarray for sensitive detection of thiram in apple using surface-enhanced Raman spectroscopy. *Food Chem.* **2021**, *343*, 128548. [CrossRef] [PubMed]
22. Xu, T.; Wang, X.; Huang, Y.; Lai, K.; Fan, Y. Rapid detection of trace methylene blue and malachite green in four fish tissues by ultra-sensitive surface-enhanced Raman spectroscopy coated with gold nanorods. *Food Control* **2019**, *106*, 106720. [CrossRef]
23. Zhao, Q.; Lu, D.; Zhang, G.; Zhang, D.; Shi, X. Recent improvements in enzyme-linked immunosorbent assays based on nanomaterials. *Talanta* **2021**, *223*, 121722. [CrossRef] [PubMed]
24. Pinzaru, S.C.; Magdas, D.A. Ag nanoparticles meet wines: SERS for wine analysis. *Food Anal. Methods* **2018**, *11*, 892–900. [CrossRef]
25. Ouyang, L.; Yao, L.; Zhou, T.; Zhu, L. Accurate SERS detection of malachite green in aquatic products on basis of graphene wrapped flexible sensor. *Anal. Chim. Acta* **2018**, *1027*, 83–91. [CrossRef]
26. ItoItoh, T.; Uchida, T.; Izu, N.; Matsubara, I.; Shin, W. Effect of Core-shell Ceria/Poly(Vinylpyrrolidone) (PVP) Nanoparticles Incorporated in Polymer Films and Their Optical Properties (2): Increasing the Refractive Index. *Materials* **2017**, *6*, 2119–2129. [CrossRef]
27. Hang, Y.; Boryczka, J.; Wu, N. Visible-light and near-infrared fluorescence and surface-enhanced Raman scattering point-of-care sensing and bio-imaging: A review. *Chem. Soc. Rev.* **2022**, *51*, 329–375.
28. Oliveira, M.J.; Rubira, R.J.; Furini, L.N.; Batagin-Neto, A.; Constantino, C.J. Detection of thiabendazole fungicide/parasiticide by SERS: Quantitative analysis and adsorption mechanism. *Appl. Surf. Sci.* **2020**, *517*, 145786. [CrossRef]
29. Zhu, J.; Zhang, F.; Li, J.-J.; Zhao, J.-W. The effect of nonhomogeneous silver coating on the plasmonic absorption of Au–Ag core–shell nanorod. *Gold Bull.* **2014**, *47*, 47–55. [CrossRef]
30. Chen, J.; Wei, Z.; Cao, X.-y. QuEChERS pretreatment combined with ultra-performance liquid chromatography–tandem mass spectrometry for the determination of four veterinary drug residues in marine products. *Food Anal. Methods* **2019**, *12*, 1055–1066. [CrossRef]
31. Ge, F.; Ga, O.L.; Peng, X.; Li, Q.; Wang, Z. Atmospheric pressure glow discharge optical emission spectrometry coupled with laser ablation for direct solid quantitative determination of Zn, Pb, and Cd in soils. *Talanta* **2020**, *218*, 121119. [CrossRef] [PubMed]
32. Sun, H.; Liu, H.; Wu, Y. A green, reusable SERS film with high sensitivity for in-situ detection of thiram in apple juice. *Appl. Surf. Sci.* **2017**, *416*, 704–709. [CrossRef]
33. Birke, R.L.; Znamenskiy, V.; Lombardi, J.R. A charge-transfer surface enhanced Raman scattering model from time-dependent density functional theory calculations on a Ag 10-pyridine complex. *J. Chem. Phys.* **2010**, *132*, 214707. [CrossRef] [PubMed]
34. Awada, C.; Dab, C.; Grimaldi, M.; Alshoaibi, A.; Ruffino, F. High optical enhancement in Au/Ag alloys and porous Au using Surface-Enhanced Raman spectroscopy technique. *Sci. Rep.* **2021**, *11*, 4714. [CrossRef]
35. Xu, N.N.; Zhang, Q.; Guo, W.; Li, Q.T.; Xu, J. Au@PVP Core-Shell Nanoparticles Used as Surface-Enhanced Raman Spectroscopic Substrate to Detect Malachite Green. *Chin. J. Anal. Chem.* **2016**, *44*, 1378–1384. [CrossRef]
36. Liu, Y.; Lei, L.; Wu, Y.; Chen, Y.; Yan, J.; Zhu, W.; Tan, X.; Wang, Q. Fabrication of sea urchin-like Au@SiO2 nanoparticles SERS substrate for the determination of malachite green in tilapia. *Vib. Spectrosc.* **2022**, *118*, 103319. [CrossRef]
37. Yue, X.; Li, Y.; Xu, S.; Li, J.; Li, M.; Jiang, L.; Jie, M.; Bai, Y. A portable smartphone-assisted ratiometric fluorescence sensor for intelligent and visual detection of malachite green. *Food Chem.* **2022**, *371*, 131164. [CrossRef]

Disclaimer/Publisher's Note: The statements, opinions and data contained in all publications are solely those of the individual author(s) and contributor(s) and not of MDPI and/or the editor(s). MDPI and/or the editor(s) disclaim responsibility for any injury to people or property resulting from any ideas, methods, instructions or products referred to in the content.

Article

Molybdenum Disulfide-Integrated Iron Organic Framework Hybrid Nanozyme-Based Aptasensor for Colorimetric Detection of Exosomes

Chao Li [1,†], Zichao Guo [2,†], Sisi Pu [1], Chaohui Zhou [1], Xi Cheng [2,*], Ren Zhao [2,*] and Nengqin Jia [1,*]

[1] The Education Ministry Key Lab of Resource Chemistry, Joint International Research Laboratory of Resource Chemistry, Ministry of Education, Shanghai Frontiers Science Center of Biomimetic Catalysis and Shanghai Key Laboratory of Rare Earth Functional Materials, College of Chemistry and Materials Science, Shanghai Normal University, Shanghai 200234, China

[2] Department of General Surgery, Ruijin Hospital, Shanghai Jiao Tong University School of Medicine, Shanghai 200025, China

* Correspondence: drchengxi@126.com (X.C.); rjzhaoren@139.com (R.Z.); nqjia@shnu.edu.cn (N.J.)

† These authors contributed equally to this work.

Abstract: Tumor-derived exosomes are considered as a potential marker in liquid biopsy for malignant tumor screening. The development of a sensitive, specific, rapid, and cost-effective detection strategy for tumor-derived exosomes is still a challenge. Herein, a visualized and easy detection method for exosomes was established based on a molybdenum disulfide nanoflower decorated iron organic framework (MoS_2-MIL-101(Fe)) hybrid nanozyme-based CD63 aptamer sensor. The CD63 aptamer, which can specifically recognize and capture tumor-derived exosomes, enhanced the peroxidase activity of the hybrid nanozyme and helped to catalyze the 3,3′,5,5′-tetramethylbenzidine (TMB)-H_2O_2 system to generate a stronger colorimetric signal, with its surface modification on the hybrid nanozyme. With the existence of exosomes, CD63 aptamer recognized and adsorbed them on the surface of the nanozyme, which rescued the enhanced peroxidase activity of the aptamer-modified nanozyme, resulting in a deep-to-moderate color change in the TMB-H_2O_2 system where the change is visible and can be monitored with ultraviolet-visible spectroscopy. In the context of optimal circumstances, the linear range of this exosome detection method is measured to be 1.6×10^4 to 1.6×10^6 particles/μL with a limit of detection as 3.37×10^3 particles/μL. Generally, a simple and accessible approach to exosome detection is constructed, and a nanozyme-based colorimetric aptamer sensor is proposed, which sheds light on novel oncological biomarker measurements in the field of biosensors.

Keywords: exosome; aptamer; nanozyme; colorimetric assay; metal organic framework

1. Introduction

Malignant tumor is one of the main causes of lethal casualty worldwide, presenting with escalating morbidity among the developing and underdeveloped countries. Besides effective therapy, early screening also plays an important role in the fight against malignant tumors, which raises a huge demand for a large volume of precise and repeatable detection [1,2]. However, the common screening strategies for malignant tumors, including positron emission tomography, magnetic resonance imaging, computed tomography, X-ray, and endoscopy, are not suitable for large-volume screening and are not accessible in primary clinics. In addition to radiological imaging examinations, tumoral biomarkers detections in liquid samples are widely used in tumor screening, diagnosis, and follow-up [3]. Although a variety of novel potential biomarkers have been identified for different cancers, methodological and technical obstacles still exist in terms of biomarker detection and limit its further applications, such as low concentration or poor stability of biomarkers in human body fluid samples, which caused attention in sensitive and stable biomarker-detecting techniques [4].

As first observed in sheep reticulocytes in the 1980s [5], exosomes are endosomal-originated extracellular vesicles with an average diameter of 100 nm. Exosomes are discovered in various human body fluids, carrying lipids, proteins, nucleic acids, and other metabolites, which were suggested to have a role in intercellular communications that maintains gene transcription and translation, cell proliferation, metabolic reprogramming, angiogenesis, immune response, cellular differentiation, and migration [6]. Emerging studies have shown that exosomes can be used as a novel, promising, and reliable biomarker for cancer screening in the field of breast cancer and pancreatic cancer [7,8]. Increasing evidence has shown that tumor-derived exosomes are involved in carcinogenesis, tumor progression, and chemoresistance by regulating intercellular communications and reshaping the tumor microenvironment. As a flourishing research topic, exosome detection is not only applicable for cancer screening, diagnosis, and follow-up but also crucial for future in-depth exploration of the exact physiological and biochemical characteristics of tumor-derived exosomes.

As a unique class of single-stranded oligonucleotides [9], aptamers fold into specific tertiary structures to serve as recognition ligands that can attach to their targets with high affinity and specificity. With its programmable, modifiable, and engineerable designs, aptamer sensors, have added a novel flourished dimension to the field of liquid biopsy. Particularly, with their characteristics of synthetic and stability, aptamer sensors show their great accessibility and economic efficiency over traditional antibodies. Functioning as recognition ligands, there are emerging studies reporting that aptamer sensors were applied for exosome detection [10,11].

At present, well-recognized methods, such as Western blot [12], nanoparticle tracking analysis (NTA) [13,14], and flow cytometry have been introduced for the identification and quantification of exosomes. However, obstacles raised by technical limitations still hinder the widespread application of exosome detection. Demands for customized instruments, specialized software, and relevant reagents limit their usability for clinical situations, especially in primary clinics. Since 2017 [15], aptamers have become widely used ligands for exosome detector construction. In terms of the signal transduction methods, aptamers sensors can be divided into the following categories: fluorescence, colorimetric, electrochemical, and luminescent, etc. The colorimetric assay is an absorbance-based quantification method that can be easily implemented without exquisite instruments, which takes the advantages of accessibility, affordability, easy operation and quick results compared to the other assays and has been used for exosome detection [15].

The rapid development of enzymes has boosted the explorations of various novel biosensors. As promising candidates for natural biological enzymes, nanozymes have attracted increasing attention because of their characteristics of stability, reasonable price, and batch production with uncompromised catalytic efficiency. Ever since Fe_3O_4 nanoparticles [16–18] were reported to have a peroxidase-like (POD-like) catalytic activity, more and more nanomaterials have been suggested to have different catalytic activities, including carbon nanohybrid materials [19,20], inorganic nanomaterials [21,22] and metal-organic skeleton materials [23,24]. Recently, the peroxidase-based colorimetric assays have been proposed for exosomes detection [25,26]. As a well-recognized colorimetric method, different peroxidase catalytic activities can oxidize substrates including 3,3′,5,5′-tetramethylbenzidine (TMB) with the existence of hydrogen peroxide (H_2O_2), exhibiting different absorbance, which can be measured quantitatively.

The metal-organic framework (MOF), as a new kind of self-assembled three-dimensional orderly coordinated polymer, is constructed by coordination bonds of organic linkers and metal ions/clusters [27], presenting the characteristics of tremendous specific surface area and large porosity. Despite these merits, owing to the presence of organic ligands and their relatively high molecular weight, MOF-based nanozymes still need improvement in the highly accessible active sites and catalytic efficiency [28–30]. Functionalized MOF-based nanozymes, and nanozymes with various exposed active units are considered very promising solutions, among which the MOF-based hybrid nanozymes constructed by hybridization regulation strategy could considerably enhance the catalytic activity of the nanozyme due to their synergistic effects of hybrid materials beneficial for the improvement of dispersion, conductivity, and specific surface area.

In terms of molybdenum disulfide (MoS_2), a transition metal dihalide compound, presented as a two-dimensional, flake, and graphene-like structure, exhibits abundant catalytic-active edges and advanced specific surface area [31]. Recently, MoS_2 nanosheets have been reported as a promising mimic of peroxidase [32]. To date, in several functional materials [33,34], MoS_2 has been adopted as the subunit to mimic natural enzymes and achieve enhanced catalytic activity. Therefore, MoS_2 is an ideal alternative for the construction of MOF-based nanozyme.

In this work, a visible exosome detection technique was developed based on an artificial, mimetic nanozyme of peroxidase, the MoS_2-MIL-101(Fe) hybrid nanozyme was constructed and was proved to possess superior peroxidase enzymatic activity. The configuration of MIL-101(Fe) provided MoS_2 with a large specific surface area, which is conducive to the absorption of substrates on the exterior of the hybrid nanozyme. Additionally, modified on the exterior of the hybrid nanozyme via electrostatic interaction, the CD63 aptamers not only specifically recognize and capture exosomes but also enhances the affinity of the hybrid nanozyme to its substrates and further improve its catalytic activity with the aptamer's single-strand DNA configuration [35,36]. Generally, with these synergistic effects, the MoS_2-MIL-101(Fe) hybrid nanozyme-based aptamer sensor is assumed to have an ideal detection limit. And a multi-purpose design of the nanozyme-based colorimetric aptamer sensor is proposed.

2. Experimental Section

2.1. Chemicals, Reagents and Cell Lines

3,3′,5,5′-tetramethylbenzidine (TMB) and terephthalic acid were purchased from Energy Chemical (Shanghai, China). Ammonium thiomolybdate (($NH_4)_2MoS_4$), ferric chloride hexahydrate ($FeCl_3·6H_2O$), ethanol (CH_3CH_2OH), hydrogen peroxide (H_2O_2), N,N′-dimethylformamide (DMF), pH4.0 acetate-sodium acetate buffer, bovine serum albumin, RMPI-1640 medium, and fetal bovine serum without exosomes (Exofree-FBS) were obtained from Shanghai Titan Technology Co., LTD (Shanghai, China). CD63 aptamer was supplied by Sangon Biotech (Shanghai, China). Cell lines HGC-7901 and LO2 were purchased from Cell Bank affiliated with the Chinese Academy of Sciences.

2.2. The Preparation of MoS_2-MIL-101(Fe)

MoS_2 nanoflowers were synthesized by the hydrothermal method as previously reported [37]. In brief, 50 mg of ($NH_4)_2MoS_4$ was dispersed in 30 mL of DMF under ultrasonic treatment for 15 min. Then, the solution was transferred into a Teflon liner and was kept at 200 °C for 10 h. After being cooled down, the raw product was washed with DMF and ethanol several times. After being dehydrated at 60 °C in a vacuum drier, MoS_2 nanoflowers were purified and collected.

MoS_2-MIL-101(Fe) nanocomposites were synthesized via heat treatment of a mixture of $FeCl_3·6H_2O$, terephthalic acid, and MoS_2 nanoflowers. In detail, under ultrasonic treatment, 20 mg of as-prepared MoS_2 nanoflowers were dispersed in 15 mL of DMF to form a homogeneous solution. Then 0.206 g of terephthalic acid and 0.675 g of $FeCl_3·6H_2O$ were dissolved in the abovementioned solution by continuous stirring. The mixture was transferred into a Teflon liner and was kept at 110 °C for 20 h. After being cooled down to room temperature, the raw product was separated, and washed by DMF and ethanol three times. Finally, after being dehydrated at 60 °C in a vacuum drier, MoS_2-MIL-101(Fe) nanocomposites were collected (Figure 1A). In addition, MIL-101(Fe) nanocomposites were synthesized through the same processes described above except for the introduction of MoS_2 nanoflowers [38].

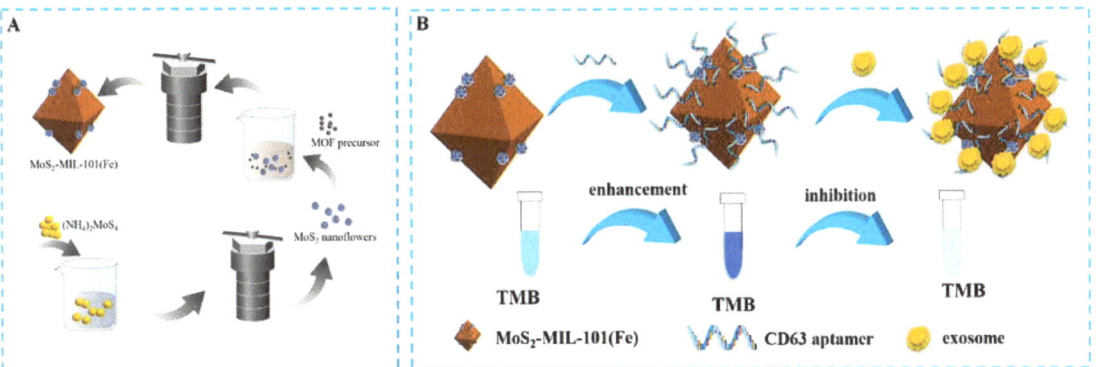

Figure 1. Schematic illustration of (**A**) the synthesis process of MoS$_2$-MIL-101(Fe) and (**B**) the detection mechanism of the proposed method for exosomes.

2.3. Cell Culture and Exosomes Preparation

The HGC-7901 cells and LO2 cells were cultured in RMPI-1640 supplemented with 10% FBS at 37 °C in 5% CO$_2$. After the cells proliferated to 80% confluence, the culture medium was replaced with RMPI-1640 supplemented with 10% Exofree-FBS. After 48 h incubation of exosome-free medium, the cell culture supernatant was collected to harvest tumor-derived exosomes, and the cells were passaged and cultured. Exosomes were isolated from the cell culture supernatant by the standard ultracentrifugation method with slight modifications [39]: (1) 1500× *g* centrifugation for 10 min to eliminate dead cells; (2) 1000× *g* for 20 min to eliminate cellular debris and the acquired supernatant was filtrated by a 0.22 µm filter; (3) 100,400× *g* for 4 h to precipitate exosomes.

2.4. Simulation of Peroxidase Activity in Nanocomposites

In the presence of a given concentration of H$_2$O$_2$ in a pH 4.0 HAc-NaAc buffer, the simulation of peroxidase activity in MoS$_2$-MIL-101(Fe) nanocomposite was evaluated by introducing MoS$_2$-MIL-101(Fe) hybrids to a TMB solution. The total volume of the reaction solution was set to 4 mL. The solution was composed of 1 mL of pH 4.0 HAc-NaAc buffer, 1 mL of MoS$_2$-MIL-101(Fe) solution (50 mg/L), 1 mL of H$_2$O$_2$ solution (10 mM), and 1 mL of TMB solution (5 mM). Next, the reaction solution was incubated at 40 °C for 5 min and the absorbance was measured by an ultraviolet-visible (UV-vis) spectrophotometer.

Kinetic experiments of MoS$_2$-MIL-101(Fe) hybrids were performed by measuring the initial rate of the reaction in the first 5 min, with one of the concentrations of H$_2$O$_2$ and TMB fixed, and the other varied. The H$_2$O$_2$ concentration was set to 4 mmol/L and the fixed TMB concentration was set to 2.5 mmol/L.

The kinetic parameters were fitted by the following equation: $V_0 = V_{max}\frac{[S]}{K_m+[S]}$. Here, K_m stands for Mi's constant, [S] for substrate concentration, V_0 for initial reaction rate, and V_{max} for maximum reaction rate. For each preset H$_2$O$_2$ and TMB concentrations, K_m and V_{max} were calculated by Hyperbola curve fit using the OriginPro 2019 (OriginLab Corporation, Northampton, MA, USA) after measuring V_0. The hyperbolic function is also the Michaelis–Menten model in enzyme kinetics, and its formula is $y = \frac{k_1 x}{k_2+x}$ corresponding to the Michaelis–Menten equation $V_0 = V_{max}\frac{[S]}{K_m+[S]}$. After being fitted, k_1 is V_{max} and k_2 is K_m [38].

2.5. Exosomes Detection

Ten microliters of CD63 aptamer (10 µM) and 200 µL MoS$_2$-MIL-101(Fe) nanocomposites (100 mg/L) were mixed and blended by vortex. Then the different solutions with different concentrations of exosomes (10 µL) were added to the mixtures and were blended by vortex. After 30 min of incubation, 100 µL TMB (2.5 mM) solution and 100 µL H$_2$O$_2$

(4 mM) solution were added to the mixtures, and then HAc-NaAc buffer was added to fill up the volume of mixtures to 1000 µL. The mixtures were incubated at 40 °C for 5 min in the dark, and then the absorbances were measured by a UV–vis spectrophotometer with a 1.0 cm quartz cell.

3. Results and Discussion

3.1. Subsection

3.1.1. The Principle and Feasibility of the Aptamer Sensor

In order to synthesize hybrid nanozymes, nanoflower-like MoS_2 materials were first synthesized by hydrothermal method. Then, the MoS_2-MIL-101(Fe) nanocomposites were constructed on the basis of MOF precursor and MoS_2 nanoflowers (Figure 1A). The design sketch of the MoS_2-MIL-101(Fe) hybrid nanozyme-based aptamer sensor is shown in Figure 1B. The chromogenic reaction was introduced to evaluate and compare the peroxidase catalytic activities of the hybrid nanozyme, as well as its precursors. MoS_2-MIL-101(Fe) nanocomposite was suggested to have a higher peroxidase activity in contrast to both MoS_2 and MIL-101(Fe) (Figure 2A), which catalyzed TMB to transform from the colorless substrates to the deep blue substances in the presence of H_2O_2. Modified by CD63 aptamers, the peroxidase catalytic activity of the MoS_2-MIL-101(Fe) was magnified, which is attributed to π-π stacking between single-stranded DNA and the substrate (Figure 2B). In detail, the bases of DNA aptamer facilitate the bindings between the substrates, especially through hydrogen bonding between DNA bases and the amino groups of TMB, as well as the nucleobase interacting between DNA bases and the benzene rings of TMB via π-π stacking, which led to the increased substrate affinity and further enhanced the catalytic activity of MoS_2-MIL-101(Fe) [35,40]. In the presence of exosomes, the specific ligand-receptor recognition between CD63 aptamers and the CD63 proteins on the exterior of exosomes confined exosomes within the external surface of MoS_2-MIL-101(Fe)@Aptamer and rescued the enhanced peroxidase catalytic activity of MoS_2-MIL-101(Fe)@Aptamer, where MoS_2-MIL-101(Fe)@Aptamer@Exosomes exhibited an even weaker peroxidase catalytic activity than MoS_2-MIL-101(Fe) (Figure 2B). With various concentrations of exosomes, the hybrid nanozyme exhibited different degrees of peroxidase catalytic activities, so as to present different absorbances of the TMB colorimetric reaction, which could be obviously visualized and measured by UV–vis spectrometer. Therefore, the exosome-detection aptamer sensor was constructed based on the integration of CD63 aptamer and the hybrid nanozyme, which was verified with the following zeta potential measurements. As shown in Figure 3, the zeta potential of MoS_2-MIL-101(Fe) is 21.0 mV, which shows that the composite material was positively charged. After incubating with the aptamer, the zeta potential of the composite material changed to 1.95 mV, which suggested that the CD63 aptamers were modified on the surface of MoS_2-MIL-101(Fe) through electrostatic interaction. In addition, while the CD63 aptamers combined with the exosomes, the overall zeta potential of MoS2-MIL-101(Fe)@Aptamer@Exosomes turned to -7.65 mV, which continued to decline; the nanocomposite was negatively charged (Figure 3).

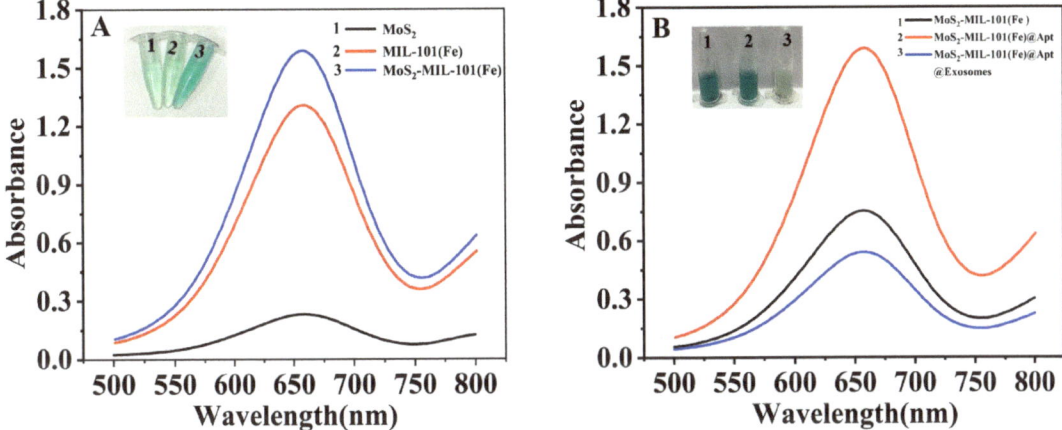

Figure 2. (**A**) Absorptions at 653 nm of different materials by UV-vis spectrophotometry, (1) MoS$_2$; (2) MIL-101(Fe); (3) MoS$_2$-MIL-101(Fe). (**B**) Absorptions at 653 nm of different materials by UV-vis spectrophotometry after the incubation of exosomes, (1) MoS$_2$-MIL-101(Fe); (2) MoS$_2$-MIL-101(Fe)@Apt; (3) MoS$_2$-MIL-101(Fe)@Apt@Exosomes.

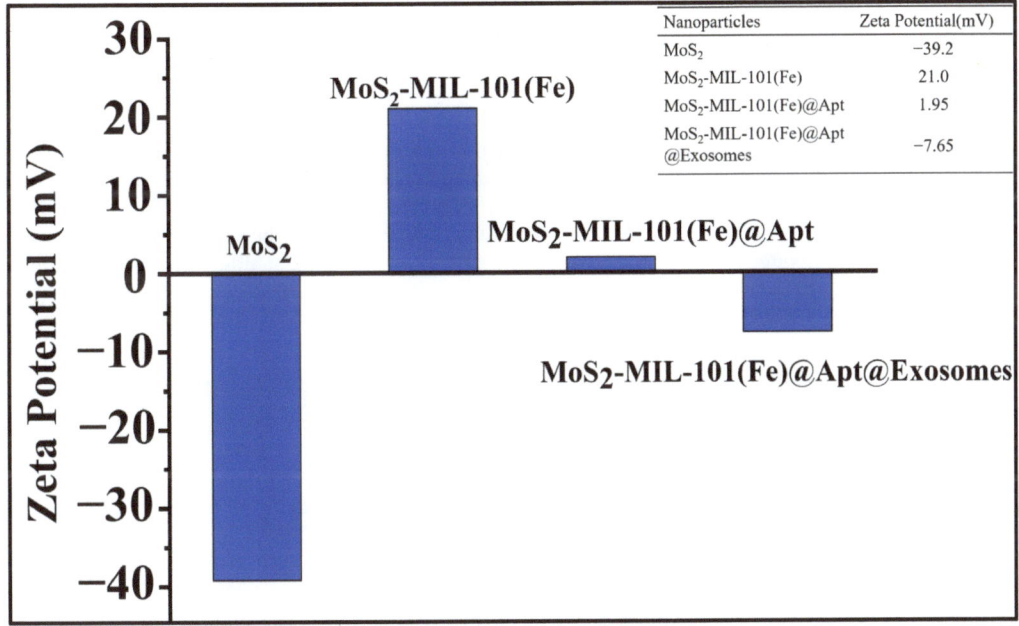

Figure 3. Zeta potential of different materials.

3.1.2. The Characterization of MoS$_2$-MIL-101(Fe) and Quantification of Exosomes

The morphology of MoS$_2$-MIL-101(Fe) nanocomposites and MoS$_2$ was demonstrated by scanning electron microscopy (SEM) and transmission electron microscopy (TEM) (Figure 4). MoS$_2$ presented a flower-like shape with a uniform size (ca. 150 nm) in diameter (Figure 4A), and MIL-101(Fe) displayed an octahedral nanostructure (Figure 4B). The successful preparation of the hybrid nanozyme was observed by TEM (Figure 4C) and further identified with X-ray diffraction (XRD) (Figure 4D), depicting a characteristic peak at 9.4° which is attributed to the (001) reflection of MoS$_2$.

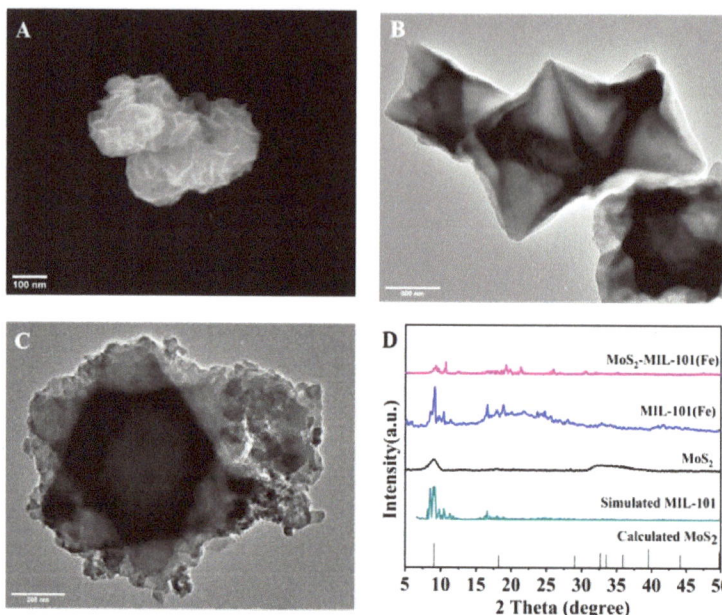

Figure 4. (**A**) The morphology of MoS$_2$ revealed by SEM; (**B**) The morphology of MIL-101(Fe) revealed by TEM; (**C**) The morphology of MoS$_2$-MIL-101(Fe) revealed by TEM; (**D**) The XRD pattern of MoS$_2$-MIL-101(Fe) and its precursors.

Exosomes were isolated by ultracentrifugation from either human gastric cancer cell line HCG-7901 cells or normal human liver cell line LO2. The morphologies of the purified exosomes were demonstrated by TEM (Figure 5A,B), which showed an average diameter of 100 nm that coincided with the previous studies. The exosome counting was carried out by NTA, and the results were further used as the standard of exosomes. (Figure 5C). The expression of typical labeled proteins [41] (transmembrane proteins CD9, CD63 and CD81) of exosomes derived from HCG-7901 was directly verified by Western Blots (WB) (Figure 5D).

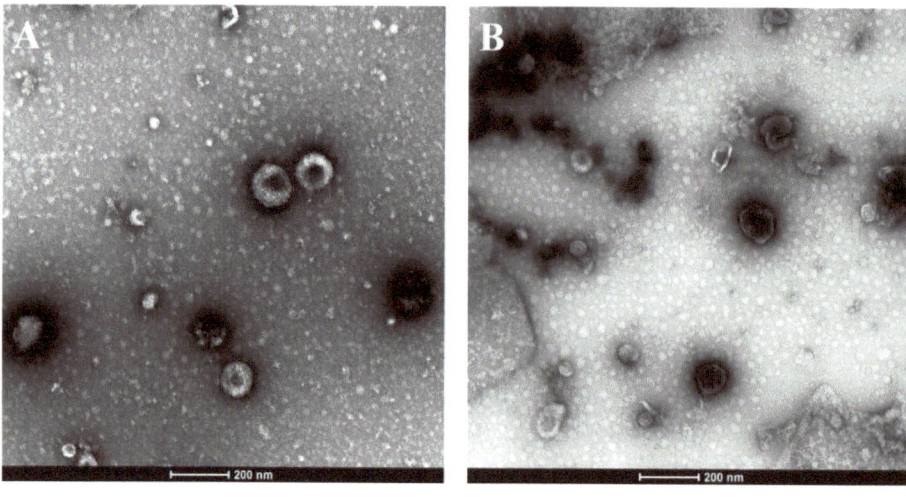

Figure 5. *Cont.*

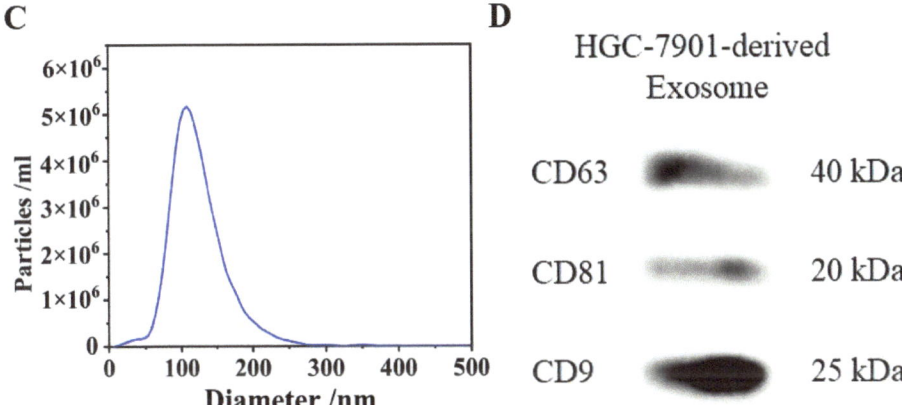

Figure 5. (**A**) The morphology of HGC-7901 cells-derived exosomes revealed by TEM; (**B**) The morphology of LO2 cells-derived exosomes revealed by TEM; (**C**) Exosome concentrations and particle sizes distribution of HGC-7901 cells; (**D**) The Western blots of HGC-7901 cells-derived exosomes.

3.1.3. Optimization of Experimental Conditions

The optimal experimental condition of the TMB reaction under MoS$_2$-MIL-101(Fe)@Aptamer was then investigated, with the total reaction volume set to 4 mL. The activity of the nanozyme active site is affected by pH. Generally, the POD-like activity of the metal-based nanozyme is more efficient in an acidic environment while its catalase-like (CAT-like) activity is more efficient in an alkaline environment, where the transferred domination between POD- and CAT-like activities were driven through different reactant decomposition pathways at different pH [42]. The peroxidase catalytic activities of MoS$_2$-MIL-101(Fe) at various pH conditions (from pH 2 to pH 8) were first investigated. The results showed that the catalytic activity was correlated with pH value, which presented the maximum activity at pH 4 (Figure S1A). Therefore, pH 4 was selected as the ideal pH value. Temperature also can regulate the catalytic activity of nanozymes. With increasing temperatures, the reactants' thermal motions in the vicinity of active sites can be further activated, thus enhancing the catalytic activities of each individual active site. Therefore, the probability of molecular collisions between nanozymes and substrates is greatly proposed, thereby lifting the reaction rate [43]. Since enzymatic catalytic activity can be affected by temperature, the catalytic activities under different temperatures were also explored, which showed that the catalytic activity reaches to maximum at 40 °C (Figure S1B). Therefore, 40 °C was selected as the optimal temperature.

Then, the effect of different CD63 aptamer concentrations (from 2 µM to 20 µM) on the absorbances at 653 nm (A653) that reflect the concentrations of oxidized TMB was investigated. The A653 gradually increased as the CD63 aptamer concentration increased from 2 µM to 10 µM, achieved the highest value at 10 µM, and then gradually fell off when the concentration was above 10 µM (Figure S1C), which suggested that the optimal concentration of CD63 aptamer was 10µM. Similarly, the effects of different TMB concentrations, different H$_2$O$_2$ concentrations, and different MoS$_2$-MIL-101(Fe) concentrations on A653 were monitored. It is reported [44] that the colorific reaction of benzidine under one-electron, or two-electron oxidation is blue or yellow, respectively, where the corresponding absorption peak is 653 nm or 450 nm. With increasing concentrations of TMB or hydrogen peroxide, the nanozyme-catalytic reaction rate, as well as the reaction rate of one-electron benzidine oxidation are lifted. Meanwhile, as a substrate, the product of one-electron benzidine oxidation also promotes two-electron oxidation, which causes the blue to yellow-green transformation of the solution system, leading to the decrease in the absorption peak at 653 nm. And it is consistent with the experimental phenomenon. The optimal conditions of these variables were shown as the followings: 2.5 mM TMB

(Figure S1D), 4 mM H_2O_2 (Figure S1E), 20 μg/mL MoS_2-MIL-101(Fe) (Figure S1F), which were chosen in the following experiments, respectively.

3.1.4. The Kinetic Properties of MoS2-MIL-101(Fe) as a Peroxidase Simulator

Next, the steady-state kinetic experiment was applied to explore the kinetic properties of MoS_2-MIL-101(Fe) as a peroxidase simulator, as well as its precursors, MIL-101(Fe) and MoS_2. The apparent kinetic parameters, K_m and V_{max}, were measured with one of the concentrations of H_2O_2 and TMB fixed, and the other varied. The hyperbolic kinetic curves of the hybrid nanozyme on TMB and H_2O_2 indicated that it followed the typical Michaelis–Menten pattern (Figure 6A,C) and possessed enzymatic kinetic properties. The lower apparent K_m value reflects the stronger affinity of the hybrid nanozyme to its substrates. The apparent K_m value of MoS_2-MIL101(Fe) to TMB is 0.12 mM. In addition, the apparent K_m value of MoS_2-MIL-101(Fe) to H_2O_2 is 0.015 mM, which was 29 times inferior to that of natural horseradish peroxidase. The results indicated that the hybrid nanozyme had advanced affinities to both TMB and H_2O_2. The double reciprocal graphs of Figures 6A and 6C correspond to Figures 6B and 6D, respectively.

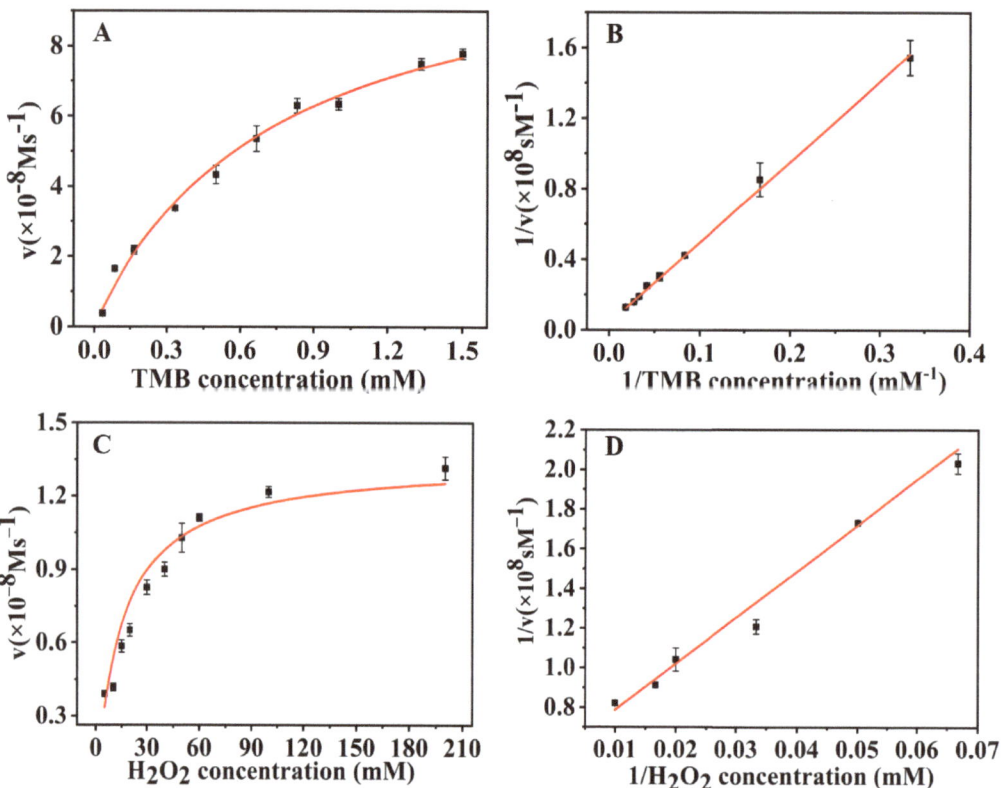

Figure 6. (**A**) Enzymatic reaction kinetics of the hybrid nanozyme on TMB; (**B**) The double reciprocal graph of enzymatic reaction kinetics of the hybrid nanozyme on TMB; (**C**) Enzymatic reaction kinetics of the hybrid nanozyme on H_2O_2; (**D**) The double reciprocal graph of enzymatic reaction kinetics of the hybrid nanozyme on H_2O_2.

3.1.5. Analytical Performance in Determination of Exosomes

Further, the absorbances at 653 nm (A653) of the TMB under diverse exosome concentrations were measured under the optimal conditions. As shown in Figure 7A, A653 decreased proportionally with the increase in exosome concentration. Correspondingly, color changes of the solutions with different exosome concentrations were obviously visible and comparable (a–h, Figure 7A). The linear range between A653 and the logarithm of exosome concentration is $1.6 \times 10^4 \sim 1.6 \times 10^6$ particles/μL, where the correlation equation is $y = -0.73081 \times \lg[c(\text{exosomes})] + 4.94426$, and the squared correlation coefficient is 0.996 (y reflects A653, and c(exosomes) reflects the exosome concentration, Figure 7B). As reported [45,46], a cancer cell can secrete more than 10^4 vesicles in 24 h in contrast to a normal epithelial cell. In the clinical approach of exosome detection [47,48], our work could reach the level of detecting tumor-derived exosomes. Designed on the basis of the 3σ method (σ is the blank standard deviation), this aptamer sensor exhibited a promising limit of detection (LOD) of 3.37×10^3 particles/μL, which is lower than that of other methods reported and is attributed to the high POD-like catalytic activity of the hybrid nanozyme (Table 1) [49–56].

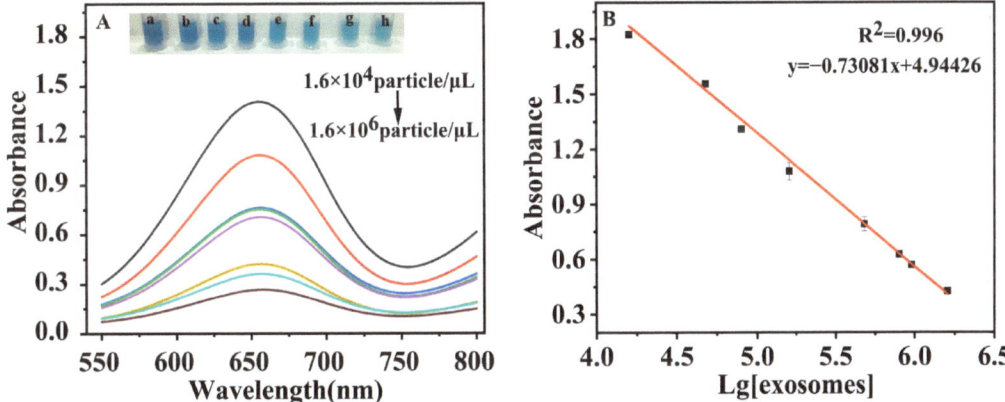

Figure 7. (**A**) Absorptions at 653 nm of the colorimetric aptasensor for detections on different exosome concentrations by UV-vis spectrophotometry (from a to h: exosome concentrations were 1.6×10^4, 4.8×10^4, 1.6×10^5, 4.8×10^5, 6.0×10^5, 9.6×10^5, 1.28×10^6, 1.6×10^6 particles/μL, respectively). The inset showed the corresponding colors of each concentration; (**B**) Linear relationship between A653 and the logarithm of exosome concentration.

Table 1. Comparisons with various reported strategies for exosome detection.

Method	Linear Range (Particles/μL)	LOD (Particles/μL)	Reference
Electrochemical (Paper-based Device)	$2.47 \times 10^5 - 2.47 \times 10^6$	7.1×10^5	[49]
Electrochemical (Au NPs)	$9 \times 10^6 - 1.4 \times 10^7$	4.5×10^6	[50]
Fluorescent (CD63-MBs)	$1.66 \times 10^3 - 1.66 \times 10^6$	4.8×10^2	[51]
Fluorescent (G-quadruplex)	$5.0 \times 10^5 - 5.0 \times 10^7$	3.4×10^5	[52]
Fluorescence (CuO NPs)	$7.5 \times 10^4 - 1.5 \times 10^7$	4.8×10^4	[53]
Colorimetric (Carbon Nanotubes)	$10^6 - 10^8$	3.94×10^4	[54]
Colorimetric (Fe$_3$O$_4$)	$4.0 \times 10^5 - 6.0 \times 10^7$	3.58×10^3	[55]
Colorimetric (Fe-MIL-88)	$1.1 \times 10^5 - 2.2 \times 10^7$	5.2×10^4	[56]
Colorimetric (CuCo$_2$O$_4$)	$5.6 \times 10^4 - 8.9 \times 10^5$	4.5×10^3	[57]
Colorimetric (ZnO)	$2.2 \times 10^5 - 2.4 \times 10^7$	2.2×10^4	[58]
Colorimetric (MoS$_2$-MIL-101(Fe))	$1.6 \times 10^4 - 1.6 \times 10^6$	3.37×10^3	This work

3.1.6. The Selectivity, Reproducibility, and Stability of the Aptamer Sensor

The specificity was a pivotal issue for developing a novel exosome-detection aptamer sensor. Heterogeneous exosomes with different CD63 protein expressions were adopted to verify the specificity of this aptamer sensor. As previously reported [36], the CD63 expressions of exosomes derived from human gastric cancer cell line HCG-7901 cells were lower than those derived from normal human liver cell line LO2. As shown in Figure 8A, the stronger signals were observed in the group of CD63-low HGC-7901 cells-derived exosomes, while the much weaker signals were observed in the group of CD63-high LO2 cells-derived exosomes with the same concentration of exosomes. In accordance with the previous studies, the results demonstrated that this aptamer sensor had a good selectivity on various exosomes based on the specificity of the CD63 aptamer.

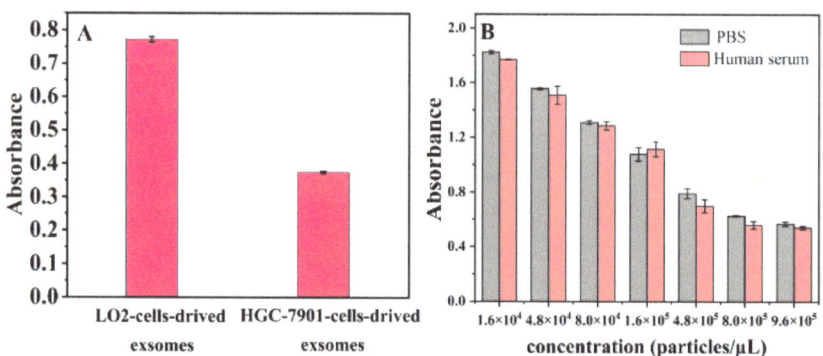

Figure 8. (**A**) The histogram of selectivity of MoS_2-MIL-101(Fe) nanozyme-based aptasensor. Each column exhibited the average value of three independent data with an error bar presented; (**B**) Comparison of absorbances between PBS and human serum with the same concentration of exosomes. Each column exhibited the average value of three independent data with an error bar presented.

The reproducibility of this aptasensor has also been explored. As shown in Figure S2, the exosome detection effectiveness of the aptasensor after 30-day storage at 4 °C exhibited a good similarity with that of synthesized on the first day as its RSD values still presented consistencies less than 5%, which indicated favorable reproducibility, stability, and the shelf life of this aptasensor.

3.1.7. Detection of Exosomes in Human Serum Sample

To explore the applicability of this hybrid nanozyme-based aptamer sensor on a clinical approach, HGC-7901-derived exosomes were added to human serum to establish artificial samples for detection. The results showed that the calculated recoveries ranged from 95% to 103% (Table S1). As shown in Figure 8B, exosome detections of the artificial samples presented results consistent with those of PBS samples with the same concentration of exosomes added. These results clearly indicated that exosome detection based on this aptasensor is accurate and suitable for detection in complex systems, providing an important tool for exosome detection in clinical applications.

4. Conclusions

In conclusion, a novel, sensitive, visible and simple approach to exosome detection is developed by means of the MoS_2-MIL-101(Fe) hybrid nanozyme-based aptasensor. Excellent linearity was obtained for exosome detection within the extent of 1.6×10^4 to 1.6×10^6 particles/µL, as well as a LOD of 3.37×10^3 particles/µL was achieved. The aptamer sensor possesses equal applicability in complex biological samples in clinical approach, which exhibited a cutting-edge economic efficiency and accessibility that can be potentially applied for portable exosome-detection devices. Nonetheless, in this study, there are limitations that need to be pointed out that the unsolved defects of nanozyme, such as the relatively low substrate selectivity and catalytic efficiency, still impede the further enhancement of the sensitivity of this aptasensor. In addition, the single CD63 DNA aptamer design of this aptasensor also hinders the further improvement of its exosome specificity.

Supplementary Materials: The following supporting information can be downloaded at: https://www.mdpi.com/article/10.3390/bios13080800/s1, Figure S1: Influences of (A) pH, (B) temperature, (C) aptamer concentration, (D) TMB concentration, (E) H2O2 concentration, (F) MoS2-MIL-101(Fe) concentration for the MoS2-MIL-101(Fe) hybrid nanozyme-based aptasensor; Figure S2: Comparison of absorbances between measurement at first-day synthesis and measurement after 30-day room-temperature storage with the same concentration of exosomes. Each column exhibited the average value of three independent data with an error bar presented; Table S1: Detection of HGC-7901-derived exosomes in human serum.

Author Contributions: Conceptualization, N.J.; methodology, C.L., Z.G., S.P., C.Z. and N.J.; software, C.L., S.P. and C.Z.; validation, C.L., Z.G. and N.J.; formal analysis, C.L.; investigation, C.L., S.P. and C.Z.; resources, N.J.; data curation, N.J.; writing—original draft preparation, C.L. and Z.G.; writing—review and editing, Z.G. and N.J.; visualization, C.L., Z.G. and N.J.; supervision, N.J., R.Z. and X.C.; project administration, N.J., R.Z. and X.C.; funding acquisition, N.J. All authors have read and agreed to the published version of the manuscript.

Funding: This research was funded by Shanghai Engineering Research Center of Green Energy Chemical Engineering (No. 18DZ2254200), and 111 Innovation and Talent Recruitment Base on Photochemical and Energy Materials (No. D18020).

Institutional Review Board Statement: The study was conducted in accordance with the Declaration of Helsinki, and approved by the Ethics Committee of Shanghai Normal University.

Informed Consent Statement: Not applicable.

Data Availability Statement: Data are available upon reasonable request from the corresponding authors.

Conflicts of Interest: The authors declare no conflict of interest.

References

1. Ginsburg, O.; Bray, F.; Coleman, M.P.; Vanderpuye, V.; Eniu, A.; Kotha, S.R.; Sarker, M.; Huong, T.T.; Allemani, C.; Dvaladze, A.; et al. The global burden of women's cancers: A grand challenge in global health. *Lancet* **2017**, *389*, 847–860. [CrossRef]
2. Zhou, B.; Xu, J.W.; Cheng, Y.G.; Gao, J.Y.; Hu, S.Y.; Wang, L.; Zhan, H.X. Early detection of pancreatic cancer: Where are we now and where are we going? *Int. J. Cancer* **2017**, *141*, 231–241. [CrossRef]
3. Wu, L.; Qu, X. Cancer biomarker detection: Recent achievements and challenges. *Chem. Soc. Rev.* **2015**, *44*, 2963–2997. [CrossRef] [PubMed]
4. Goossens, N.; Nakagawa, S.; Sun, X.; Hoshida, Y. Cancer biomarker discovery and validation. *Transl. Cancer Res.* **2015**, *4*, 256–269.
5. Xiong, H.; Huang, Z.; Yang, Z.; Lin, Q.; Yang, B.; Fang, X.; Liu, B.; Chen, H.; Kong, J. Recent Progress in Detection and Profiling of Cancer Cell-Derived Exosomes. *Small* **2021**, *17*, e2007971. [CrossRef]
6. Kalluri, R.; LeBleu, V.S. The biology, function, and biomedical applications of exosomes. *Science* **2020**, *367*, eaau6977. [CrossRef]
7. Wang, M.; Ji, S.; Shao, G.; Zhang, J.; Zhao, K.; Wang, Z.; Wu, A. Effect of exosome biomarkers for diagnosis and prognosis of breast cancer patients. *Clin. Transl. Oncol.* **2018**, *20*, 906–911. [CrossRef] [PubMed]
8. Nilsson, J.; Skog, J.; Nordstrand, A.; Baranov, V.; Mincheva-Nilsson, L.; Breakefield, X.O.; Widmark, A. Prostate cancer-derived urine exosomes: A novel approach to biomarkers for prostate cancer. *Br. J. Cancer* **2009**, *100*, 1603–1607. [CrossRef]
9. Wu, L.; Wang, Y.; Xu, X.; Liu, Y.; Lin, B.; Zhang, M.; Zhang, J.; Wan, S.; Yang, C.; Tan, W. Aptamer-Based Detection of Circulating Targets for Precision Medicine. *Chem. Rev.* **2021**, *121*, 12035–12105. [CrossRef]
10. Wang, T.; Chen, C.; Larcher, L.M.; Barrero, R.A.; Veedu, R.N. Three decades of nucleic acid aptamer technologies: Lessons learned, progress and opportunities on aptamer development. *Biotechnol. Adv.* **2019**, *37*, 28–50. [CrossRef] [PubMed]
11. Zhu, C.; Yang, G.; Ghulam, M.; Li, L.; Qu, F. Evolution of multi-functional capillary electrophoresis for high-efficiency selection of aptamers. *Biotechnol. Adv.* **2019**, *37*, 107432. [CrossRef]
12. Zhou, H.; Yuen, P.; Pisitkun, T.; Gonzales, P.; Yasuda, H.; Dear, J.; Gross, P.; Knepper, M.; Star, R. Collection, storage, preservation, and normalization of human urinary exosomes for biomarker discovery. *Kidney Int.* **2006**, *69*, 1471–1476. [CrossRef] [PubMed]
13. Oosthuyzen, W.; Sime, N.E.L.; Ivy, J.R.; Turtle, E.J.; Street, J.M.; Pound, J.; Bath, L.E.; Webb, D.J.; Gregory, C.D.; Bailey, M.; et al. Quantification of human urinary exosomes by nanoparticle tracking analysis. *J. Physiol.* **2013**, *591*, 5833–5842. [CrossRef]
14. Vestad, B.; Llorente, A.; Neurauter, A.; Phuyal, S.; Kierulf, B.; Kierulf, P.; Skotland, T.; Sandvig, K.; Haug, K.B.F.; Øvstebø, R. Size and concentration analyses of extracellular vesicles by nanoparticle tracking analysis: A variation study. *J. Extracell. Vesicles* **2017**, *6*, 1344087. [CrossRef]
15. Zhu, C.; Li, L.; Wang, Z.; Irfan, M.; Qu, F. Recent advances of aptasensors for exosomes detection. *Biosens. Bioelectron.* **2020**, *160*, 112213. [CrossRef]

16. Ju, J.; Chen, Y.; Liu, Z.; Huang, C.; Li, Y.; Kong, D.; Shen, W.; Tang, S. Modification and application of Fe3O4 nanozymes in analytical chemistry: A review. *Chin. Chem. Lett.* **2023**, *34*, 107820. [CrossRef]
17. Wu, J.; Wang, X.; Wang, Q.; Lou, Z.; Li, S.; Zhu, Y.; Qin, L.; Wei, H. Nanomaterials with enzyme-like characteristics (nanozymes): Next-generation artificial enzymes (II). *Chem. Soc. Rev.* **2019**, *48*, 1004–1076. [CrossRef] [PubMed]
18. Gao, L.; Zhuang, J.; Nie, L.; Zhang, J.; Zhang, Y.; Gu, N.; Wang, T.; Feng, J.; Yang, D.; Perrett, S.; et al. Intrinsic peroxidase-like activity of ferromagnetic nanoparticles. *Nat. Nanotechnol.* **2007**, *2*, 577–583. [CrossRef] [PubMed]
19. Ahmed, S.R.; Takemeura, K.; Li, T.C.; Kitamoto, N.; Tanaka, T.; Suzuki, T.; Park, E.Y. Size-controlled preparation of peroxidase-like graphene-gold nanoparticle hybrids for the visible detection of norovirus-like particles. *Biosens. Bioelectron.* **2017**, *87*, 558–565. [CrossRef]
20. Jin, G.H.; Ko, E.; Kim, M.K.; Tran, V.-K.; Son, S.E.; Geng, Y.; Hur, W.; Seong, G.H. Graphene oxide-gold nanozyme for highly sensitive electrochemical detection of hydrogen peroxide. *Sens. Actuators B Chem.* **2018**, *274*, 201–209. [CrossRef]
21. Zhang, Y.; Wang, Y.-N.; Sun, X.-T.; Chen, L.; Xu, Z.-R. Boron nitride nanosheet/CuS nanocomposites as mimetic peroxidase for sensitive colorimetric detection of cholesterol. *Sens. Actuators B Chem.* **2017**, *246*, 118–126. [CrossRef]
22. Kuo, P.C.; Lien, C.W.; Mao, J.Y.; Unnikrishnan, B.; Chang, H.T.; Lin, H.J.; Huang, C.C. Detection of urinary spermine by using silver-gold/silver chloride nanozymes. *Anal. Chim. Acta* **2018**, *1009*, 89–97. [CrossRef] [PubMed]
23. Bhardwaj, N.; Bhardwaj, S.K.; Mehta, J.; Kim, K.H.; Deep, A. MOF-Bacteriophage Biosensor for Highly Sensitive and Specific Detection of Staphylococcus aureus. *ACS Appl. Mater. Interfaces* **2017**, *9*, 33589–33598. [CrossRef] [PubMed]
24. Zhong, X.; Xia, H.; Huang, W.; Li, Z.; Jiang, Y. Biomimetic metal-organic frameworks mediated hybrid multi-enzyme mimic for tandem catalysis. *Chem. Eng. J.* **2020**, *381*, 122758. [CrossRef]
25. Wang, Y.M.; Liu, J.W.; Adkins, G.B.; Shen, W.; Trinh, M.P.; Duan, L.Y.; Jiang, J.H.; Zhong, W. Enhancement of the Intrinsic Peroxidase-Like Activity of Graphitic Carbon Nitride Nanosheets by ssDNAs and Its Application for Detection of Exosomes. *Anal. Chem.* **2017**, *89*, 12327–12333. [CrossRef]
26. Xia, Y.; Liu, M.; Wang, L.; Yan, A.; He, W.; Chen, M.; Lan, J.; Xu, J.; Guan, L.; Chen, J. A visible and colorimetric aptasensor based on DNA-capped single-walled carbon nanotubes for detection of exosomes. *Biosens. Bioelectron.* **2017**, *92*, 8–15. [CrossRef]
27. Liang, W.; Wied, P.; Carraro, F.; Sumby, C.J.; Nidetzky, B.; Tsung, C.K.; Falcaro, P.; Doonan, C.J. Metal-Organic Framework-Based Enzyme Biocomposites. *Chem. Rev.* **2021**, *121*, 1077–1129. [CrossRef]
28. Boyjoo, Y.; Wang, M.; Pareek, V.K.; Liu, J.; Jaroniec, M. Synthesis and applications of porous non-silica metal oxide submicrospheres. *Chem. Soc. Rev.* **2016**, *45*, 6013–6047. [CrossRef]
29. Huang, Y.; Ren, J.; Qu, X. Nanozymes: Classification, Catalytic Mechanisms, Activity Regulation, and Applications. *Chem. Rev.* **2019**, *119*, 4357–4412. [CrossRef]
30. He, Y.; Zhou, W.; Qian, G.; Chen, B. Methane storage in metal-organic frameworks. *Chem. Soc. Rev.* **2014**, *43*, 5657–5678. [CrossRef]
31. Cai, S.; Han, Q.; Qi, C.; Lian, Z.; Jia, X.; Yang, R.; Wang, C. Pt74Ag26 nanoparticle-decorated ultrathin MoS2 nanosheets as novel peroxidase mimics for highly selective colorimetric detection of H2O2 and glucose. *Nanoscale* **2016**, *8*, 3685–3693. [CrossRef]
32. Lin, T.; Zhong, L.; Guo, L.; Fu, F.; Chen, G. Seeing diabetes: Visual detection of glucose based on the intrinsic peroxidase-like activity of MoS2 nanosheets. *Nanoscale* **2014**, *6*, 11856–11862. [CrossRef]
33. Peng, D.; Yang, Y.; Que, M.; Ding, Y.; Wu, H.; Deng, X.; He, Q.; Ma, X.; Li, X.; Qiu, H. Partially oxidized MoS(2) nanosheets with high water-solubility to enhance the peroxidase-mimic activity for sensitive detection of glutathione. *Anal. Chim. Acta* **2023**, *1250*, 340968. [CrossRef] [PubMed]
34. Abdolmohammad-Zadeh, H.; Ahmadian, F. A chemiluminescence biosensor based on the peroxidase-like property of molybdenum disulfide/zirconium metal-organic framework nanocomposite for diazinon monitoring. *Anal. Chim. Acta* **2023**, *1253*, 341055. [CrossRef] [PubMed]
35. Li, S.; Zhao, X.; Yu, X.; Wan, Y.; Yin, M.; Zhang, W.; Cao, B.; Wang, H. Fe(3)O(4) Nanozymes with Aptamer-Tuned Catalysis for Selective Colorimetric Analysis of ATP in Blood. *Anal. Chem.* **2019**, *91*, 14737–14742. [CrossRef] [PubMed]
36. Liu, M.X.; Zhang, H.; Zhang, X.W.; Chen, S.; Yu, Y.L.; Wang, J.H. Nanozyme Sensor Array Plus Solvent-Mediated Signal Amplification Strategy for Ultrasensitive Ratiometric Fluorescence Detection of Exosomal Proteins and Cancer Identification. *Anal. Chem.* **2021**, *93*, 9002–9010. [CrossRef]
37. Chen, G.; Dong, W.F.; Deng, Y.H.; Li, B.L.; Li, X.L.; Luo, H.Q.; Li, N.B. Nanodots of transition metal (Mo and W) disulfides grown on NiNi Prussian blue analogue nanoplates for efficient hydrogen production. *Chem. Commun.* **2018**, *54*, 11044–11047. [CrossRef]
38. Dong, W.; Chen, G.; Hu, X.; Zhang, X.; Shi, W.; Fu, Z. Molybdenum disulfides nanoflowers anchoring iron-based metal organic framework: A synergetic catalyst with superior peroxidase-mimicking activity for biosensing. *Sens. Actuators B Chem.* **2020**, *305*, 127530. [CrossRef]
39. Clark, D.J.; Fondrie, W.E.; Liao, Z.; Hanson, P.I.; Fulton, A.; Mao, L.; Yang, A.J. Redefining the Breast Cancer Exosome Proteome by Tandem Mass Tag Quantitative Proteomics and Multivariate Cluster Analysis. *Anal. Chem.* **2015**, *87*, 10462–10469. [CrossRef]
40. Wang, L.; Zhou, H.; Hu, H.; Wang, Q.; Chen, X. Regulation Mechanism of ssDNA Aptamer in Nanozymes and Application of Nanozyme-Based Aptasensors in Food Safety. *Foods* **2022**, *11*, 544. [CrossRef] [PubMed]
41. Yu, D.; Li, Y.; Wang, M.; Gu, J.; Xu, W.; Cai, H.; Fang, X.; Zhang, X. Exosomes as a new frontier of cancer liquid biopsy. *Mol. Cancer* **2022**, *21*, 56. [CrossRef] [PubMed]
42. Li, J.; Liu, W.; Wu, X.; Gao, X. Mechanism of pH-switchable peroxidase and catalase-like activities of gold, silver, platinum and palladium. *Biomaterials* **2015**, *48*, 37–44. [CrossRef] [PubMed]

43. Xiong, X.; Tang, Y.; Xu, C.; Huang, Y.; Wang, Y.; Fu, L.; Lin, C.; Zhou, D.; Lin, Y. High Carbonization Temperature to Trigger Enzyme Mimicking Activities of Silk-Derived Nanosheets. *Small* **2020**, *16*, e2004129. [CrossRef] [PubMed]
44. Zhang, X.; Yang, Q.; Lang, Y.; Jiang, X.; Wu, P. Rationale of 3,3′,5,5′-Tetramethylbenzidine as the Chromogenic Substrate in Colorimetric Analysis. *Anal. Chem.* **2020**, *92*, 12400–12406. [CrossRef]
45. Balaj, L.; Lessard, R.; Dai, L.; Cho, Y.J.; Pomeroy, S.L.; Breakefield, X.O.; Skog, J. Tumour microvesicles contain retrotransposon elements and amplified oncogene sequences. *Nat. Commun.* **2011**, *2*, 180. [CrossRef]
46. Ding, L.; Liu, L.-E.; He, L.; Effah, C.Y.; Yang, R.; Ouyang, D.; Jian, N.; Liu, X.; Wu, Y.; Qu, L. Magnetic-Nanowaxberry-Based Simultaneous Detection of Exosome and Exosomal Proteins for the Intelligent Diagnosis of Cancer. *Anal. Chem.* **2021**, *93*, 15200–15208. [CrossRef]
47. Yu, W.; Hurley, J.; Roberts, D.; Chakrabortty, S.K.; Enderle, D.; Noerholm, M.; Breakefield, X.O.; Skog, J.K. Exosome-based liquid biopsies in cancer: Opportunities and challenges. *Ann. Oncol.* **2021**, *32*, 466–477. [CrossRef]
48. Choi, D.; Montermini, L.; Jeong, H.; Sharma, S.; Meehan, B.; Rak, J. Mapping Subpopulations of Cancer Cell-Derived Extracellular Vesicles and Particles by Nano-Flow Cytometry. *ACS Nano* **2019**, *13*, 10499–10511. [CrossRef] [PubMed]
49. Kasetsirikul, S.; Tran, K.T.; Clack, K.; Soda, N.; Shiddiky, M.J.A.; Nguyen, N.T. Low-cost electrochemical paper-based device for exosome detection. *Analyst* **2022**, *147*, 3732–3740. [CrossRef]
50. Oliveira-Rodriguez, M.; Serrano-Pertierra, E.; Garcia, A.C.; Lopez-Martin, S.; Yanez-Mo, M.; Cernuda-Morollon, E.; Blanco-Lopez, M.C. Point-of-care detection of extracellular vesicles: Sensitivity optimization and multiple-target detection. *Biosens. Bioelectron.* **2017**, *87*, 38–45. [CrossRef]
51. Pan, Y.; Wang, L.; Deng, Y.; Wang, M.; Peng, Y.; Yang, J.; Li, G. A simple and sensitive method for exosome detection based on steric hindrance-controlled signal amplification. *Chem. Commun.* **2020**, *56*, 13768–13771. [CrossRef]
52. Chen, J.; Meng, H.M.; An, Y.; Geng, X.; Zhao, K.; Qu, L.; Li, Z. Structure-switching aptamer triggering hybridization displacement reaction for label-free detection of exosomes. *Talanta* **2020**, *209*, 120510. [CrossRef] [PubMed]
53. He, F.; Wang, J.; Yin, B.C.; Ye, B.C. Quantification of Exosome Based on a Copper-Mediated Signal Amplification Strategy. *Anal. Chem.* **2018**, *90*, 8072–8079. [CrossRef]
54. Kuang, J.; Fu, Z.; Sun, X.; Lin, C.; Yang, S.; Xu, J.; Zhang, M.; Zhang, H.; Ning, F.; Hu, P. A colorimetric aptasensor based on a hemin/EpCAM aptamer DNAzyme for sensitive exosome detection. *Analyst* **2022**, *147*, 5054–5061. [CrossRef] [PubMed]
55. Chen, J.; Xu, Y.; Lu, Y.; Xing, W. Isolation and Visible Detection of Tumor-Derived Exosomes from Plasma. *Anal. Chem.* **2018**, *90*, 14207–14215. [CrossRef]
56. Ding, Z.; Lu, Y.; Wei, Y.; Song, D.; Xu, Z.; Fang, J. DNA-Engineered iron-based metal-organic framework bio-interface for rapid visual determination of exosomes. *J. Colloid. Interface Sci.* **2022**, *612*, 424–433. [CrossRef] [PubMed]
57. Zhang, Y.; Su, Q.; Song, D.; Fan, J.; Xu, Z. Label-free detection of exosomes based on ssDNA-modulated oxidase-mimicking activity of $CuCo_2O_4$ nanorods. *Anal. Chim. Acta* **2021**, *1145*, 9–16. [CrossRef]
58. Chen, Z.; Cheng, S.B.; Cao, P.; Qiu, Q.F.; Chen, Y.; Xie, M.; Xu, Y.; Huang, W.H. Detection of exosomes by ZnO nanowires coated three-dimensional scaffold chip device. *Biosens. Bioelectron.* **2018**, *122*, 211–216. [CrossRef]

Disclaimer/Publisher's Note: The statements, opinions and data contained in all publications are solely those of the individual author(s) and contributor(s) and not of MDPI and/or the editor(s). MDPI and/or the editor(s) disclaim responsibility for any injury to people or property resulting from any ideas, methods, instructions or products referred to in the content.

Article

A Novel Dual Bacteria-Imprinted Polymer Sensor for Highly Selective and Rapid Detection of Pathogenic Bacteria

Xiaoli Xu [1,†], Xiaohui Lin [1,†], Lingling Wang [1], Yixin Ma [1], Tao Sun [1] and Xiaojun Bian [1,2,3,*]

1. College of Food Science and Technology, Shanghai Ocean University, Shanghai 201306, China
2. Laboratory of Quality and Safety Risk Assessment for Aquatic Product on Storage and Preservation (Shanghai), Ministry of Agriculture, Shanghai 201306, China
3. Shanghai Engineering Research Center of Aquatic-Product Processing & Preservation, Shanghai 201306, China
* Correspondence: xjbian@shou.edu.cn; Fax: +86-21-61900753
† These authors contributed equally to this work.

Abstract: The rapid, sensitive, and selective detection of pathogenic bacteria is of utmost importance in ensuring food safety and preventing the spread of infectious diseases. Here, we present a novel, reusable, and cost-effective impedimetric sensor based on a dual bacteria-imprinted polymer (DBIP) for the specific detection of *Escherichia coli* O157:H7 and *Staphylococcus aureus*. The DBIP sensor stands out with its remarkably short fabrication time of just 20 min, achieved through the efficient electro-polymerization of o-phenylenediamine monomer in the presence of dual bacterial templates, followed by in-situ template removal. The key structural feature of the DBIP sensor lies in the cavity-free imprinting sites, indicative of a thin layer of bacterial surface imprinting. This facilitates rapid rebinding of the target bacteria within a mere 15 min, while the sensing interface regenerates in just 10 min, enhancing the sensor's overall efficiency. A notable advantage of the DBIP sensor is its exceptional selectivity, capable of distinguishing the target bacteria from closely related bacterial strains, including different serotypes. Moreover, the sensor exhibits high sensitivity, showcasing a low detection limit of approximately 9 CFU mL^{-1}. The sensor's reusability further enhances its cost-effectiveness, reducing the need for frequent sensor replacements. The practicality of the DBIP sensor was demonstrated in the analysis of real apple juice samples, yielding good recoveries. The integration of quick fabrication, high selectivity, rapid response, sensitivity, and reusability makes the DBIP sensor a promising solution for monitoring pathogenic bacteria, playing a crucial role in ensuring food safety and safeguarding public health.

Keywords: dual bacteria-imprinted polymer; electrochemical sensor; bacterial detection; molecularly imprinted polymer; o-phenylenediamine; *Escherichia coli* O157:H7; *Staphylococcus aureus*

1. Introduction

In recent times, global food safety issues arising from pathogenic bacteria have sparked considerable concern due to a rise in severe food poisoning cases, particularly among vulnerable populations such as children, the elderly, and immunocompromised individuals [1,2]. Among the plethora of pathogenic bacteria, *Escherichia coli* O157:H7 (*E. coli* O157:H7) and *Staphylococcus aureus* (*S. aureus*) have gained notoriety for their role in causing severe foodborne illnesses and infections [3,4]. *E. coli* O157:H7 infection can lead to distressing symptoms, including abdominal cramps, bloody diarrhea, vomiting, and fever [5,6]. In more severe instances, it may progress to hemolytic uremic syndrome (HUS), which poses serious risks, particularly to young children and the elderly, often resulting in kidney failure and other complications [7]. On the other hand, *S. aureus* can cause a wide range of infections, from minor skin issues, such as boils and abscesses, to life-threatening conditions, such as pneumonia, bloodstream infections (bacteremia), and surgical site infections [8,9]. Of particular concern with *S. aureus* is its ability to develop resistance to multiple antibiotics, making effective treatment challenging [9]. Consequently, the development of an efficient

and reliable strategy for detecting these pathogens is of utmost importance to safeguard food safety and public health.

Conventional culture-based methods have long been considered the gold standard for bacterial detection [10]. However, these methods often involve tedious procedures, including bacterial culturing steps and subsequent biochemical or serological tests, leading to time-consuming and labor-intensive processes [11]. While polymerase chain reaction (PCR) and enzyme-linked immunosorbent assay (ELISA) offer relatively rapid alternatives, they still require several hours to generate results [12–16]. Additionally, these methods are constrained by their high cost, complex operations, and reliance on prior knowledge of target sequences or the availability of specific antibodies/antigens [17]. Furthermore, their limited ability to distinguish closely related bacterial strains within a species poses challenges, as certain strains may share highly conserved DNA sequences, or antibodies/antigens may target common epitopes present across strains. As a consequence, there is a pressing demand for more rapid, user-friendly, cost-effective, and specific detection strategies for pathogenic bacteria.

Electrochemical sensors have emerged as promising tools for pathogen detection due to their rapid response, user-friendly nature, affordability, and potential for miniaturization [18]. Notably, these sensors offer a significant advantage by directly detecting whole bacterial cells without the need for time-consuming procedures such as cell lysis, nucleic acid extraction, or signal amplification. In the development of electrochemical sensors, the selection of appropriate receptors and their effective attachment to the transducer surface (e.g., glass carbon, gold, etc.) play a critical role. Various receptors, including antibodies, aptamers, phages, and carbohydrates, can be employed to target bacteria [19–23]. Among these, antibodies are the most commonly used recognition elements due to their exceptional selectivity and binding affinity. However, antibodies come with certain limitations, including their reliance on animal production, high expenses, and susceptibility to harsh conditions such as high temperatures, salt concentrations, strong acids or bases, and organic solvents [24,25]. Another challenge with using antibodies is the potential for denaturation or conformational changes when immobilized on the transducer surface through adsorption or covalent coupling [26].

Molecularly imprinted polymers (MIPs) are synthetic receptors that can be tailored to have precise binding sites that match a specific template, often representing the target analyte of interest [27,28]. Compared to antibodies, MIPs offer the benefits of simple preparation, cost-effectiveness, and enhanced physical and chemical stability, making them potential substitutes for natural antibodies [29,30]. While bacteria-imprinted polymers (BIPs) have shown promise in identifying single types of bacteria using a single bacteria template [31,32], real-life scenarios often involve co-contamination, with multiple species or strains of bacteria present simultaneously [33]. Therefore, it becomes essential to develop BIPs with multiple recognition sites capable of capturing multiple types of bacteria concurrently. However, the research on imprinting multiple-template bacteria remains limited, with only a few existing studies available [34,35]. Existing studies face certain limitations, such as lengthy preparation times for BIPs, often exceeding 48 h, and the necessity for additional measures, including drive dielectrophoretic or machine learning assistance. These prolonged preparation times can hinder the application of BIPs in situations requiring quick results or time-sensitive experiments. Additionally, the incorporation of extra measures can introduce complexity to the experimental setup and demand expertise in specific domains, thereby limiting their applicability in certain areas.

Building upon the preceding description, this study presents a novel approach for the highly selective screening of pathogenic bacteria by constructing a simple and robust electrochemical detection system based on a dual bacteria-imprinted polymer (DBIP) with double recognition sites. We demonstrate the efficacy of the DBIP sensor using two prominent pathogens, *E. coli* O157:H7 and *S. aureus*, as examples. The fabrication of the DBIP involves a facile and in situ electro-polymerization-based imprinting process, which results in the direct formation of highly specific binding sites tailored to the target bacteria on the electrode surface. The subsequent recognition and capture of the target bacteria induced detectable changes in the electrochemical impedance signal, enabling quantitative analysis. A key advantage of the electro-polymerization technique is its ability to control the deposition thickness, ensuring a thin layer of bacterial surface imprinting. This feature facilitates more rapid rebinding and unbinding of the target bacteria, leading to a faster recognition and regeneration process. By combining the unique advantages of MIPs with the inherent sensitivity and versatility of electrochemical techniques, this proposed sensor holds promise for highly selective, rapid, and sensitive detection of pathogens. The performance of the DBIP sensor was comprehensively evaluated, focusing on critical aspects such as selectivity, sensitivity, reusability, and practical applicability. Notably, we achieved a remarkable reduction in the preparation and recognition times for the DBIP, requiring only 20 and 10 min, respectively. The outcomes of this investigation hold the potential to advance the development of detection platforms for rapid and reliable identification of pathogenic bacteria, thereby enhancing food safety and public health surveillance.

2. Experimental Section

2.1. Materials and Reagents

The bacterial strains used in the experiment included *Staphylococcus aureus* (*S. aureus* ATCC 27661), *Escherichia coli* O157:H7 (*E. coli* O157:H7, ATCC 43889), *Escherichia coli* O6 (*E. coli* O6, ATCC 25922), and *Streptococcus hemolyticus* (*S. hemolyticus* ATCC 21059). LB liquid medium, trypsin soy broth, nutrient agar, Baird-Parker agar, egg-yolk tellurite emulsion, acetic acid (HAc), and potassium chloride (KCl) were purchased from Sangon Biotech (Shanghai, China). Lysozyme, cetyltrimethylammonium bromide (CTAB), and dimethyl sulfoxide (DMSO) were obtained from BBI Life Sciences Corporation (Shanghai, China). *o*-phenylenediamine (*o*PD) was obtained from TCI (Shanghai, China). Milli-Q grade (>18 MΩ) water was used throughout the experiment.

2.2. Apparatus and Measurements

All electrochemical experiments were conducted using a CHI 660E workstation with a standard three-electrode system. The glass carbon electrode (GCE, 3 mm in diameter), platinum sheet, and saturated calomel electrode (SCE) serve as the working, auxiliary, and reference electrodes, respectively. The cyclic voltammetry (CV) measurements were recorded in 0.1 M KCl solution containing 5 mM $K_3[Fe(CN)_6]$. The electrochemical impedance spectroscopy (EIS) was performed in 0.1 M KCl solution containing 1 mM $K_3[Fe(CN)_6]$ and 1 mM $K_4[Fe(CN)_6]$ by applying an open circuit voltage over a frequency range of 0.1 to 100,000 Hz with an amplitude of 5 mV.

2.3. Bacterial Cultivation

E. coli O157:H7 and *E. coli* O6 were cultured individually in LB liquid medium at 37 °C overnight with continuous agitation at 200 rpm. Similarly, *S. aureus* and *S. hemolyticus* were cultured separately in trypsin soy broth medium at 37 °C overnight under continuous agitation at 200 rpm. Enumeration of bacterial colonies was performed using the plate count technique. For subsequent experiments, the bacteria were rendered nonviable by treating the cultures with formaldehyde at a 1:100 ratio. Throughout the entirety of the experimental procedures, the formaldehyde-treated inactivated bacterial cultures were utilized. Subsequently, the bacterial cultures were centrifuged at 10,000 rpm for 3 min to pellet the bacterial cells, which were then subjected to two rounds of washing to remove the residual culture medium. Following the removal of the culture medium, the bacterial pellet was resuspended in a specific volume of 0.01 M sterile phosphate-buffered solution at pH 7.4. This resuspended bacterial suspension underwent a series of 10-fold serial dilutions to generate a range of dilutions with varying concentration gradients, ranging from 10 to 10^6 CFU mL^{-1}. These prepared dilutions were subsequently employed in the experimental protocols.

2.4. Preparation of the DBIP-Modified Electrode

Before use, the GCE was polished with 0.3–0.05 μm of alumina aqueous slurry until it had a shiny appearance. Then, the polished GCE was immersed in acetate buffer solution (0.1 M, pH 5.8) containing *o*PD (5 mM) and double bacterial template of *E. coli* O157:H7 and *S. aureus* (both at 10^8 CFU mL^{-1}), and CV was carried out under gentle stirring for 15 cycles with a potential range of −0.05 to 0.95 V vs. SCE and a scan rate of 0.05 V s^{-1} [36]. To elute the bacterial template, the modified electrode was soaked in CTAB/HAc solution (1 mM CTAB dispersed in 36% HAc) at 37 °C for 10 min under constant shaking (400 rpm). The fabricated modified electrodes before and after template removal were named P*o*PD+dual bacteria/GCE and DBIP/GCE, respectively. A non-imprinted polymer (NIP)-modified electrode (NIP/GCE) was prepared using the same steps but without adding the bacterial template.

2.5. Detection of E. coli O157:H7 and S. aureus

For capture, the freshly prepared DBIP/GCE was incubated with 250 μL of phosphate-buffered solution (0.01 M, pH 7.4) containing a specific concentration of bacteria at 37 °C for 15 min under constant shaking (300 rpm). The fabricated modified electrode was denoted as DBIP-(*E. coli* O157:H7+*S. aureus*)/GCE. Following this, the DBIP-(*E. coli* O157:H7+*S. aureus*)/GCE was washed with deionized water and analyzed by EIS under the conditions mentioned above.

2.6. Optimization of Experimental Conditions

To obtain better sensing performances, several parameters involved in the DBIP preparation (concentration of monomer and bacterial template, polymerization cycles, and conditions for template elution) and bacterial recognition (time, pH, and oscillation speed) were systematically optimized. The selection of optimum eluents and elution time for template removal was guided by the degree of reduction in charge transfer resistance (R_{ct}). A greater reduction in impedance indicated more effective template removal. Other optimal conditions were chosen based on the EIS response ($\Delta R/R$) towards a mixture of *E. coli* O157:H7 and *S. aureus*, each at a concentration of 10^5 CFU mL^{-1}. The $\Delta R/R$ was calculated using the following formula:

$$\Delta R/R = (R_{cta} - R_{ctb})/R_{ctb}$$

Here, R_{ctb} and R_{cta} represent the values of R_{ct} before and after capturing the target bacterial template, respectively.

2.7. Real Sample

To assess the sensor's suitability for real-world applications, we selected a sample of apple juice purchased from a local supermarket as our test specimen. We diluted the apple juice 100 times and then introduced different amounts of *E. coli* O157:H7 or *S. aureus* into separate portions, achieving final concentrations ranging from 10^2 to 10^4 CFU mL^{-1}. As a control, we conducted additional experiments using a phosphate-buffered solution (0.01 M, pH 7.4) instead of the bacteria. Subsequent incubation with the DBIP/GCE and EIS detection followed the same procedures used for the pure bacterial solution. Each group underwent at least three parallel experiments to ensure reliable and consistent results.

3. Results and Discussion

3.1. Fabrication of DBIP-Modified Electrode for Bacterial Capture and Detection

Scheme 1 illustrates the preparation of the dual-template bacteria-imprinted polymer (DBIP) designed to simultaneously capture *E. coli* O157:H7 and *S. aureus*. The DBIP was efficiently fabricated on a GCE surface using cyclic voltammetry electro-copolymerization of *o*-phenylenediamine (*o*PD) monomer along with dual bacterial templates (*E. coli* O157:H7 and *S. aureus*). Subsequently, in-situ elution of the templates was performed for 10 min.

The entire preparation process of the DBIP sensor is remarkably swift, taking just 20 min, which is much faster than the majority of reported electrochemical sensors designed for pathogen detection.

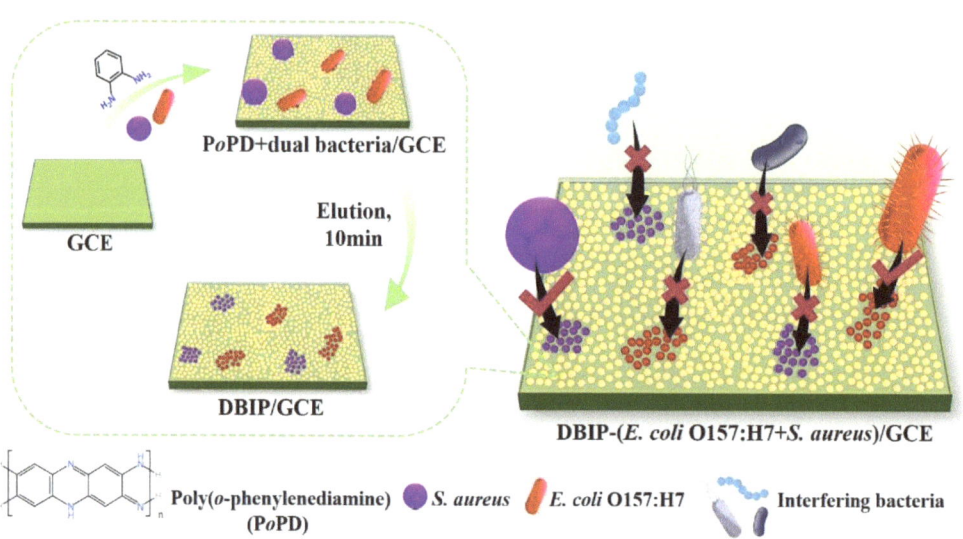

Scheme 1. Preparation of dual bacteria-imprinted polymer (DBIP) for simultaneous capture of *E. coli* O157:H7 and *S. aureus*.

The resulting DBIP on the electrode surface exhibits rapid and selective recognition of the target bacteria from complex bacterial mixtures in a mere 15 min. This specificity arises from the formation of imprinted sites within the polymer, which are tailored to complement the size, shape, and chemical characteristics of the target bacteria. Considering the low conductivity of bacteria, the target *E. coli* O157:H7 or *S. aureus* captured by the DBIP can be individually analyzed using the EIS technique.

The developed DBIP-modified electrode provides a promising platform for the capture and detection of specific bacteria, paving the way for potential applications in various fields, including food safety monitoring and biomedical diagnostics.

3.2. Characterization of the DBIP Fabrication

The morphological changes of the electrode surface during the fabrication of the DBIP were investigated using scanning electron microscopy (SEM). In the absence of bacteria, during the electro-polymerization of *o*PD, only a few irregular aggregates were observed on the electrode surface (Figure 1A). However, in the presence of the dual bacterial template (*E. coli* O157:H7 and *S. aureus*), numerous well-defined rod-shaped and globule-shaped bacteria were observed to be embedded within the polymer matrix on the electrode surface (Figure 1B). Further magnification provided a clearer view of the two distinct types of bacteria, with the rod-shaped bacteria measuring approximately 2 μm in length and 0.6 μm in width, and the globule-shaped bacteria having a diameter of about 0.8 μm (Figure 1C). These dimensions align well with the reported sizes of *E. coli* O157:H7 and *S. aureus*, respectively [32,37]. The presence of the dual bacterial template within the P*o*PD matrices was confirmed by these results. However, after eluting the copolymerized film for 10 min, almost no bacteria were found on the electrode surface (Figure 1D). This observation indicates that the bacteria were successfully removed from the P*o*PD matrices during the elution process. The DBIP, featuring a cavity-free structural trait, provides enhanced accessibility, thereby enabling the rapid recognition of the target bacteria [38]. Following the recognition of the dual target bacteria, namely *E. coli* O157:H7 and *S. aureus*, within a rapid timeframe of 15 min (Figure 1E), a subsequent observation revealed the reappearance of numerous rod-shaped and globule-shaped bacteria reappearing on the electrode surface. From the further magnified image (Figure 1F), the distinct morphologies of the two bacterial types become more apparent.

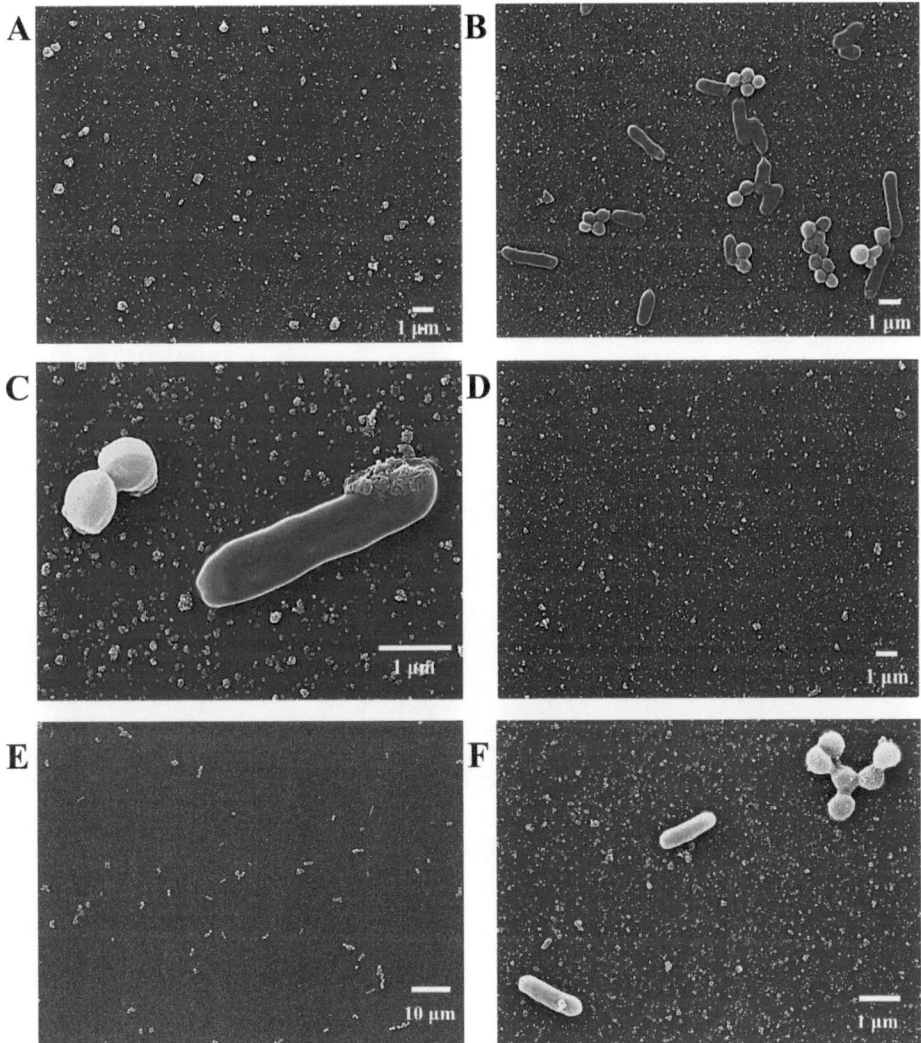

Figure 1. SEM images showing the morphological changes in the electrode surface at various stages during the preparation of the DBIP. (**A**) PoPD/GCE, (**B,C**) PoPD+dual bacteria/GCE with 5000- and 20,000-fold magnification, (**D**) DBIP/GCE, (**E,F**) DBIP-(*E. coli* O157:H7+*S. aureus*)/GCE with 1000- and 10,000-fold magnification.

3.3. Electrochemical Characterization of the DBIP-Based Sensor

The DBIP fabrication processes and the recognition response were thoroughly characterized using EIS and CV. Figure 2A–C represent the EIS Nyquist plots of different modified GCEs. The bare GCE displayed an almost straight line (Figure 2C, solid square), indicating a mass diffusion-limited electron transfer process [39]. The impedance behavior was high when the dual bacterial template was electro-polymerized onto the GCE surface with *o*PD monomer, owing to the non-conductive nature of both bacteria and PoPD. However, after elution with CTAB/HAc for 10 min, the impedance substantially decreased to around 400 Ω for the modified electrode (DBIP/GCE) (Figure 2B, hollow square), confirming the successful removal of the bacterial template.

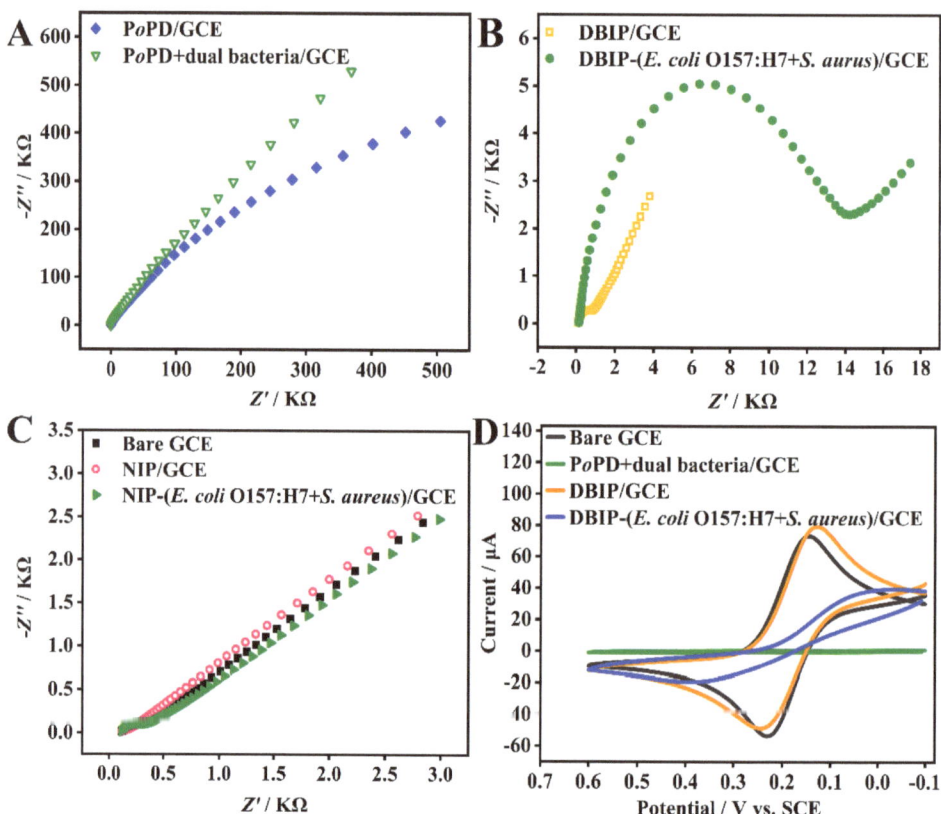

Figure 2. Electrochemical characterization of the DBIP fabrication and recognition. (**A**–**C**) EIS Nyquist plots and (**D**) CV curves of different modified electrodes as indicated.

The fabricated DBIP, acting as the recognition element, efficiently captured the target *E. coli* O157:H7 and *S. aureus*, each at a concentration of 10^5 CFU mL^{-1}. The semicircular diameter of the modified electrode (Figure 2B, solid circle) was much larger than that of the DBIP. Conversely, the non-imprinted polymer (NIP) showed almost no recognition response toward the template bacteria (Figure 2C). These results indicate that the DBIP can specifically recognize the target bacteria due to the imprinted sites formed on the polymer matrices with complementary physical and chemical characteristics to the template bacteria.

These findings were further corroborated by the CV analysis shown in Figure 2D. The bare GCE showed a pair of distinct redox peaks attributed to the redox behavior of the electroactive ion pair $[Fe(CN)_6]^{3-/4-}$. However, after the electro-copolymerization of oPD and the dual bacterial template, the redox peak disappeared. Upon elution, the fabricated DBIP exhibited an evident redox peak, confirming the successful removal of the dual bacterial template from the PoPD matrices. Subsequently, when the DBIP/GCE was exposed to a phosphate-buffered solution (0.01 M, pH 7.4) containing the dual template bacteria, the redox peak current decreased, and the ΔE_p (potential separation between the cathodic and anodic peaks) increased significantly. These changes indicated that the target bacteria were captured by the DBIP, leading to hindered electron transfer of $[Fe(CN)_6]^{3-/4-}$ over the GCE surface.

In summary, the electrochemical characterization using EIS and CV demonstrates the successful fabrication of the DBIP sensor and its efficient recognition of the target bacteria through specific imprinted sites, showcasing its potential for selective bacterial capture and detection applications.

3.4. Optimization of Experimental Conditions

The achievement of optimal sensing performance necessitates the optimization of various parameters related to the fabrication and recognition processes of the DBIP. The first step involved the optimization of electro-polymerization conditions, specifically focusing on the concentrations of oPD monomer and the dual bacterial template, as well as the number of polymerization cycles. Higher concentrations of oPD or the bacterial template led to an increase in the EIS response up to a certain point, beyond which the response declined. After careful evaluation (Figure 3A,B), the optimal concentrations of oPD and the bacterial template were determined to be 5 mM and 10^8 CFU mL^{-1}, respectively. The polymerization cycle was also critical, affecting the thickness of the imprinted layer and the generation of sufficient imprinted sites. After extensive testing (Figure 3C), 15 cycles of polymerization demonstrated the highest recognition response.

The second step focused on optimizing template elution conditions, including the choice of eluents and elution time. Comparative assessment of three eluents, namely DMSO, lysozyme (10 mg mL^{-1}), and CTAB/HAc, revealed CTAB/HAc to be the most effective eluent (Figure 3D). Subsequently, the elution time with CTAB/HAc was fine-tuned, with 10 min being the optimal duration for the successful removal of the dual bacterial template (Figure 3E).

To further enhance the sensing performance, the recognition conditions were refined, encompassing recognition time, pH, and oscillation speed. Extensive investigations (Figure 3F–H) identified the optimal recognition time as 15 min, the preferred pH value as 7.4, and the optimal oscillation speed as 300 rpm.

3.5. Quantitative Detection of E. coli O157:H7 and S. aureus

Under the optimized experimental conditions, we conducted separate quantitative detection of E. coli O157:H7 and S. aureus using the DBIP-based sensor. To achieve this, 10-fold serially diluted samples of E. coli O157:H7 and S. aureus (ranging from 10^1 to 10^6 CFU mL^{-1}) were individually incubated with the DBIP/GCE for 15 min. The subsequent quantitative detection was performed using the EIS strategy. Each sample was subjected to at least three parallel experiments to ensure reliability.

Figure 4A,C display the EIS Nyquist plots of the DBIP-based sensor for the separate detection of E. coli O157:H7 and S. aureus. As the bacterial concentration increased from 0 to 10^6 CFU mL^{-1}, the impedances also increased. This behavior can be attributed to the partial blocking of electron transfer between the redox probe and the electrode surface caused by the captured bacterial cells on the DBIP/GCE, leading to an increase in film impedance [31]. Figure 4B,D present the corresponding calibration curves of the EIS response plotted against the logarithmic concentration of E. coli O157:H7 and S. aureus, respectively. Over a wide concentration range from 10 to 10^6 CFU mL^{-1}, both calibration curves demonstrated excellent linearity between $\Delta R/R$ (Ω) and the logarithmic concentration of E. coli O157:H7 or S. aureus. The linear regression equations were expressed as $\Delta R/R$ (Ω) = 4.13 lg $C_{E.coli\ O157:H7}$ − 2.54 (R^2 = 0.995), and $\Delta R/R$ (Ω) = 4.10 lg $C_{S.\ aureus}$ − 2.32 (R^2 = 0.996), respectively. Based on the $3\sigma/S$ rule [17], the calculated detection limits for E. coli O157:H7 and S. aureus were found to be 9.4 and 9.5 CFU mL^{-1}, respectively. These results indicate that the DBIP-based sensor exhibits remarkable sensitivity in quantitatively detecting both E. coli O157:H7 and S. aureus over a broad concentration range, making it a promising tool for rapid bacterial detection in various applications.

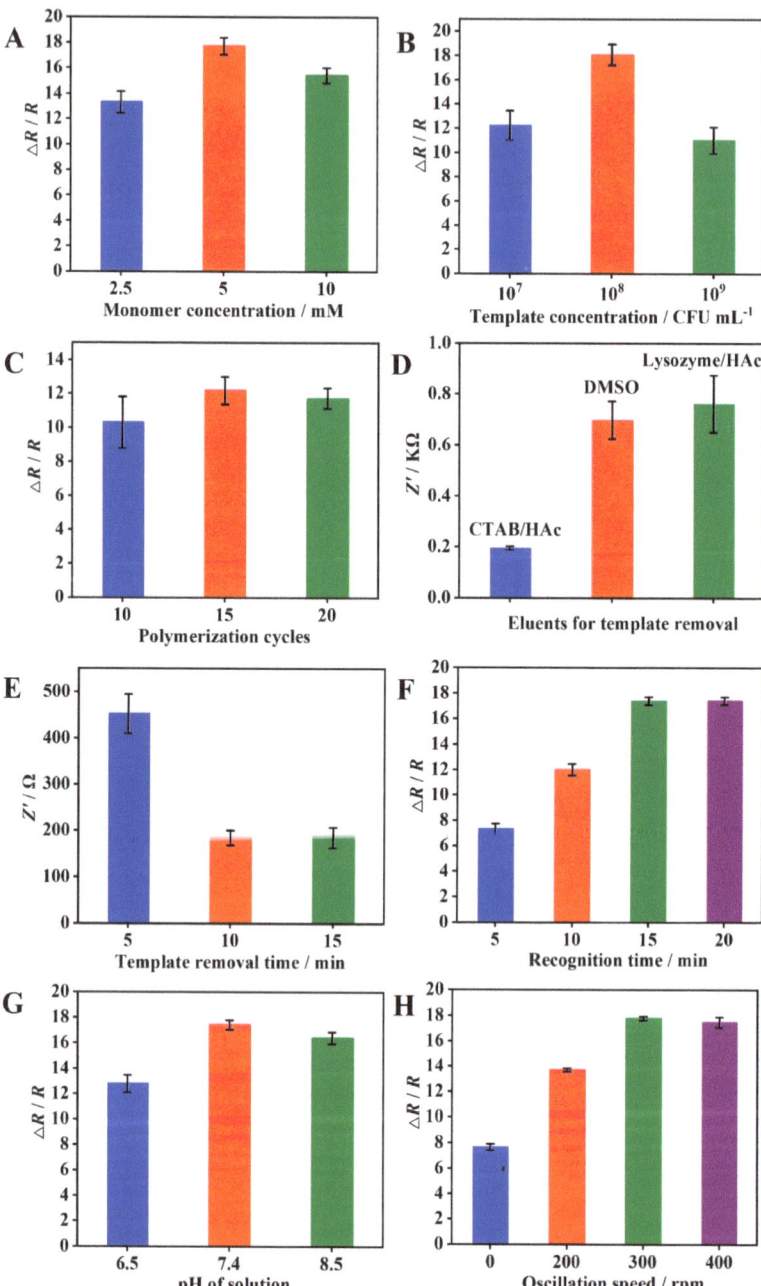

Figure 3. Optimization of parameters for the DBIP fabrication and bacterial recognition: (**A**–**B**) concentration of monomer and bacterial template during electro-polymerization; (**C**) number of polymerization cycles; (**D**–**E**) different eluents and elution time for removal of bacterial template; (**F**–**H**) recognition time, pH condition, and oscillation speed during the recognition process.

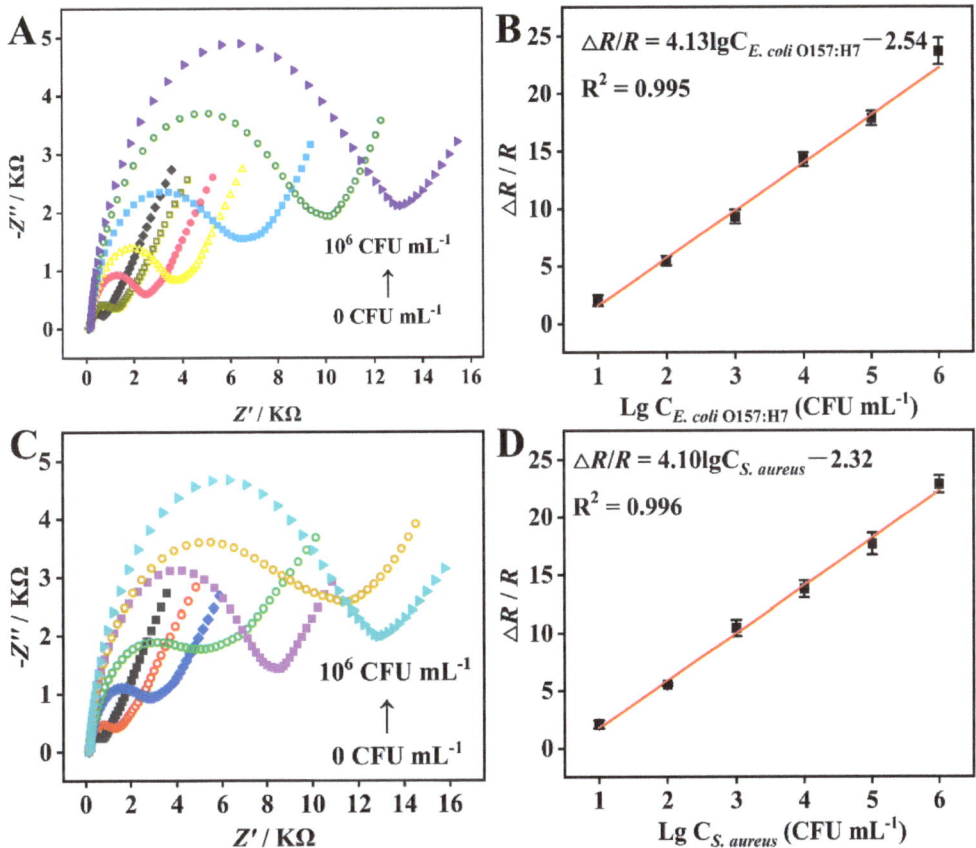

Figure 4. (**A,C**) EIS Nyquist plots and (**B,D**) corresponding calibration curves obtained from the DBIP sensor for the separate quantitative detection of *E. coli* O157:H7 and *S. aureus* over a range of concentrations from 0 to 10^6 CFU mL^{-1}.

3.6. Selectivity of the DBIP Sensor

The sensor's selectivity for bacteria is a crucial factor that enhances its potential applicability in diverse fields, including food safety, environmental monitoring, and clinical diagnostics. To assess the selectivity of the DBIP with dual recognition sites, we chose *E. coli* O6 and *S. hemolyticus* as potential interferences. These strains were selected due to their close resemblance to *E. coli* O157:H7 and *S. aureus*, respectively. As shown in Figure 5, the EIS response for single *E. coli* O157:H7 and *S. aureus* was significantly higher than that for the blank phosphate-buffered solution (0.01 M, pH 7.4), indicating clear and distinct detection of these target bacteria. Remarkably, the response to dual bacteria (both *E. coli* O157:H7 and *S. aureus*) was even more pronounced than that of the single bacterium, demonstrating an enhanced recognition response when detecting both strains simultaneously. Notably, the EIS recognition response to either single *E. coli* O157:H7 and *S. aureus*, or the dual bacteria, remained unaffected by the presence of closely related strains, such as *E. coli* O6 and/or *S. hemolyticus*. These findings demonstrate the outstanding selectivity of the DBIP sensor.

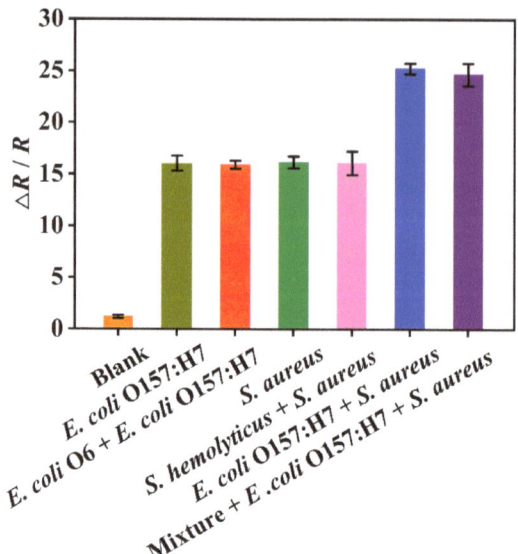

Figure 5. Selectivity of the DBIP sensor for detection of *E. coli* O157:H7 and *S. aureus*. Mixture refers to a mixed bacterial solution of *E. coli* O6 and *S. hemolyticus*. Each bacterium in the different samples was kept at the same concentration of 10^5 CFU mL^{-1}.

3.7. Reusability of the DBIP-Based Sensor

The DBIP-based sensing interface exhibited efficient regeneration capability within a short duration of 15 min, achieved by removing the rebound bacteria using CTAB/HAc, as previously mentioned. Figure 6 illustrates the successful and clean removal of rebound bacteria, and remarkably, after being recycled four times, the sensor's response signal remained at 64.4% of its initial value. This finding reveals the excellent reusability of the DBIP-based sensor, significantly reducing the need for frequent sensor replacements or replenishments. As a result, this reusability feature offers notable advantages in terms of cost and time savings. The sensor's ability to maintain its detection performance even after multiple recycling cycles enhances its practicality and sustainability, making it a promising choice for various applications in fields such as food safety, environmental monitoring, and clinical diagnostics.

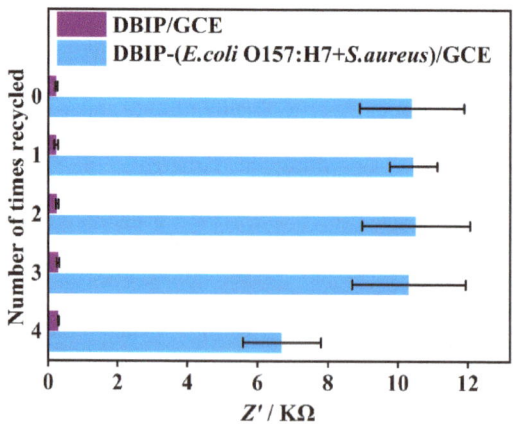

Figure 6. Reusability of the DBIP sensor for bacteria detection.

3.8. Detection of E. coli O157:H7 and S. aureus in Real Samples

To validate the practical applicability of the DBIP-based sensor in detecting *E. coli* O157:H7 and *S. aureus* in real samples, we utilized apple juice samples as a representative example. The obtained results are summarized in Table 1. The recovery rates for *E. coli* O157:H7 and *S. aureus* ranged from 86.86% to 98.40% and 81.36% to 100.58%, respectively. These findings demonstrate the sensor's effectiveness in detecting the target bacteria in actual samples. The high recovery rates highlight the sensor's reliability and accuracy in real-world applications.

Table 1. Detection of *E. coli* O157:H7 and *S. aureus* in apple juice (n = 3).

Sample	Bacteria	Original (CFU mL^{-1})	Added (CFU mL^{-1})	Found (Mean ± SD/CFU mL^{-1})	Recovery (%)
Apple juice	*E. coli* O157:H7	Not found	10^2	$(9.84 \pm 0.82) \times 10^1$	98.4
			10^3	$(9.80 \pm 1.35) \times 10^2$	98.0
			10^4	$(8.69 \pm 1.16) \times 10^3$	86.9
	S. aureus		10^2	$(9.36 \pm 0.98) \times 10^1$	93.6
			10^3	$(8.14 \pm 0.80) \times 10^2$	81.4
			10^4	$(1.01 \pm 0.02) \times 10^4$	101

4. Conclusions

In summary, we have successfully developed a simple, and reusable impedimetric sensor based on a novel dual bacteria-imprinted polymer for highly selective, rapid, and sensitive detection of pathogenic bacteria. The remarkable fabrication time of just 20 min renders the DBIP sensor an expedient and practical tool for real-world applications. The outstanding selectivity exhibited by the DBIP sensor in distinguishing between bacterial serotypes and closely related strains can be attributed to the formation of complementary binding sites that precisely match the unique features of the target bacteria. The rapid recognition and regeneration process of the DBIP sensor is made possible by the presence of cavity-free imprinted sites on its surface. This unique structural feature enables enhanced mass transfer and binding kinetics, resulting in faster and more efficient binding and unbinding of the target bacteria. The impressive combination of quick fabrication, superior selectivity, rapid response, sensitivity, and reusability underscores the great potential of DBIP-based sensors as versatile and cost-effective solutions for monitoring pathogenic bacteria across diverse fields such as food safety, environmental monitoring, and clinical diagnostics. Nevertheless, the DBIP sensor does exhibit certain limitations. Notably, it currently lacks the capability to conduct simultaneous quantitative analyses of two types of bacteria.

Author Contributions: Conceptualization, X.B., X.X., X.L. and L.W.; methodology, X.B. and Y.M.; investigation, L.W. and T.S.; writing—original draft preparation, X.X. and X.L.; writing—review and editing, X.B.; visualization, T.S.; supervision, X.B.; project administration, X.B.; funding acquisition, X.B. All authors have read and agreed to the published version of the manuscript.

Funding: This research was funded by the Program of Shanghai Academic Research Leader (21XD1401200), and Chenguang Program supported by the Shanghai Education Development Foundation and the Shanghai Municipal Education Commission (15CG54).

Institutional Review Board Statement: Not applicable.

Informed Consent Statement: Not applicable.

Acknowledgments: We gratefully acknowledge financial support from the Program of Shanghai Academic Research Leader (21XD1401200), and Chenguang Program supported by the Shanghai Education Development Foundation and the Shanghai Municipal Education Commission (15CG54).

Conflicts of Interest: The authors declare no conflict of interest.

References

1. Gu, R.H.; Duan, Y.X.; Li, Y.X.; Luo, Z.W. Fiber-Optic-Based Biosensor as an Innovative Technology for Point-of-Care Testing Detection of Foodborne Pathogenic Bacteria to Defend Food and Agricultural Product Safety. *J. Agric. Food Chem.* **2023**, *71*, 10982–10988. [CrossRef]
2. Nguyen, T.T.Q.; Gu, M.B. An ultrasensitive electrochemical aptasensor using Tyramide-assisted enzyme multiplication for the detection of Staphylococcus aureus. *Biosens. Bioelectron.* **2023**, *228*, 115199. [CrossRef] [PubMed]
3. Zhou, Y.Q.; Li, Z.Y.; Huang, J.J.; Wu, Y.X.; Mao, X.Y.; Tan, Y.Z.; Liu, H.; Ma, D.X.; Li, X.; Wang, X.Y. Development of a phage-based electrochemical biosensor for detection of Escherichia coli O157: H7 GXEC-N07. *Bioelectrochemistry* **2023**, *150*, 108345. [CrossRef] [PubMed]
4. Wan, Y.L.; Wang, X.W.; Yang, L.; Li, Q.H.; Zheng, X.T.; Bai, T.Y.; Wang, X. Antibacterial Activity of Juglone Revealed in a Wound Model of Staphylococcus aureus Infection. *Int. J. Mol. Sci.* **2023**, *24*, 3931. [CrossRef]
5. Dai, G.; Li, Y.; Li, Z.; Zhang, J.W.; Geng, X.; Zhang, F.; Wang, Q.J.; He, P.A. Zirconium-Based Metal-Organic Framework and Ti3C2Tx Nanosheet- Based Faraday Cage-Type Electrochemical Aptasensor for Escherichia coli Detection. *ACS Appl. Nano Mater.* **2022**, *5*, 9201–9208. [CrossRef]
6. Li, S.; Konoval, H.M.; Marecek, S.; Lathrop, A.A.; Feng, S.; Pokharel, S. Control of Escherichia coli O157:H7 using lytic bacteriophage and lactic acid on marinated and tenderized raw pork loins. *Meat Sci.* **2023**, *196*, 109030. [CrossRef]
7. Kolodziejek, A.M.; Minnich, S.A.; Hovde, C.J. Escherichia coli O157:H7 virulence factors and the ruminant reservoir. *Curr. Opin. Infect. Dis.* **2022**, *35*, 205–214. [CrossRef]
8. Kong, W.J.; Xiong, J.; Yue, H.; Fu, Z.F. Sandwich Fluorimetric Method for Specific Detection of Staphylococcus aureus Based on Antibiotic-Affinity Strategy. *Anal. Chem.* **2015**, *87*, 9864–9868. [CrossRef]
9. Ali, M.M.; Silva, R.; White, D.; Mohammadi, S.; Li, Y.F.; Capretta, A.; Brennan, J.D. A Lateral Flow Test for Staphylococcus aureus in Nasal Mucus Using a New DNAzyme as the Recognition Element. *Angew. Chem.* **2022**, *61*, e202112346. [CrossRef]
10. Foddai, A.C.G.; Grant, I.R. Methods for detection of viable foodborne pathogens: Current state-of-art and future prospects. *Appl. Microbiol. Biot.* **2020**, *104*, 4281–4288. [CrossRef]
11. Wang, J.F.; Wu, X.Z.; Wang, C.W.; Shao, N.S.; Dong, P.T.; Xiao, R.; Wang, S.Q. Magnetically Assisted Surface-Enhanced Raman Spectroscopy for the Detection of Staphylococcus aureus Based on Aptamer Recognition. *ACS Appl. Mater. Interfaces* **2015**, *7*, 20919–20929. [CrossRef] [PubMed]
12. Macori, G.; McCarthy, S.C.; Burgess, C.M.; Fanning, S.; Duffy, G. A quantitative real time PCR assay to detect and enumerate Escherichia coli O157 and O26 serogroups in sheep recto-anal swabs. *J. Microbiol. Methods* **2019**, *165*, 105703. [CrossRef]
13. Athamanolap, P.; Hsieh, K.; O'Keefe, C.M.; Zhang, Y.; Yang, S.; Wang, T.H. Nanoarray Digital Polymerase Chain Reaction with High-Resolution Melt for Enabling Broad Bacteria Identification and Pheno-Molecular Antimicrobial Susceptibility Test. *Anal. Chem.* **2019**, *91*, 12784–12792. [CrossRef] [PubMed]
14. Wu, B.Y.; Hu, J.S.; Li, Y. Development of an ultra sensitive single-tube nested PCR assay for rapid detection of Campylobacter jejuni in ground chicken. *Food Microbiol.* **2022**, *106*, 104052. [CrossRef]
15. Cui, Y.F.; Wang, H.W.; Guo, F.F.; Cao, X.Y.; Wang, X.; Zeng, X.M.; Cui, G.L.; Lin, J.; Xu, F.Z. Monoclonal antibody-based indirect competitive ELISA for quantitative detection of Enterobacteriaceae siderophore enterobactin. *Food Chem.* **2022**, *391*, 133241. [CrossRef]
16. Yin, W.; Zhu, L.; Xa, H.; Tang, Q.; Ma, Y.X.; Chou, S.H.; He, J. Bio-hybrid nanoarchitectonics of nanoflower-based ELISA method for the detection of Staphylococcus aureus. *Sens. Actuators B-Chem.* **2022**, *366*, 132005. [CrossRef]
17. Wang, L.L.; Lin, X.H.; Liu, T.; Zhang, Z.H.; Kong, J.; Yu, H.; Yan, J.; Luan, D.L.; Zhao, Y.; Bian, X.J. Reusable and universal impedimetric sensing platform for the rapid and sensitive detection of pathogenic bacteria based on bacteria-imprinted polythiophene film. *Analyst* **2022**, *147*, 4433–4441. [CrossRef]
18. Simoska, O.; Stevenson, K.J. Electrochemical sensors for rapid diagnosis of pathogens in real time. *Analyst* **2019**, *144*, 6461–6478. [CrossRef]
19. Pankratov, D.; Bendixen, M.; Shipovskov, S.; Gosewinkel, U.; Ferapontova, E.E. Cellulase-Linked Immunomagnetic Microbial Assay on Electrodes: Specific and Sensitive Detection of a Single Bacterial Cell. *Anal. Chem.* **2020**, *92*, 12451–12459. [CrossRef] [PubMed]
20. Wang, S.H.; Zhu, X.L.; Meng, Q.Y.; Zheng, P.M.; Zhang, J.; He, Z.W.; Jiang, H.Y. Gold interdigitated micro-immunosensor based on Mn-MOF-74 for the detection of Listeria monocytogens. *Biosens. Bioelectron.* **2021**, *183*, 113186. [CrossRef]
21. Farooq, U.; Yang, Q.L.; Ullah, M.W.; Wang, S.Q. Bacterial biosensing: Recent advances in phage-based bioassays and biosensors. *Biosens. Bioelectron.* **2018**, *118*, 204–216. [CrossRef]
22. Bulard, E.; Bouchet-Spinelli, A.; Chaud, P.; Roget, A.; Calemczuk, R.; Fort, S.; Livache, T. Carbohydrates as New Probes for the Identification of Closely Related Escherichia coli Strains Using Surface Plasmon Resonance Imaging. *Anal. Chem.* **2015**, *87*, 1804–1811. [CrossRef] [PubMed]
23. Hui, Y.Y.; Peng, H.S.; Zhang, F.X.; Zhang, L.; Liu, Y.F.; Jia, R.; Song, Y.X.; Wang, B.N. An ultrasensitive sandwich-type electrochemical aptasensor using silver nanoparticle/titanium carbide nanocomposites for the determination of Staphylococcus aureus in milk. *Microchim. Acta* **2022**, *189*, 276. [CrossRef]

24. Lee, S.; Kang, T.W.; Hwang, I.J.; Kim, H.I.; Jeon, S.J.; Yim, D.; Choi, C.; Son, W.; Kim, H.; Yang, C.S.; et al. Transition-Metal Dichalcogenide Artificial Antibodies with Multivalent Polymeric Recognition Phases for Rapid Detection and Inactivation of Pathogens. *J. Am. Chem. Soc.* **2021**, *143*, 14635–14645. [CrossRef]
25. Xu, J.J.; Miao, H.H.; Wang, J.X.; Pan, G.Q. Molecularly Imprinted Synthetic Antibodies: From Chemical Design to Biomedical Applications. *Small* **2020**, *16*, 6644. [CrossRef]
26. Amiri, M.; Bezaatpour, A.; Jafari, H.; Boukherroub, R.; Szunerits, S. Electrochemical Methodologies for the Detection of Pathogens. *ACS Sens.* **2018**, *3*, 1069–1086. [CrossRef]
27. Pan, J.; Chen, W.; Ma, Y.; Pan, G. Molecularly imprinted polymers as receptor mimics for selective cell recognition. *Chem. Soc. Rev.* **2018**, *47*, 5574–5587. [CrossRef] [PubMed]
28. Yu, L.; Shen, Y.; Chen, L.; Zhang, Q.; Hu, X.; Xu, Q. Molecularly imprinted ultrasensitive cholesterol photoelectrochemical sensor based on perfluorinated organics functionalization and hollow carbon spheres anchored organic-inorganic perovskite. *Biosens. Bioelectron.* **2023**, *237*, 115496. [CrossRef] [PubMed]
29. Xing, R.R.; Wen, Y.R.; Dong, Y.R.; Wang, Y.J.; Zhang, Q.; Liu, Z. Dual Molecularly Imprinted Polymer-Based Plasmonic Immunosandwich Assay for the Specific and Sensitive Detection of Protein Biomarkers. *Anal. Chem.* **2019**, *91*, 9993–10000. [CrossRef]
30. Xu, C.-Y.; Ning, K.-P.; Wang, Z.; Yao, Y.; Xu, Q.; Hu, X.-Y. Flexible Electrochemical Platform Coupled with In Situ Prepared Synthetic Receptors for Sensitive Detection of Bisphenol A. *Biosensors* **2022**, *12*, 1076. [CrossRef] [PubMed]
31. Wu, J.; Wang, R.; Lu, Y.; Jia, M.; Yan, J.; Bian, X. Facile preparation of a bacteria imprinted artificial receptor for highly selective bacterial recognition and label-free impedimetric detection. *Anal. Chem.* **2019**, *91*, 1027–1033. [CrossRef] [PubMed]
32. Lin, X.; Liu, P.P.; Yan, J.; Luan, D.; Sun, T.; Bian, X. Dual Synthetic Receptor-Based Sandwich Electrochemical Sensor for Highly Selective and Ultrasensitive Detection of Pathogenic Bacteria at the Single-Cell Level. *Anal. Chem.* **2023**, *95*, 5561–5567. [CrossRef]
33. Yang, M.Y.; Liu, X.B.; Luo, Y.G.; Pearlstein, A.J.; Wang, S.L.; Dillow, H.; Reed, K.; Jia, Z.; Sharma, A.; Zhou, B.; et al. Machine learning-enabled non-destructive paper chromogenic array detection of multiplexed viable pathogens on food. *Nat. Food* **2021**, *2*, 110–117. [CrossRef]
34. Wang, C.; Hao, T.T.; Wang, Z.L.; Lin, H.; Wei, W.T.; Hu, Y.F.; Wang, S.; Shi, X.Z.; Guo, Z.Y. Machine learning-assisted cell-imprinted electrochemical impedance sensor for qualitative and quantitative analysis of three bacteria. *Sens. Actuators B-Chem.* **2023**, *384*, 133672. [CrossRef]
35. Tokonami, S.; Shimizu, E.; Tamura, M.; Iida, T. Mechanism in External Field-mediated Trapping of Bacteria Sensitive to Nanoscale Surface Chemical Structure. *Sci. Rep.* **2017**, *7*, 16651. [CrossRef] [PubMed]
36. Kazemi, R.; Potts, E.I.; Dick, J.E. Quantifying Interferent Effects on Molecularly Imprinted Polymer Sensors for Per- and Polyfluoroalkyl Substances (PFAS). *Anal. Chem.* **2020**, *92*, 10597–10605. [CrossRef]
37. Shan, X.L.; Yamauchi, T.; Yamamoto, Y.; Niyomdecha, S.; Ishiki, K.; Le, D.Q.; Shiigi, H.; Nagaoka, T. Spontaneous and specific binding of enterohemorrhagic Escherichia coli to overoxidized polypyrrole-coated microspheres. *Chem. Commun.* **2017**, *53*, 3890–3893. [CrossRef] [PubMed]
38. Haupt, K. Molecularly imprinted polymers: The next generation. *Anal. Chem.* **2003**, *75*, 377A–383A. [CrossRef]
39. Dong, S.B.; Zhao, R.T.; Zhu, J.G.; Lu, X.; Li, Y.; Qiu, S.F.; Jia, L.L.; Jiao, X.; Song, S.P.; Fan, C.H.; et al. Electrochemical DNA Biosensor Based on a Tetrahedral Nanostructure Probe for the Detection of Avian Influenza A (H7N9) Virus. *ACS Appl. Mater. Interfaces* **2015**, *7*, 8834–8842. [CrossRef]

Disclaimer/Publisher's Note: The statements, opinions and data contained in all publications are solely those of the individual author(s) and contributor(s) and not of MDPI and/or the editor(s). MDPI and/or the editor(s) disclaim responsibility for any injury to people or property resulting from any ideas, methods, instructions or products referred to in the content.

Communication

The Development of a Specific Nanofiber Bioreceptor for Detection of *Escherichia coli* and *Staphylococcus aureus* from Air

Leontýna Varvařovská [1,*], Petr Kudrna [1], Bruno Sopko [2,3] and Taťána Jarošíková [1]

[1] Department of Natural Sciences, Faculty of Biomedical Engineering, Czech Technical University in Prague, 272 01 Kladno, Czech Republic; kudrnpet@fbmi.cvut.cz (P.K.); jarostat@fbmi.cvut.cz (T.J.)
[2] Laboratory of Advanced Biomaterials, University Centre for Energy Efficient Buildings, Czech Technical University in Prague, 273 43 Buštěhrad, Czech Republic; bruno.sopko@lfmotol.cuni.cz
[3] Department of Medical Chemistry and Biomedical Biochemistry, Second Faculty of Medicine, Charles University, 150 00 Prague, Czech Republic
* Correspondence: varvaleo@fbmi.cvut.cz

Abstract: Polluted air and the presence of numerous airborne pathogens affect our daily lives. The sensitive and fast detection of pollutants and pathogens is crucial for environmental monitoring and effective medical diagnostics. Compared to conventional detection methods (PCR, ELISA, metabolic tests, etc.), biosensors bring a very attractive possibility to detect chemicals and organic particles with the mentioned reliability and sensitivity in real time. Moreover, by integrating nanomaterials into the biosensor structure, it is possible to increase the sensitivity and specificity of the device significantly. However, air quality monitoring could be more problematic even with such devices. The greatest challenge with conservative and sensing methods for detecting organic matter such as bacteria is the need to use liquid samples, which slows down the detection procedure and makes it more difficult. In this work, we present the development of a polyacrylonitrile nanofiber bioreceptor functionalized with antibodies against bacterial antigens for the specific interception of bacterial cells directly from the air. We tested the presented novel nanofiber bioreceptor using a unique air filtration system we had previously created. The prepared antibody-functionalized nanofiber membranes for air filtration and pathogen detection (with model organisms *E. coli* and *S. aureus*) show a statistically significant increase in bacterial interception compared to unmodified nanofibers. Creating such a bioreceptor could lead to the development of an inexpensive, fast, sensitive, and incredibly selective bionanosensor for detecting bacterial polluted air in commercial premises or medical facilities.

Keywords: nanofibers; nanofiber biosensor; immuno-nanosensor; bacterial detection

1. Introduction

Nowadays, the rapid and eminent development of biomedicine and environmental monitoring is mainly due to the possibility of easy, fast, precise, and sensitive diagnostics and detection [1–3]. For such a development, sensors are the tools of great interest. In addition, combined with bioactive molecules (antibodies, enzymes, nucleic acids, etc.), (bio)sensors allow for the reliable detection of different biological and chemical markers. The main attractivity of biosensors stands especially on particular and sensitive biological interactions between analytes and the recognition bioactive element of the sensor (so-called bioreceptor) [1]. The most common biosensors commercially used are glucometers—sensors for glucose monitoring in blood [4,5]. However, in addition to monitoring and detecting glucose and other chemical analytes and biomarkers (hormones, enzymes, lipids, etc.), fast and so-called online detection of pathogens is also a significant priority.

Biosensors have become an exciting alternative to pathogen detection in microbiology and epidemiology. Today, the most common methods for determining bacteria and viruses are ELISA, PCR, and metabolic tests [6–9]. However, biosensors reduce costs (in some cases up to 96% [9], but on average, around 40% [10]) and time (from hours with PCR

Citation: Varvařovská, L.; Kudrna, P.; Sopko, B.; Jarošíková, T. The Development of a Specific Nanofiber Bioreceptor for Detection of *Escherichia coli* and *Staphylococcus aureus* from Air. *Biosensors* **2024**, *14*, 234. https://doi.org/10.3390/bios14050234

Received: 31 March 2024
Revised: 24 April 2024
Accepted: 3 May 2024
Published: 8 May 2024

Copyright: © 2024 by the authors. Licensee MDPI, Basel, Switzerland. This article is an open access article distributed under the terms and conditions of the Creative Commons Attribution (CC BY) license (https://creativecommons.org/licenses/by/4.0/).

to units to tens of minutes with biosensors [11]). Among other things, device sensitivity can be increased by incorporating nanomaterials [12–18] into the biosensor system, and it is possible to achieve LoD in fM concentration [12,13]. This increase in sensitivity is secured mainly using nanofibers. Their characteristic structure with the immense number of pores [19–21] creates an enormous active surface that can be modified, enriched, or functionalized [22,23].

Functionalization is a process of immobilizing bioactive molecules in the matrix structure [24]. Nanofibers modified by this process are the subject of recent studies. Whether it is functionalization with nucleic acids (such as DNA immobilization for the detection of Salmonella [25]) or antibodies (specific antibodies against Pseudomonas aeruginosa [26], Helicobacter pylori [27], or Streptococcus agalactiae [28]), low detection limits in the units of $CFU \cdot mL^{-1}$, high sensitivity, and fast response characterize these biosensors. These mentioned studies are dedicated to pathogen detection from liquid samples. However, many pathogens are transmitted through the air, and in addition to causing respiratory diseases, they also cause nosocomial infections. Pathogen detection directly from the air is becoming an attractive and desired method for the environmental monitoring of polluted air. Although many different biosensors exist, their use for detecting analytes from air faces challenges in bioreceptor preservation [29–32]. Nevertheless, pathogen monitoring in the air could help prevent respiratory disease epidemics or the emergence of nosocomial infections in operating rooms, intensive units, and hospitals in general.

The main goal of this work was to prepare antibody-functionalized nanofibers as bioreceptors for the interception and detection of selected bacterial organisms. In this work, we present the needleless electrospinning process of polyacrylonitrile nanofiber fabrication; the process of their functionalization; and finally, the evaluation of the bioreceptor's bacterial interception effectiveness through optical density measurement. This work is directly linked to the conference paper from the EHB 2023 conference but expands the mentioned paper with more detailed methodology and new results (supplemented results of detecting *E. coli* and added new results of detecting *S. aureus*) [33].

2. Materials and Methods

Considering nanofibers' characteristic structure, a mechanically and chemically durable synthetic polymer material with the possibility of functionalization had to be chosen to prepare desirable filtration membranes. The immobilization of proper bioactive molecules (antibodies) secured the functionalization of nanofiber membranes. For the required application, specific antibodies were selected as a biorecognition element for detecting the model bacteria (*Escherichia coli* and *Staphylococcus aureus*). After preparing and characterizing the prepared bioreceptor, functionalized nanofiber membranes were tested in the laboratory.

2.1. Materials

Polymer polyacrylonitrile (PAN) was purchased from Sigma-Aldrich (USA) to fabricate electrospun nanofibers. This polymer was chosen due to its mechanical and chemical endurance and the possibility of surface functionalization. The functionalization of PAN nanofibers was performed by the immobilization of specific antibodies. For the interception of Gram-negative model bacteria, Rabbit polyclonal IgG anti-*Escherichia coli* antibodies (4329–4906) were purchased from Bio-Rad (USA). Anti-*Staphylococcus aureus* LTA antibodies (SAB4200883-100UL) from Sigma-Aldrich (USA) were immobilized to nanofibers to detect the Gram-positive model bacteria *Staphylococcus aureus*.

The University of Chemistry and Technology, Prague, provided Gram-negative model bacteria *Escherichia coli* reference strains (O26:B6, *E. coli* DBM 3125—collection CCM 3988). The Institute of Medical Biochemistry and Laboratory Diagnostics, First Faculty of Medicine, Charles University in Prague, provided Gram-positive bacteria *Staphylococcus aureus* (STAV) strains.

2.2. Nanofiber Fabrication, Modification, and Characterization

For the biosensor matrix, polyacrylonitrile nanofibers were fabricated using the electrospinning method. Electrospinning uses the charge polymer solution under a high-voltage electric field to prepare ultrafine fibers with diameters of hundreds of nanometers [34]. Electrospun nanofibers are characterized by extremely high surface-to-volume ratio, high porosity, low weight, and excellent mechanical and chemical properties. Nevertheless, all the properties can be customized by adequately selecting a polymer solution and setting the process parameters of the fabrication method [19,35,36].

Polyacrylonitrile (PAN) polymer is suitable for preparing fine nanofibers with excellent mechanical and chemical stability. Fibers fabricated from polyacrylonitrile are ideal for filtration and the creation of biosensor matrices (mats). These fibers are also suited for surface functionalization by immobilizing bioactive molecules [37,38].

To fabricate suitable nanofibers, the powder of polyacrylonitrile was mixed with N, N-dimethylformamide (DMF) and homogenized for 2 h at 35 °C. Electrospun PAN nanofibers were fabricated (roller electrospinning—Figure 1) using Nanospider NS 1WS500U (Elmarco, Liberec, Czech Republic). The process parameters are shown in Table 1.

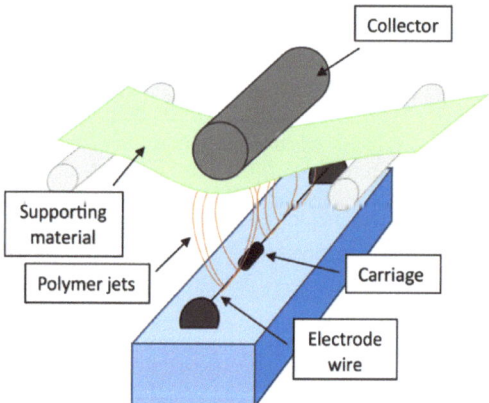

Figure 1. Setup of Nanospider device for fabrication of electrospun nanofibers [33].

Table 1. Set process parameters of the electrospinning (with the deviation given by the Nanospider NS 1WS500U device) [33].

Fabrication Parameters	Values
Solution	PAN + DMF
Solution concentration [%]	15
Diameter of the wire electrode [mm]	0.2
Distance between electrodes [cm]	25
Temperature [°C]	20
Relative humidity [%]	20
Voltage [kV]	50–90

After fabrication, samples of nanofibers were gilded and characterized through the scanning electron microscope Vega3 SB (Tescan, Brno, Czech Republic).

The created nanofibers were later modified and functionalized. PAN nanofibers' surface modification (reduction) ensures the formation of functional groups suitable for bonding bioactive molecules [39]. Specific antibodies against bacteria *E. coli* and *S. aureus* were then covalently immobilized in the structure of PAN nanofibers. The concentration of bonded antibodies was determined by infrared spectroscopy IRAFfinity-1 (Shimadzu, Kyoto, Japan), and the absorbance of 1685 cm^{-1}, characteristic of the peptide bond, was used. The calibration curve was determined using avidin and measuring the remaining protein in the solution after immobilization [40,41].

The functionalized nanofibers were prepared and preserved in a saline buffer with sodium azide. Samples preserved this way were stored in the fridge. Previous testing shows that preserved functionalized nanofiber membranes can be stored in the fridge for at least 2 months without changing the antibody activity.

2.3. Bacterial Cultivation

Both model organisms—*E. coli* and *S. aureus*—were cultured on a solid agar medium prepared from 2.5 g of yeast extract, 2.5 g of peptone, 1.125 g of NaCl, and 5 g of agar. Individual media components were mixed in 250 mL of distilled water, homogenized, heated, and sterilized before being poured into the Petri dishes.

From the reference strains, a single colony of bacteria was transferred to the agar medium using the streak plate method; passaged bacteria were cultured at 37 °C in the incubator (mini-incubator ICT 18, FALC Instruments, Treviglio BG, Italy). *E. coli* was incubated for 21 h and *S. aureus* for 24 h to achieve adequately grown bacterial colonies.

2.4. Testing of the Nanofiber Bioreceptor

A unique pump system was designed to test the detection effectivity of the functionalized nanofibers. The created system consists of a mechanical pump enabling the filtration of the air sample through the nanofiber membrane in a sealed chamber. A sample container with a volume of 1.5 l is connected directly to the sealed chamber. The whole pump system is closed and provided with filters and thus does not allow bacteria to escape from the experimental setup. Moreover, this unique pump system was designed to maintain suitable conditions for the immobilized antibodies by continually humidifying filtered air. The detailed layout (Figure 2) and function of the air filtration system are presented in the original paper from 2024 [42].

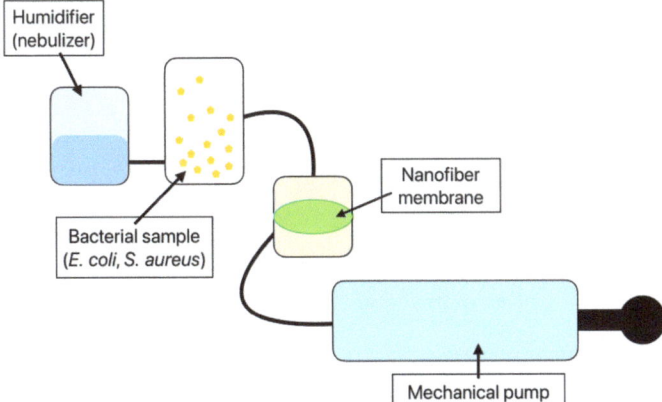

Figure 2. The layout of the air filtration system consisting of a mechanical pump, a 1.5 l sample container, a sealed container with a nanofiber membrane, and a humidifier sustaining the proper environment for the antibody immobilized to the nanofiber structure [42].

Nanofiber membranes were tested as a bioreceptor for the interception of bacterial cells. Before use, membranes were washed in distilled water so the saline buffer and preservative residues would not affect the detection. After washing, the nanofiber membrane was evenly spread to the holder in the sealed chamber. The volume of contaminated air in the sample container was then filtered through the functionalized nanofiber membranes using the mechanical pump. After the filtration, membranes were cleansed for 10 s in 1× PBS buffer to wash out bacterial cells that did not bind to the antibodies.

The functionalized PAN nanofibers as bioreceptors were tested in different conditions, namely dry air filtration and air filtration with additional humidification of the nanofiber membranes.

Nanofiber membranes were transferred to the liquid growth medium and incubated at 37 °C for 21 h (*E. coli*) or 24 h (*S. aureus*). After the incubation, 1 mL of homogenized bacterial suspension was transferred to the spectrophotometric cuvette. The bacterial suspensions' optical density (wavelength 600 nm) was measured through the spectrophotometer UV-3600 (Shimadzu, Kyoto, Japan) to evaluate the number of captured bacteria. The parameters of the used spectrophotometer are shown in Table 2.

Table 2. Spectrophotometer hardware parameters [33].

Hardware Parameters	Values
Wavelength range [nm]	185–3300
Wavelength accuracy for UV and VIS [nm]	±0.2
Wavelength accuracy for IR [nm]	±0.8
Photometric range [Abs]	−6–6
Photometric accuracy [Abs] for 1 Abs	±0.003
Photometric accuracy [Abs] for 0.5 Abs	±0.002

2.5. Data Analysis and Evaluation of Bioreceptor Effectivity

The bioreceptor effectivity evaluation dataset consists of 144 measurements for *E. coli* and 90 measurements for *S. aureus*. For both model organisms, three types of samples were used: functionalized nanofibers FNn for humid air filtration, FNs for dry air filtration, and unmodified nanofibers NN for humid air filtration. Using a series of samples ensured the reproducibility and repeatability of the experiments. The individual series were compared with each other, and the comparison was evaluated.

For *E. coli*, 24 nanofiber membranes (8 for each type) were used. A series of 15 nanofiber membranes were tested through air filtration polluted by the model organism *S. aureus*. After the air filtration through the membranes and membrane incubation, bacterial suspensions were created, and the optical density (OD_{600}) was measured (spectrophotometer UV-3600, Shimadzu, Kyoto, Japan).

The optical densities dataset consists of six measured values for each nanofiber sample, ranging from OD_{600} of 0.164 to 1.677 for *E. coli* and OD_{600} of 0.456 to 1.132 for *S. aureus*. From these values, the mean and the median were calculated and then compared for each type of nanofiber membrane (FNn, FNs, and NN). In addition, the statistical significance ($p = 0.05$) of the obtained data was determined through the *t*-test.

R software with an EZR plug-in was used to analyze the data and graphically represent the results [43].

3. Results

3.1. Preparation and Characterization of PAN Nanofibers

PAN nanofibers were prepared using the roller electrospinning method (needleless electrospinning) and functionalized by immobilizing the specific antibodies. Due to the high surface-to-volume ratio, even a small part of the functionalized nanofiber obtains many antibodies. The final concentration of antibodies was determined by IR spectroscopy to be 108 ± 12 μM/g.

The structure of PAN nanofibers was characterized through SEM. Predominantly regular fibers with a mean diameter between 500 and 900 nm were observed (Figure 3).

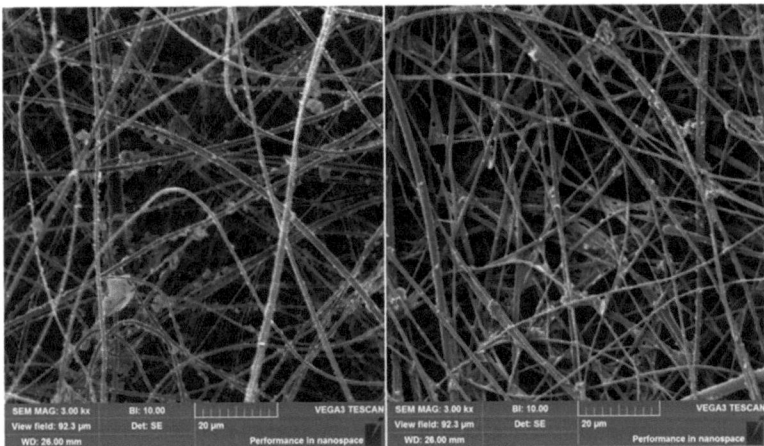

Figure 3. Surface-modified (**left**) and anti-*E. coli*-functionalized (**right**) PAN nanofibers.

From the prepared nanofibers, circle membranes with a diameter of 1.5 cm were cut. Due to the use of a 3D-printed stand for the nanofiber membranes, the real functional diameter was limited to 1 cm (the part through which the air was filtered).

The nanofiber membranes were stored in a saline buffer, so the antibody was preserved. For longer preservation, sodium azide was added to the saline buffer. Before their use as filters, nanofiber membranes were washed from chemical residues and preservatives with distilled water.

3.2. The Detection of Bacteria and Evaluation of Bioreceptor Effectiveness

The effectiveness of the bacterial interception by nanofiber membrane was evaluated through the optical density of created bacterial suspensions. The obtained results are divided according to the detected model organisms in the following subsections:

3.2.1. Detection of *Escherichia coli*

To detect *E. coli* bacteria from sufficiently humid air (an average of 60%), unmodified and anti-*E. coli* PAN nanofiber targets were used and compared (Figure 4). In addition, filtration under different conditions was tested. To determine the extent of the proper environment, anti-*E. coli* PAN nanofibers were used to detect bacteria during humid air and dry air filtration, and the bacterial interception was compared (Figure 4). The measurements were divided into eight series always consisting of the three samples (FNn, FNs, and NN).

For better clarity, Figure 5 compares the filter effectiveness between unmodified and functionalized (FNn/NN and FNs/NN) nanofiber membranes and the two used filtration methods under different conditions (FNn/FNs).

3.2.2. Detection of *Staphylococcus aureus*

As explained previously for bacteria *E. coli*, two experiments were performed for Gram-positive bacteria *Staphylococcus aureus*. Functionalized anti-*S. aureus* PAN nanofibers (FNn) were compared to the unmodified ones (NN). In addition, a comparison of the bacterial interception of the functionalized nanofibers under different conditions was performed. The estimated optical densities of both experiments are shown in Figure 6. Five series consisting of the three nanofiber samples (FNn, FNs, and NN) were evaluated.

A more detailed comparison of the interception effectivity is shown in Figure 7.

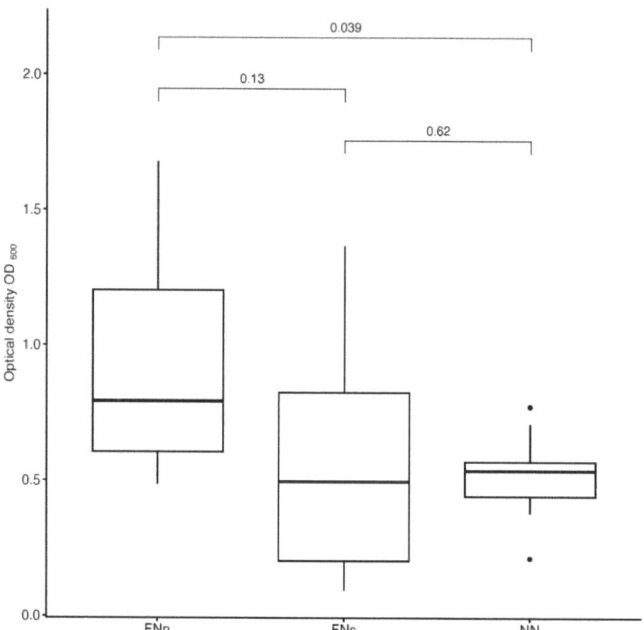

Figure 4. Comparison of bacterial suspensions' optical densities OD_{600} created from *E. coli* cells captured into the nanofiber structure during humid air and dry air filtration. In the figure, FNn (anti-*E. coli* PAN nanofibers) and NN (unmodified PAN nanofibers) show the data obtained during humid air filtration and FNs (anti-*E. coli* PAN nanofibers) during dry air filtration. The numbers above the boxplots show the *p*-values.

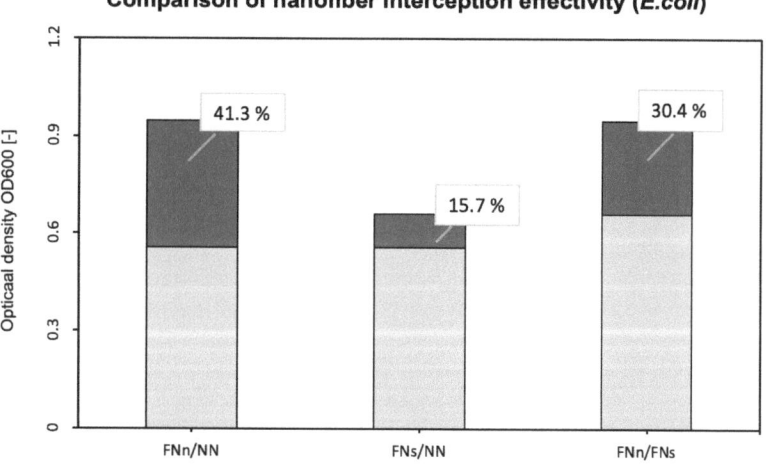

Figure 5. Comparison of the interception effectivity for functionalized and unmodified nanofibers and two types of filtrations. The dark part and percentages show the increase in the effectivity of functionalized nanofibers (FNn and FNs) compared to unmodified nanofibers NN (FNn/NN for humid air filtration and FNs/NN for dry air filtration). The third column shows the increase in interception effectivity of functionalized nanofibers during humid air filtration (FNn) compared to dry air filtration (FNs).

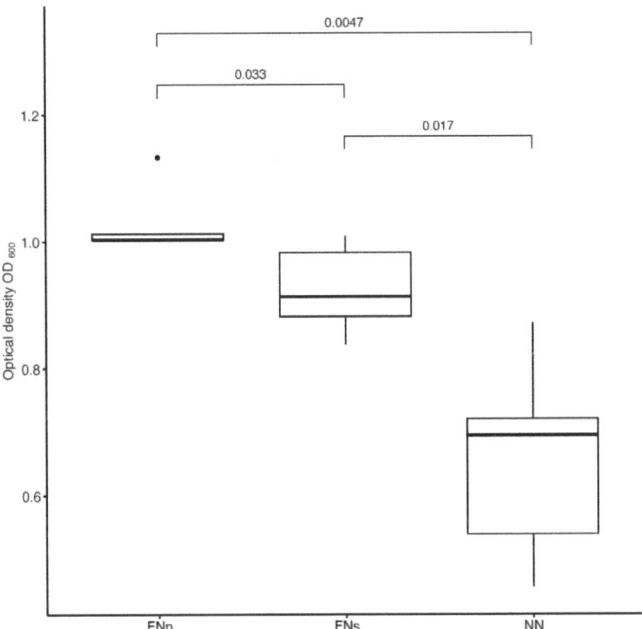

Figure 6. Comparison of bacterial suspensions' optical densities OD_{600} created from *S. aureus* cells captured into the nanofiber structure during humid air and dry air filtration. In the figure, FNn (anti-*S. aureus* PAN nanofibers) and NN (unmodified PAN nanofibers) show the data obtained during humid air filtration, and FNs (anti-*S. aureus* PAN nanofibers) during dry air filtration. The numbers above the boxplots show the corresponding *p*-values.

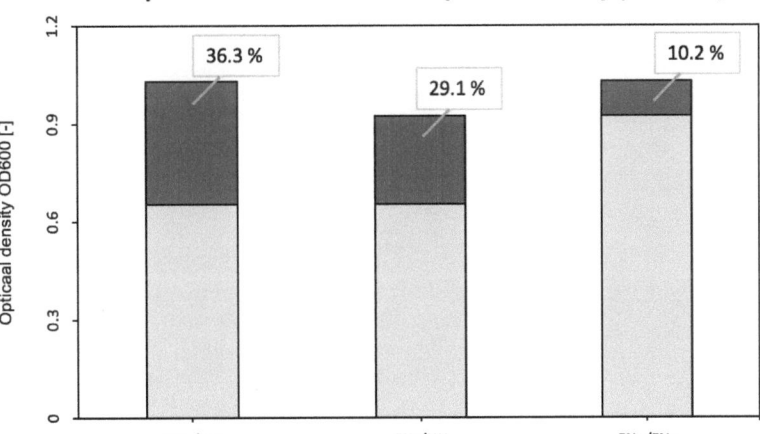

Figure 7. Comparison of the interception effectivity for functionalized and unmodified nanofibers and two types of filtrations. FNn/NN shows the difference in the interception effectivity of the functionalized and unmodified nanofibers during humid air filtration. FNs/NN shows the same difference but during dry air filtration. The FNn/FNs column then shows the increase in interception effectivity of functionalized nanofibers during humid air filtration (FNn) compared to dry air filtration (FNs).

4. Discussion

This work presents the creation and the bacterial interception effectivity evaluation of a novel immunoreceptor based on antibody-functionalized PAN nanofibers. To detect airborne bacteria (*E. coli* and *S. aureus*) directly from the air, electrospun nanofibers with great mechanical and chemical durability were used as filtration membranes. PAN nanofibers were selected due to their exceptional filtration ability and the possibility of surface functionalization. Although electrospun PAN nanofiber membranes are capable of bacterial interception itself and with great effectivity (up to 99%) [44], antibody-functionalized nanofibers capture bacterial cells with specific biochemical bonds (antigen-antibody reaction). In the case of nanofiber bioreceptors, the mechanical interception of bacterial cells is undesirable due to the rapid clogging of the filtration membranes. In comparison with a previous study dealing with the filtration effectivity of PAN nanofibers [44], the area density of functionalized membranes for bacterial detection was reduced to 2.5 g/m^2, so the mechanical interception was suppressed.

As mentioned earlier, PAN nanofibers were functionalized by immobilizing specific antibodies against *E. coli* and *S. aureus*. Bioactive molecules, such as antibodies, used as a biosensing layer of biosensors are dependent on stable and specific conditions (temperature, pH, humidity, and electrostatic repulsion). When detecting antigens directly from the air, humidity is the most challenging condition to maintain. Without additional moisturization, immobilized antibodies lose their bioactivity [45,46]. For this reason, bacterial detection, whether using conservative methods (ELISA, PCR, etc.) or (bio)sensors, is performed in liquid samples (water, body fluids, food, etc.) [26,29,30,47,48]. Airborne samples, thus, must undergo post-collection processing [31,49–51]. However, with the use of a previously designed air filtration system [42], the presented nanofiber bioreceptor was used and tested for the detection of bacterial cells directly from the air. This system humidifies nanofiber membranes during air filtration and protects immobilized antibodies from desiccations and, thus, inactivation (Figures 4 and 6) [42].

To evaluate the bioreceptor effectiveness, bacterial interception through unmodified and functionalized nanofibers was compared. The increase in the optical density of bacterial suspensions (around 41 % for *E. coli* and 36 % for *S. aureus*, as seen in Figures 5 and 7) belonging to the functionalized nanofiber membranes testing shows the effectivity of immobilized antibodies (the specific binding reaction of the bioreceptor). For both model organisms, the increase in interception effectivity due to the antibodies' activity was found to be statistically significant at the significance level of $p < 0.05$. Thus, in comparison with other mentioned nanofiber biosensors for bacterial detection [26–28], the designed nanofiber bioreceptor combines both biosensing and filtration abilities. In further research, a combination of such a bioreceptor with a proper transducer could be a pioneering alternative for fast, sensitive, and continual environment monitoring presented in recent years [52–55].

As in other studies [56–58], PAN nanofibers have been proven to be membranes with extraordinary air filtration abilities. After enrichment by metal particles (TiO_2, ZnO, Ag, etc.) [57] or bioactive molecules (enzymes and antibodies), PAN membranes show additional abilities, such as antibacterial [57] or biosensing activity, in relation to bacteria. Presented antibody-functionalized PAN nanofibers, thus, show great potential as a novel sensitive bioreceptor for detecting Gram-negative and Gram-positive bacteria such as *E. coli* and *S. aureus*.

5. Conclusions

Herein, we presented the preparation and use of the novel antibody-functionalized PAN nanofibers as bioreceptors for bacterial detection from the air. To detect model bacterial organisms *E. coli* and *S. aureus*, PAN nanofiber membranes were fabricated through the needleless electrospinning process and later functionalized by immobilizing corresponding antibodies. The specific structure of electrospun nanofibers enables the use of the membranes for air filtration. In addition, antibody functionalization significantly increases the bacterial interception effectivity of the membrane (on average about 40%) and facilitates the formation of special biochemical bonds with detected antigens (bacteria). In combination with the system for air filtration presented in previous work, the designed antibody-functionalized PAN nanofiber bioreceptor enables reliable, specific, and sensitive detection of Gram-negative and Gram-positive bacteria directly from the air and without inactivation and disintegration of the immobilized bioactive layer. Our finding opens the door for the development of a novel solution for continual environment monitoring. In addition, further studies will focus on combining the presented bioreceptor with a suitable electrode and the development of an ultrasensitive biosensor for bacterial detection.

Author Contributions: Methodology, L.V., B.S.; validation and data analysis, L.V. and B.S.; writing—original draft preparation, L.V.; writing—review and editing, L.V., B.S., P.K. and T.J.; visualization, L.V.; supervision, P.K. and T.J. All authors have read and agreed to the published version of the manuscript.

Funding: This research was funded by the Student Grant Competition of CTU (SGS22/199/OHK4/3T/17) provided by Czech Technical University in Prague, Czech Republic.

Institutional Review Board Statement: Not applicable.

Informed Consent Statement: Not applicable.

Data Availability Statement: The data presented in this study are available upon request.

Acknowledgments: We thank the Department of Natural Sciences, FBME, CTU in Prague, and UCEEB, CTU in Prague, for providing the laboratories for our experiments. We would also like to acknowledge the doc. Dana Gášková and Tomáš Bartl from the Faculty of Mathematics and Physics, Charles University, and Evžen Amler from the Second Faculty of Medicine, Charles University, for the advice and all help on the study.

Conflicts of Interest: The authors declare no conflicts of interest.

References

1. Bhatia, D.; Paul, S.; Acharjee, T.; Ramachairy, S.S. Biosensors and their widespread impact on human health. *Sens. Int.* **2024**, *5*, 100257. [CrossRef]
2. Kim, E.R.; Joe, C.; Mitchell, R.J.; Gu, M.B. Biosensors for healthcare: Current and future perspectives. *Trends Biotechnol.* **2023**, *41*, 374–395. [CrossRef] [PubMed]
3. Murzin, D.; Mapps, D.J.; Levada, K.; Belyaev, V.; Omelyanchik, A.; Panina, L.; Rodionova, V. Ultrasensitive Magnetic Field Sensors for Biomedical Applications. *Sensors* **2020**, *20*, 1569. [CrossRef] [PubMed]
4. Fiedorova, K.; Augustynek, M.; Kubicek, J.; Kudrna, P.; Bibbo, D. Review of present method of glucose from human blood and body fluids assessment. *Biosens. Bioelectron.* **2022**, *211*, 114348. [CrossRef]
5. Yoon, J.-Y. *Introduction to Biosensors: From Electric Circuits to Immunosensors*, 2nd ed.; Springer: New York, NY, USA, 2016; ISBN 978-1-4419-6021-4.
6. Yanagihara, K.; Kitagawa, Y.; Tomonaga, M.; Tsukasaki, K.; Kohno, S.; Seki, M.; Sugimoto, H.; Shimazu, T.; Tasaki, O.; Matsushima, A.; et al. Evaluation of pathogen detection from clinical samples by real-time polymerase chain reaction using a sepsis pathogen DNA detection kit. *Crit. Care* **2010**, *14*, 159. [CrossRef]
7. Wolk, D.; Mitchell, S.; Patel, R. Principles Of Molecular Microbiology Testing Methods. *Infect. Dis. Clin. N. Am.* **2001**, *15*, 1157–1204. [CrossRef]
8. Váradi, L.; Luo, J.L.; Hibbs, D.E.; Perry, J.D.; Anderson, R.J.; Orenga, S.; Groundwater, P.W. Methods for the detection and identification of pathogenic bacteria: Past, present, and future. *R. Soc. Chem.* **2017**, *46*, 4818–4832. [CrossRef] [PubMed]
9. Alahi, M.E.; Mukhopadhyay, S.C. Detection Methodologies for Pathogen and Toxins: A Review. *Sensors* **2017**, *17*, 1885. [CrossRef]
10. Myatt, C.J.; Delaney, M.; Todorof, K.; Heil, J. Low-Cost, Multiplexed Biosensor for Disease Diagnosis. In Proceedings of the SPIE Proceedings, Frontiers in Pathogen Detection: From Nanosensors to Systems, San Jose, CA, USA, 24–29 January 2009; Volume 1767, p. 716703. [CrossRef]
11. Malhotra, S.; Pham, D.S.; Lau, M.P.H.; Nguyen, A.H.; Cao, H. A Low-Cost, 3D-Printed Biosensor for Rapid Detection of Escherichia coli. *Sensors* **2022**, *22*, 2382. [CrossRef]
12. Fernando, L.M. Nanobiosensors for Detection of Pathogens. In Proceedings of the 16th Engineering Research and Development for Technology Conference, Pasay, Philippines, 25 October 2019.
13. Song, M.; Yang, M.; Hao, J. Pathogenic Virus Detection by Optical Nanobiosensors. *Cell Rep. Phys. Sci.* **2021**, *2*, 100288. [CrossRef]
14. Yang, L.; Li, Y.; Fang, F.; Li, L.; Yan, Z.; Zhang, L.; Sun, Q. Highly Sensitive and Miniature Microfiber-Based Ultrasound Sensor for Photoacoustic Tomography. *Opto-Electron. Adv.* **2022**, *5*, 200076. [CrossRef]

15. Yu, W.; Yao, N.; Pan, J.; Fang, W.; Li, X.; Tong, L.; Zhang, L. Highly Sensitive and Fast Response Strain Sensor Based on Evanescently Coupled Micro/Nanofibers. *Opto-Electron. Adv.* **2022**, *5*, 210101. [CrossRef]
16. Eivazzadeh-Keihan, R.; Noruzi, E.B.; Chidar, E.; Jafari, M.; Davoodi, F.; Kashtiaray, A.; Gorab, M.G.; Hashemi, S.M.; Javanshir, S.; Cohan, R.A.; et al. Applications of Carbon-Based Conductive Nanomaterials in Biosensors. *Chem. Eng. J.* **2022**, *442*, 136183. [CrossRef]
17. Štukovnik, Z.; Fuchs-Godec, R.; Bren, U. Nanomaterials and Their Recent Applications in Impedimetric Biosensing. *Biosensors* **2023**, *13*, 899. [CrossRef] [PubMed]
18. Dong, T.; Pires, N.M.M.; Yang, Z.; Jiang, Z. Advances in Electrochemical Biosensor Based on Nanomaterials for Protein Biomarker Detection in Saliva. *Adv. Sci.* **2023**, *10*, 2205429. [CrossRef] [PubMed]
19. Bayrak, E. Nanofibers: Production, Characterization, and Tissue Engineering Applications. In *21st Century Nanostructured Materials—Physics, Chemistry, Classification, and Emerging Application in Industry, Biomedicine, and Agriculture*; IntechOpen: London, UK, 2022. [CrossRef]
20. Xue, J.; Xie, J.; Liu, W.; Xia, Y. Electrospun nanofibers: New concepts, materials, and applications. *Acc. Chem. Res.* **2017**, *50*, 1976–1987. [CrossRef] [PubMed]
21. Al-Abduljabbar, A.; Farooq, I. Electrospun Polymer Nanofibers: Processing, Properties, and Applications. *Polymers* **2023**, *15*, 65. [CrossRef]
22. Chakrapani, G.; Ramakrishna, S.; Zare, M. Functionalization of electrospun nanofiber for biomedical application. *J. Appl. Polym. Sci.* **2023**, *140*, e53906. [CrossRef]
23. Pashchenko, A.; Stuchlíková, S.; Varvařovská, L.; Firment, P.; Staňková, L.; Nečasová, A.; Filipejová, Z.; Urbanová, L.; Jarošíková, T.; Nečas, A.; et al. Smart Nanofibres For Specific And Ultrasensitive Nanobiosensors And Drug Delivery System. *Acta Vet. Brno* **2022**, *91*, 163–170. [CrossRef]
24. Kulkarni, D.; Musale, S.; Panzade, P.; Paiva-Santos, A.C.; Sonwane, P.; Madibone, M.; Choundhe, P.; Giram, P.; Cavalu, S. Surface Functionalization of Nanofibers: The Multifaceted Approach for Advanced Biomedical Applications. *Nanomaterials* **2022**, *12*, 3899. [CrossRef]
25. Gokce, Z.G.; Akalin, P.; Kok, F.N.; Sarac, A.S. Impedimetric DNA Biosensor Based On Polyurethane/Poly(M-Anthranilic Acid) Nanofibers. *Sens. Actuators B Chem.* **2018**, *254*, 719–726. [CrossRef]
26. Sarabaegi, M.; Roushani, M.; Hosseini, H. Hollow Carbon Nanocapsules-Based Nitrogen-Doped Carbon Nanofibers With Rosary-Like Structure As A High Surface Substrate For Impedimetric Detection Of Pseudomonas Aeruginosa. *Talanta* **2021**, *223*, 121700. [CrossRef]
27. Sarabaegi, M.; Roushani, M.; Hosseini, H.; Saedi, Z.; Lemraski, E.G. A Novel Ultrasensitive Biosensor Based On Nico-Mof Nanostructure And Confined To Flexible Carbon Nanofibers With High-Surface Skeleton To Rapidly Detect Helicobacter Pylori. *Mater. Sci. Semicond. Process.* **2021**, *139*, 106351. [CrossRef]
28. Ghasemi, R.; Mirahmadi-Zare, S.Z.; Allafchian, A.; Behmanesh, M. Fast fluorescent screening assay and dual electrochemical sensing of bacterial infection agent (streptococcus agalactiae) based on fluorescent-immune nanofibers. *Sens. Actuators B. Chem.* **2022**, *352*, 130968. [CrossRef]
29. Rajamanickam, S.; Yoon Lee, N. Recent advances in airborne pathogen detection using optical and electrochemical biosensors. *Anal. Chim. Acta* **2022**, *1234*, 340297. [CrossRef]
30. Al-Taie, A.; Han, X.; Williams, C.M.; Abdulwhhab, M.; Abbott, A.P.; Goddard, A.; Wegrzyn, M.; Garton, N.J.; Barer, M.R.; Pan, J. 3-D printed polyvinyl alcohol matrix for detection of airborne pathogens in respiratory bacterial infections. *Microbiol. Res.* **2020**, *241*, 126587. [CrossRef]
31. Bhardwaj, S.K.; Bhardwaj, N.; Kumar, V.; Bhatt, D.; Azzouz, A.; Bhaumik, J.; Kim, K.-H.; Deep, A. Recent progress in nanomaterial-based sensing of airborne viral and bacterial pathogens. *Environ. Int.* **2021**, *146*, 106183. [CrossRef]
32. Triadó-Margarit, X.; Cáliz, J.; Casamayor, E.O. A long-term atmospheric baseline for intercontinental exchange of airborne pathogens. *Environ. Int.* **2022**, *158*, 106916. [CrossRef] [PubMed]
33. Varvařovská, L.; Kudrna, P.; Jarošíková, T. The development of a specific nanofiber bioreceptor for bacterial detection. In *Advances in Digital Health and Medical Bioengineering, Proceedings of the 11th International Conference on E-Health and Bioengineering (EHB-2023), Bucharest, Romania, 9–10 November 2023*; Springer Nature Publishing AG: Cham, Switzerland, 2024.
34. Ramakrishna, S. *An Introduction to Electrospinning and Nanofibers*; World Scientific: Hackensack, NJ, USA, 2005. [CrossRef]
35. Lim, C.T. Nanofiber Technology: Current Status and Emerging Developments. *Prog. Polym. Sci.* **2017**, *70*, 1–17. [CrossRef]
36. Mercante, L.A.; Pavinatto, A.; Pereira, T.S.; Migliorini, F.L.; dos Santos, D.M.; Correa, D.S. Nanofibers interfaces for biosensing: Design and applications. *Sens. Actuators Rep.* **2021**, *3*, 100048. [CrossRef]
37. Lasenko, I.; Grauda, D.; Butkauskas, D.; Sanchaniya, J.V.; Viluma-Gudmona, A.; Lusis, V. Testing the physical and mechanical properties of polyacrylonitrile nanofibers reinforced with succinite and silicon dioxide nanoparticles. *Textiles* **2022**, *2*, 162–173. [CrossRef]
38. Sanchaniya, J.V.; Kanukuntla, K. Morphology and mechanical properties of PAN nanofiber Mat. *J. Phys. Conf. Ser.* **2022**, *2423*, 012018. [CrossRef]
39. Senthil, R.; Sumathi, V.; Tamilselvi, A.; Kavukcu, S.B.; Aruni, A.W. Functionalized electrospun nanofibers for high efficiency removal of particulate matter. *Sci. Rep.* **2022**, *12*, 8411. [CrossRef] [PubMed]

40. Haris, P.I. Infrared Spectroscopy of Protein Structure. In *Encyclopedia of Biophysics*; Springer: Berlin/Heidelberg, Germany, 2013; ISBN 978-3-642-16712-6.
41. Tatulian, S.A. FTIR Analysis of Proteins and Protein-Membrane Interactions. *Methods Mol. Biol.* **2019**, *2003*, 281–325. [CrossRef]
42. Varvařovská, L.; Sopko, B.; Gášková, D.; Bartl, T.; Amler, E.; Jarošíková, T. Surface-Functionalized PAN Nanofiber Membranes for the Sensitive Detection of Airborne Specific Markers. *PeerJ* **2024**. accepted.
43. Kanda, Y. Investigation of the freely available easy-to-use software 'EZR' for medical statistics. *Bone Marrow Transplant.* **2013**, *48*, 452–458. [CrossRef]
44. Varvařovská, L.; Sopko, B.; Divín, R.; Pashschenko, A.; Fedačko, J.; Sabo, J.; Nečas, A.; Amler, E.; Jarošíková, T. Bacteria trapping effectivity on nanofibre membrane in liquids is exponentially dependent on the surface density. *Acta Vet. Brno* **2023**, *92*, 435–441. [CrossRef]
45. Wang, J.; Yiu, B.; Obermeyer, J.; Filipe, C.D.M.; Brennan, J.D.; Pelton, R. Effects of Temperature and Relative Humidity on the Stability of Paper-Immobilized Antibodies. *Biomacromolecules* **2012**, *13*, 559–564. [CrossRef]
46. Slocik, J.M.; Dennis, P.B.; Kuang, Z.; Pelton, A.; Naik, R.R. Creation of stable water-free antibody based protein liquids. *Commun. Mater.* **2021**, *2*, 118. [CrossRef]
47. Zhou, Y.; Liu, Y.; Zhang, M.; Feng, Z.; Yu, D.-G.; Wang, K. Electrospun Nanofiber Membranes for Air Filtration: A Review. *Nanomaterials* **2022**, *12*, 1077. [CrossRef]
48. Ventura, B.D.; Cennamo, M.; Minopoli, A.; Campanile, R.; Censi, S.B.; Terracciano, D.; Portella, G.; Velotta, R. Colorimetric test for fast detection of SARS-CoV-2 in nasal and throat swabs. *ACS Sens.* **2020**, *5*, 3043–3048. [CrossRef]
49. Ménard-Moyon, C.; Bianco, A.; Kalantar-Zadeh, K. Two-dimensional material-based biosensors for virus detection. *ACS Sens.* **2020**, *5*, 3739–3769. [CrossRef]
50. Prieto-Simón, B.; Bandaru, N.M.; Saint, C.; Voelcker, N.H. Tailored carbon nanotube immunosensors for the detection of microbial contamination. *Biosens. Bioelectron.* **2015**, *67*, 642–648. [CrossRef]
51. Liu, Q.; Zhang, X.; Yao, Y.; Jing, W.; Liu, S.; Sui, G. A novel microfluidic module for rapid detection of airborne waterborne pathogens. *Sens. Actuators B Chem.* **2018**, *258*, 1138–1145. [CrossRef]
52. Eltzov, E.; Pavluchov, V.; Burstin, M.; Marks, R.S. Creation of fiber optic based biosensor for air toxicity monitoring. *Sens. Actuators B Chem.* **2011**, *155*, 859–867. [CrossRef]
53. Kim, H.-J.; Park, S.J.; Park, C.S.; Le, T.-H.; Lee, S.H.; Ha, T.H.; Kim, H.-I.; Kim, J.; Lee, C.-S.; Yoon, H.; et al. Surface-modified polymer nanofiber membrane for high-efficiency microdust capturing. *Chem. Eng. J.* **2018**, *339*, 204–213. [CrossRef]
54. Shuvo, S.N.; Gomez, A.M.U.; Mishra, A.; Chen, W.Y.; Dongare, A.M.; Stanciu, L.A. Sulfur-doped titanium carbide MXenes for room-temperature gas sensing. *ACS Sens.* **2020**, *5*, 2915–2924. [CrossRef]
55. Deng, Y.; Lu, T.; Cui, J.; Samal, S.K.; Xiong, R.; Huang, C. Bio-based electrospun nanofiber as building block for a novel eco-friendly air filtration membrane: A review. *Sep. Purif. Technol.* **2021**, *277*, 119623. [CrossRef]
56. Zhu, M.; Han, J.; Wang, F.; Shao, W.; Xiong, R.; Zhang, Q.; Pan, H.; Yang, Y.; Samal, S.K.; Zhang, F.; et al. Electrospun nanofiber membranes for effective air filtration. *Macromol. Mater. Eng.* **2016**, *302*, 1600353. [CrossRef]
57. Bortolassi, A.C.C.; Guerra, V.G.; Aguiar, M.L.; Soussan, L.; Cornu, D.; Miele, P.; Bechelany, M. Composites Based on Nanoparticle and Pan Electrospun Nanofiber Membranes for Air Filtration and Bacterial Removal. *Nanomaterials* **2019**, *9*, 1740. [CrossRef]
58. Fahimirad, S.; Fahimirad, Z.; Sillanpää, M. Efficient removal of water bacteria and viruses using electrospun nanofibers. *Sci. Total Environ.* **2021**, *751*, 141673. [CrossRef] [PubMed]

Disclaimer/Publisher's Note: The statements, opinions and data contained in all publications are solely those of the individual author(s) and contributor(s) and not of MDPI and/or the editor(s). MDPI and/or the editor(s) disclaim responsibility for any injury to people or property resulting from any ideas, methods, instructions or products referred to in the content.

Article

Parallel Monitoring of Glucose, Free Amino Acids, and Vitamin C in Fruits Using a High-Throughput Paper-Based Sensor Modified with Poly(carboxybetaine acrylamide)

Xinru Yin [1,†], Cheng Zhao [1,2,†], Yong Zhao [1,*] and Yongheng Zhu [1,*]

1. College of Food Science and Technology, Shanghai Ocean University, Shanghai 201306, China; m210311019@st.shou.edu.cn (X.Y.); allenz_1222@foxmail.com (C.Z.)
2. Henan Railway Food Safety Management Engineering Technology Research Center, Zhengzhou Railway Vocational & Technical College, Zhengzhou 451460, China
* Correspondence: yzhao@shou.edu.cn (Y.Z.); yh-zhu@shou.edu.cn (Y.Z.); Tel.: +86-15692165928 (Y.Z.); +86-15000137862 (Y.Z.)
† These authors contributed equally to this work.

Abstract: Herein, a cost-effective and portable microfluidic paper-based sensor is proposed for the simultaneous and rapid detection of glucose, free amino acids, and vitamin C in fruit. The device was constructed by embedding a poly(carboxybetaine acrylamide) (pCBAA)-modified cellulose paper chip within a hydrophobic acrylic plate. We successfully showcased the capabilities of a filter paper-based microfluidic sensor for the detection of fruit nutrients using three distinct colorimetric analyses. Within a single paper chip, we simultaneously detected glucose, free amino acids, and vitamin C in the vivid hues of cyan blue, purple, and Turnbull's blue, respectively, in three distinctive detection zones. Notably, we employed more stable silver nanoparticles for glucose detection, replacing the traditional peroxidase approach. The detection limits for glucose reached a low level of 0.049 mmol/L. Meanwhile, the detection limits for free amino acids and vitamin C were found to be 0.236 mmol/L and 0.125 mmol/L, respectively. The feasibility of the proposed sensor was validated in 13 different practical fruit samples using spectrophotometry. Cellulose paper utilizes capillary action to process trace fluids in tiny channels, and combined with pCBAA, which has superior hydrophilicity and anti-pollution properties, it greatly improves the sensitivity and practicality of paper-based sensors. Therefore, the paper-based colorimetric device is expected to provide technical support for the nutritional value assessment of fruits in the field of rapid detection.

Keywords: microfluidic paper-based sensor; nanoparticles; glucose; amino acid; vitamin C; colorimetric detection

Citation: Yin, X.; Zhao, C.; Zhao, Y.; Zhu, Y. Parallel Monitoring of Glucose, Free Amino Acids, and Vitamin C in Fruits Using a High-Throughput Paper-Based Sensor Modified with Poly(carboxybetaine acrylamide). *Biosensors* **2023**, *13*, 1001. https://doi.org/10.3390/bios13121001

Received: 24 October 2023
Revised: 18 November 2023
Accepted: 20 November 2023
Published: 28 November 2023

Copyright: © 2023 by the authors. Licensee MDPI, Basel, Switzerland. This article is an open access article distributed under the terms and conditions of the Creative Commons Attribution (CC BY) license (https://creativecommons.org/licenses/by/4.0/).

1. Introduction

Fruits, as a kind of highly nutritious food, are considered to be a significant source of carbohydrates, vitamins, amino acids, and dietary fiber in the human diet [1,2]. Consequently, they have assumed a vital role in the dietary guidelines of numerous countries, including China [3]. Gu et al. [4] demonstrated that an increased intake of fruits and vegetables was associated with a reduction in total mortality among Chinese adults. Meanwhile, there is a stronger inverse correlation between fruit intake and mortality. Yuan et al. [5] revealed that the consumption of fruits and vegetables had the potential to counteract the development of hypertension resulting from high fat intake. The abundance of nutrients found in fruits caters to various nutritional requirements of the human body. For instance, glucose intake from fruits can be readily absorbed to provide energy [6], while amino acids play a significant role in regulating bodily functions and enhancing immunity [7]. Moreover, vitamin C, known for its potent antioxidant properties, aids in eliminating excessive free radicals from the body and reducing the risk of cancer. Considering these factors, the nutritional composition of fruits has emerged as a vital criterion for consumers to consider. The detection of these nutrients has been achieved through various methods such as high-performance liquid chromatography (HPLC) [8,9], near-infrared (NIR) spectroscopy [10,11], fluorescence techniques [12–14], and liquid chromatography–tandem mass spectrometry (LC/MS) [15,16]. Although these methods offer high sensitivity and accuracy, they involve multiple intricate steps, costly equipment, and skilled personnel, rendering them unsuitable for rapid on-field fruit detection.

Miniaturization, integration, and ease of manipulation are key advantages that have propelled microfluidic devices into the spotlight within the field of application and detection [17,18]. These devices, through the incorporation of grooves or microchannels engraved onto silicon or polymer layers, effectively control fluid direction and reaction, allowing for the integration of multiple reaction steps [19]. In a significant breakthrough, Whitesides' laboratory reported the development of the first easy-fabricated microfluidic paper-based device (µPAD) setup for chemical analysis in 2007 [20]. The unique feature of this pioneering work was the use of filter paper as a substrate to create a hydrophobic/hydrophilic channel on the paper without any pump or external energy source, relying on capillary action for unpowered fluid transport. This significant breakthrough has made researchers realize that paper served as an outstanding substrate in contemporary applications that prioritize cost-effectiveness, high throughput, and portability [21]. Then, in the realm of paper-based microfluidic devices, various advancements were made in the rapid detection of fruit nutrients. For instance, Akhmad Sabarudin's research group utilized paper-based platforms to determine the average vitamin C content in four different fruits by leveraging redox reactions [22]. In a similar vein, Siriwan Teepoo et al. constructed a hydrophobic channel with polylactic acid solution to detect sugar levels in sugarcane juice [23], as well as vitamin C levels in beverages [24]. Furthermore, Manas Ranjan Gartia et al. [25] implemented wax printing on Whatman Paper Grade No.1 to analyze the composition and content of fruit juices. Nevertheless, these studies failed to address the limitations of paper in terms of analysis and detection, such as its limited color development effect and inadequate anti-pollution ability. Consequently, it is necessary to incorporate additional materials to adjust and enhance the sensitivity and stability of the paper substrate.

Zwitterionic polymers have exhibited exceptional properties including high hydrophilicity, long-term durability, resistance to fouling, and environmental stability [26]. Due to their exceptional properties, zwitterionic polymers are grafted onto diverse inorganic/organic surfaces using "graft-from" and "graft-to" strategies, post-zwitterionization, and surface grafting copolymerization methods to fulfill specific application requirements [27]. The negatively charged membranes of Wang et al. were modified by a positively charged zwitterion copolymer through surface adsorption and a cross-linking reaction, which significantly improved the hydrophilicity and antifouling properties of the film surface [28]. Fu et al. enhanced the microfluidic sensor device by incorporating a superhydrophilic zwitterionic polymer. The prepared microfluidic sweat sensor presented exceptional wettability and excellent infusion capacity after modification [29].

In this research, a microfluidic paper-based detection platform was constructed via using an acrylic plate as a hydrophobic channel and grafting cellulose filter (CF) with the zwitterionic polymer pCBAA to create a hydrophilic channel. By integrating colorimetry, this platform enabled rapid and simultaneous identification and quantification of three nutrients in fruits with high throughput and low cost. The three analytes underwent enzyme catalysis, ninhydrin reaction, and redox reactions individually within three detection zones on the paper base, resulting in the exhibition of three distinct colors. The concentration of each analyte was then calculated by evaluating the average color intensity of the corresponding detection zone using Image J 1.51 software. Additionally, the pCBAA-µPAD successfully determined glucose, free amino acids, and vitamin C in various fruit samples. The reliability was verified through spectral analysis simultaneously.

2. Materials and Methods

2.1. Materials and Instruments

Whatman filter paper No. 1 (150 mm diameter) was purchased from Whatman International Ltd. (Shanghai, China). Glucose, vitamin C, and leucine were provided by Sinopharm Chemical Reagent Co., Ltd. (Shanghai, China). β-propiolactone (95%), cuprous(I) bromide (CuBr, 98%), 11-hydroxy-1-undecanethiol ($C_{11}H_{24}OS$), 2-bromoisobutyryl bromide (BIBB, 98%), 2,2′-bipyridine (BPY), bromoisobutyryl bromide ($C_4H_6Br_2O$), triethylamine (TEA, 99%), tetrahydrofuran (THF, HPLC grade), chitosan (deacetylation degree > 92%), glucose oxidase (GOx, ≥ 100 U mg^{-1}), ninhydrin, silver nitrate, and 3,3,5,5-tetramethylbenzidine (TMB) were purchased from Sigma-Aldrich (St. Louis, MO, USA). Ethanol (CH_3CH_2OH), hydrochloric acid (HCl), ferric chloride ($FeCl_3$), sodium borohydride ($NaBH_4$), potassium ferricyanide ($K_3[Fe(CN)_6]$), acetone (CH_3COCH_3), acetic acid, and sodium acetate were obtained from Aladdin Industrial Inc. (Shanghai, China). Phosphate-buffered saline (PBS) was provided by Solarbio (Beijing, China). All chemical reagents utilized were compounded by ultrapure water without further purification.

2.2. Instruments

The XPS spectra were identified by X-ray photoelectron spectroscopy (XPS, Thermo Scientific EscaLab 250Xi, Waltham, MA, USA). All FT-IR measurement was performed on an FT-IR spectrometer (Thermo Scientific, Waltham, MA, USA). Transmission electron microscopy (TEM) images were captured on a Tecnai G2 T20 electron microscope at 200 kV. The UV–vis spectrum and absorbance were measured using a UV–vis absorption spectrophotometer (Shimadzu Corporation, UV-vis-2550, Kyoto, Japan).

2.3. Design and Fabrication of the pCBAA-µPAD

The snowflake-shaped pattern, which served as the fundamental design for the µPAD, was created using AutoCAD. The detection zones of this pattern were strategically positioned at the six vertices of a regular hexagon. Each detection area was comprised of a circle with a diameter of 5 mm. Figure 1a illustrate the pCBAA-µPAD, which consisted of two components. Firstly, a hydrophilic paper substrate and a hydrophobic spacer. In the first step, a laser cutting machine was used to carve multiple snowflake-shaped grooves on an acrylic plate composed of poly(methylmethacrylate) (PMMA). This PMMA acrylic plate naturally served as a hydrophobic spacer for the paper-based microfluidic device. Subsequently, in the second step, snowflake-shaped patterns of the same size were laser cut on Whatman filter paper No. 1. Through the surface-initiated atom transfer radical polymerization (SI-ATRP) technique, the synthesized CBAA was successfully grafted onto the cellulose surface, resulting in a hydrophilic paper substrate. The detailed process of the grafting reaction is presented in Figure S1. Finally, the hydrophobic plate and hydrophilic paper substrate were assembled to create a high-throughput pCBAA-µPAD.

2.4. Synthesis of Chitosan-Stabilized Silver Nanoparticles

The preparation of chitosan-stabilized silver nanoparticles (Ch-Ag NPs) was carried out following the method previously described by Yang et al. [30]. As presented in Figure 1b, chitosan (2 mg/mL) was dissolved in a 1% acetic acid solution (200 mL) and stirred for 6 h, followed by filtration through a 0.22 µm microporous filter. Then, 5 mL of 0.01 mol/L silver nitrate solution was mixed with 127.5 mL of the chitosan solution and stirred for 30 min. Then, 2.5 mL of 0.1 mol/L $NaBH_4$ solution was rapidly added, and the mixture was further stirred for 90 min. The resulting Ch-Ag NPs exhibited a concentration of 40 mg/L (as Ag) and were stored in a refrigerator at 4 °C away from light.

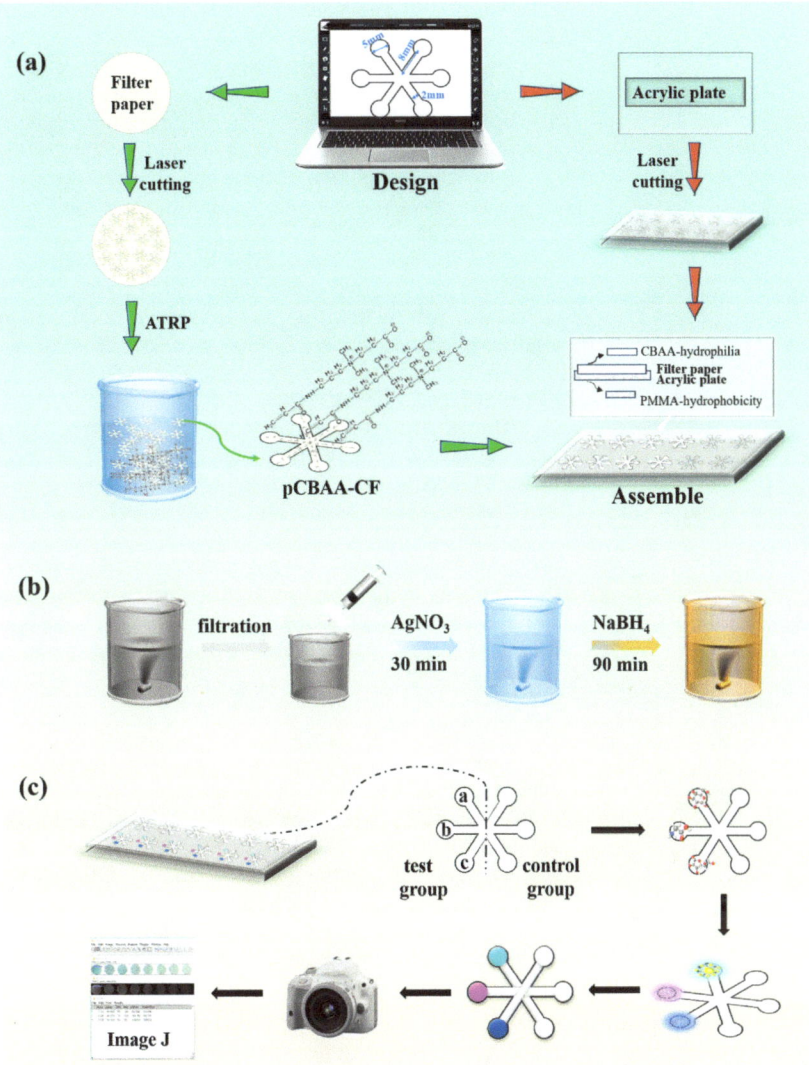

Figure 1. (**a**) Design and production process using a functional poly(carboxybetaine acrylamide)-coated μPAD; (**b**) synthesis of chitosan-stabilized silver nanoparticles; (**c**) the reaction scheme of the pCBAA-μPAD for glucose, vitamin C, and amino acid detection.

2.5. Simultaneous Colorimetric Detection of Glucose, Free Amino Acids, and Vitamin C

The chromogenic agent in glucose detection was prepared using acetic acid–sodium acetate solution, TMB solution, and Ch-Ag NPs. As shown in Figure 1c, in the detection area a, 1 μL of glucose standard solution and 1 μL of glucose oxidase were first added, followed by 1μL of the configured color developer mixture. The sealing film covered the surface of the sensor and was placed in the oven at 37 °C until the color development completed. Several effective parameters, such as the proportion of each solution, concentration of Ch-Ag NPs, pH value, and reaction time were investigated. For the detection of free amino acids, 1 μL of leucine standard solution, 1 μL of acetic acid buffer solution and 1 μL of ninhydrin were added to the detection area b successively and placed in the oven at 60 °C. The parameters of pH value, concentration, temperature, and time were optimized experimentally. The reactive chromogenic agent of vitamin C was composed of $FeCl_3$, $K_3[Fe(CN)_6]$, and HCl. Next, 1 μL of vitamin C standard solution and 1 μL of chromogenic agent were added to the detection area c successively, and a series of adjustments were made to the concentration and proportion of each solution in the mixed chromogenic agent. The optimal detection scheme after optimization of the three target detection objects is described in detail in the Supplementary Material.

After the completion of three colorimetric reactions, image acquisition (presented in the Materials and Methods section of the Supplementary Material) was performed promptly. The acquired images were analyzed and compared for color intensity using Image J 1.51 software. This analysis established a linear relationship between the concentrations of glucose, free amino acids, and vitamin C and the average relative intensity of grayscale. This relationship was subsequently utilized for qualitative and quantitative analysis of actual samples.

2.6. Sample Preparation

Thirteen fruit varieties were purchased from the local market and used as test objects to assess the performance of the sensor. The fruits were subjected to pretreatment based on the Chinese National Standard system (GB 5009.8-2016 and GB 5009.86-2016). The fruit was washed, dried, peeled, and cored, then cut into small pieces. Approximately 30 g of the fruit sample was weighed for homogenization. The homogenate was subjected to ultrasonic extraction in an ice bath for 5 min, followed by centrifugation at 6000 rpm for 15 min to obtain the supernatant. The extracted supernatant was further filtered using a sterile microporous ultrafiltration membrane (0.22 μm) and stored in a refrigerator at 4 °C for future use. All fruit samples used in the experiment were freshly purchased.

3. Results

3.1. FT-IR and XPS Analysis

The grafting results on the paper substrate were confirmed using FT-IR and XPS analysis. As depicted in Figure 2a, the bare-CF spectrum exhibited characteristic peaks of cellulose paper, including the broad stretching vibration of -OH at 3337 cm^{-1}, the skeletal vibration of -CH_2- at 2900 cm^{-1}, and the skeletal stretching vibrations of C-O at 1055 cm^{-1} and 1026 cm^{-1} [31]. In the FT-IR spectrum of pCBAA-CF, characteristic absorption peaks of pCBAA were observed in addition to the presence of characteristic peaks of cellulose paper. These included the vibrational stretching of the C=O bond at 1662 cm^{-1} [32] and the asymmetric stretching of the COO^- group at 1585 cm^{-1} [33].

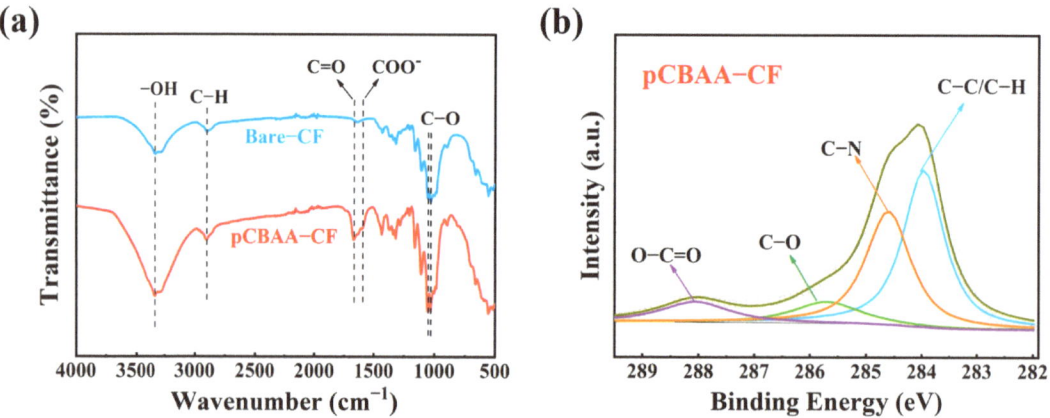

Figure 2. (**a**) The FT−IR spectra of bare−CF and pCBAA−CF; (**b**) representative XPS high−resolution C1s spectra of pCBAA−CF.

As shown in Figure S2, pCBAA-CF revealed the presence of carbon and oxygen elements, similar to bare-CF, as well as the detection of nitrogen element specific to pCBAA. The high-resolution XPS C 1s spectrum of bare-CF was fitted with two peaks centered at 284.6 and 286.3 eV, corresponding to C-C/C-H and C-O bonds [31], respectively (Figure S3). In Figure 2b, besides the binding energy observed in bare-CF, pCBAA-CF displayed two additional signals at 285.2 and 288.6 eV, respectively assigned to C-N and O-C=O groups [34]. Thus, all above results confirm that the pCBAA was successfully grafted onto bare-CF using the ATRP polymerization method.

3.2. Properties Characterization of pCBAA-μPAD

To verify the hydrophilic performance of pCBAA-μPADs, an equal amount of red ink was simultaneously dropped onto the surface of pre-modified and post-modified paper substrates. Then, the liquid flow time on both surfaces was compared. As shown in Figure 3a, the red ink on pCBAA-CF completely covered the entire paper platform within 20 s, while on the unmodified bare-CF, it did not reach the full coverage. This demonstrated that pCBAA significantly increased the hydrophilicity of the paper substrate, accelerating the capillary action of the detection liquid and thus shortening the detection time. The colorimetric sensing performance of the pCBAA-μPAD is further revealed in Figure 3b. While the concentrations of glucose, amino acids, and vitamin C increased, the corresponding color depth and grayscale values gradually increased. It was confirmed that pCBAA-functionalized paper-based platforms exhibited significant detection performance for various target nutrients.

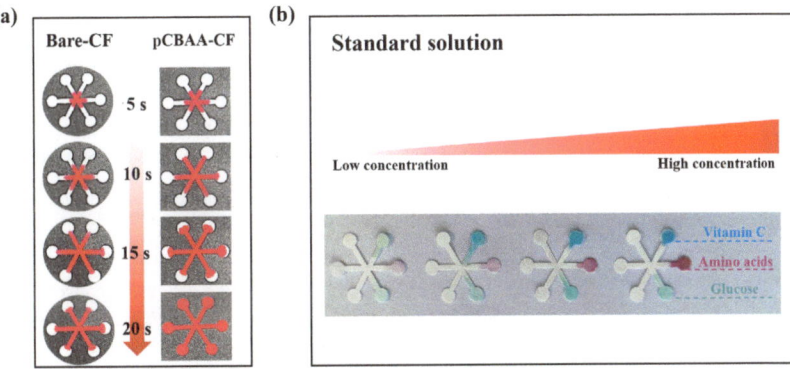

Figure 3. (**a**) Time dependence of red ink flow rate on bare-CF and pCBAA-CF; (**b**) gradient sensing performance of glucose, amino acids, and vitamin C on pCBAA-CF.

3.3. Detection of Glucose

The detection of glucose was achieved through the catalytic activity of chitosan–silver nanoparticle (Ch-Ag NP) peroxidase. Glucose was first decomposed by glucose oxidase into hydrogen peroxide (H_2O_2) and gluconic acid. H_2O_2 molecules adsorb on Ch-Ag NP surfaces, generating hydroxyl radicals (·OH). Hydroxyl radicals reacted with colorless TMB, resulting in the formation of the TMB diamine oxidized in a blue form, as shown in Figure 4a [35,36]. Ch-Ag NPs exhibited excellent catalytic ability in the presence of H_2O_2, and the presence of chitosan increased the stability and storage of silver nanoparticles [37]. In order to demonstrate the successful combination of chitosan and silver nanoparticles, a comparison of FT-IR analysis was conducted for chitosan and Ch-Ag NPs. As shown in Figure 4b, characteristic peaks of chitosan appeared in Ch-Ag NPs, with the peak around 1649 cm^{-1} attributed to the stretching vibration of the C=O bond (amide I), and the peak at 1559 cm^{-1} was enhanced with the addition of chitosan [38]. The synthesized Ch-Ag NPs were spherical with a diameter of approximately 12 nm (Figure S4).

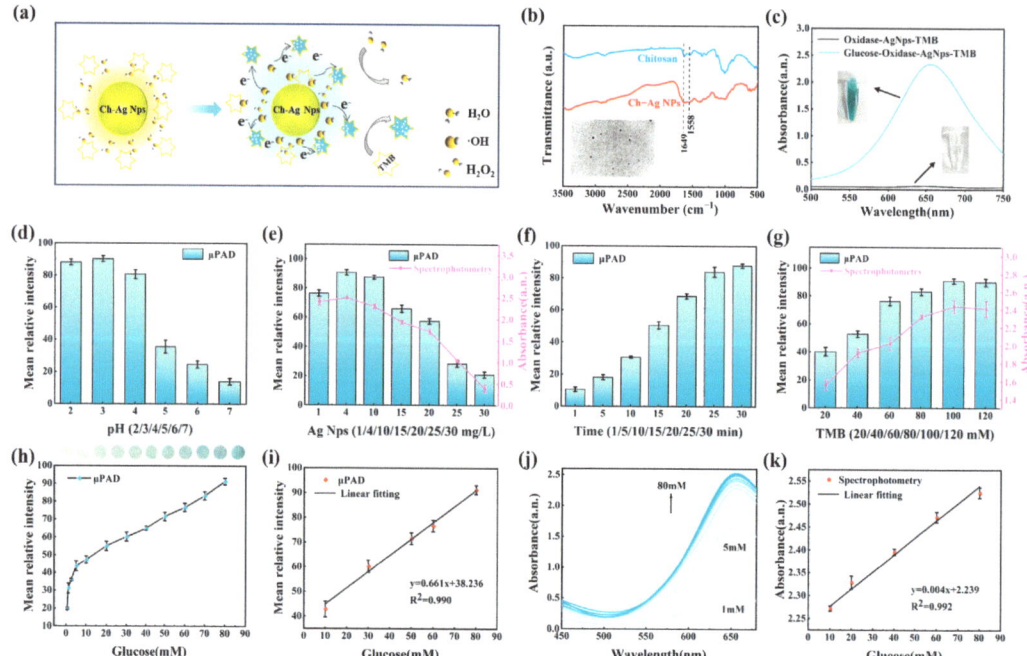

Figure 4. (**a**) Principle of glucose detection using Ch-Ag NPs; (**b**) FTIR spectrum of Ch-Ag NPs and chitosan; (**c**) UV absorbance of experimental group and blank control group; (**d**) effect of pH on detection method; (**e**) effect of Ch-Ag NP concentration on enzymatic activity; (**f**) effect of time on reaction progress; (**g**) effect of TMB concentration on colorimetric reaction; (**h**) effect of different glucose concentrations on average grayscale intensity; (**i**) linear relationship between glucose concentrations and average grayscale intensity; (**j**) effect of different glucose concentrations on UV absorbance; (**k**) correlation between glucose concentration and UV absorbance.

The cyan blue product showed a maximum absorption peak at 650 nm. In contrast, no significant color change was observed in the blank reaction system without the addition of glucose (Figure 4c). The influence of different pH environments on catalytic reaction was firstly investigated through grayscale intensity analysis. Ch-Ag NPs presented strong activity within a relatively wide pH range of 2–4. Then, pH = 3 was selected as the optimal reaction environment for subsequent catalytic reaction kinetics (Figure 4d). Subsequently, the effect of Ch-Ag NP concentration on peroxidase activity was explored. As observed in Figure 4e, the grayscale intensity was significantly higher when Ch-Ag NPs were diluted in acetic acid buffer at concentrations of 4–10 mg/L. Considering the characteristic absorption at 650 nm, Ch-Ag NPs exhibited the strongest enzymatic activity at a concentration of 4 mg/L. The catalytic reaction was time-dependent, as shown in Figure 4f, where the production of oxidized TMB (oxTMB) and the corresponding grayscale intensity increased with prolonged reaction time. After 30 min, the average grayscale intensity of oxTMB tended to stabilize. Similarly, the concentration of TMB also affected the colorimetric reaction. TMB concentrations between 60 and 120 mmol/L demonstrated higher grayscale intensity and UV absorbance (Figure 4g). A TMB concentration of 100 mmol/L was chosen for the preparation of the reaction color reagent. Under the optimal conditions, as revealed in Figure 4h,i, there was a good linear relationship (y = 0.661x + 38.236, R^2 = 0.990) between glucose concentration and the average grayscale intensity of the pCBAA-µPAD within the range of 10–80 mmol/L. Meanwhile, the detection limit (*LOD*) was as low as 0.049 mmol/L.

$$LOD = \frac{3\sigma}{S} \quad (1)$$

The value of *LOD* is calculated using Equation (1), where σ represents the standard deviation of the blank sample ($n = 11$), and *S* refers to the slope of the fitted standard curve. A comparative experiment using spectrophotometry showed a good linear equation (y = 0.004x + 2.239, R^2 = 0.992) for the data collected at the characteristic peak of 650 nm (Figure 4j,k). These results demonstrated that there was good synchronization between the two methods. Thus, the pCBAA-μPAD offers the advantages of a microfluidic device, providing a more convenient and rapid approach for glucose detection.

3.4. Detection of Free Amino Acids

The reaction principle between free amino acids and ninhydrin was depicted in Figure 5a. Amino acids produced carbon dioxide, ammonia, and aldehyde by oxidation, while ninhydrin hydrate was reduced to its reduced form. Subsequently, the generated ammonia and reductive ninhydrin with another molecule of ninhydrin hydrated, forming a purple compound [39]. This purple compound exhibited maximum absorbance at 570 nm, and the UV absorbance of the blank control group without amino acids remained almost unchanged (Figure 5b). To evaluate the selectivity of ninhydrin for amino acid detection on the pCBAA-μPAD, potential interfering compounds such as glucose, sucrose, fructose, and vitamin C were introduced in the sample zone. All experiments were conducted with the same concentration of 30 mmol/L. The results in Figure 5c show that there was no significant change in the grayscale intensities of all interfering compounds, while the average intensity of amino acids was considerably high. Consequently, the developed pCBAA-μPAD presented high specificity for free amino acids.

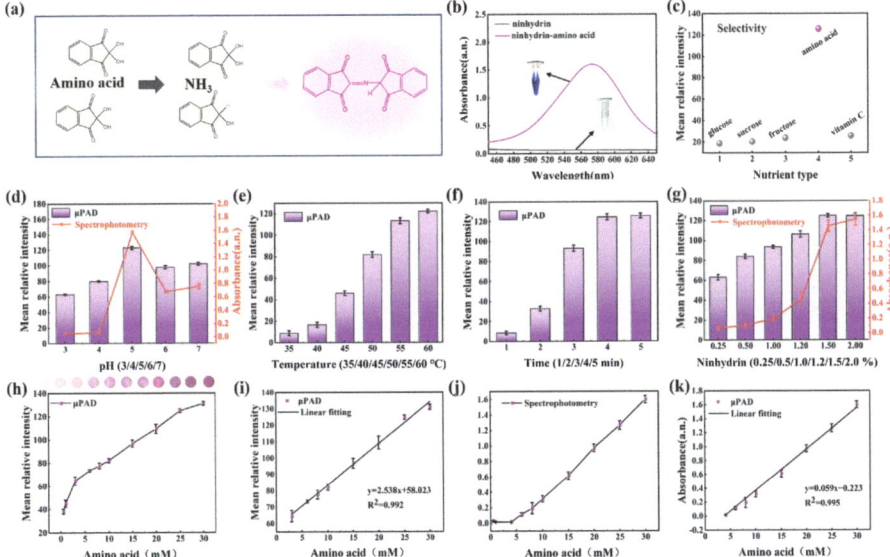

Figure 5. (**a**) Principle of free amino acid detection; (**b**) UV absorbance of experimental group and blank control group; (**c**) selective testing of detection method; (**d**) effect of pH on detection method; (**e**) effect of temperature on color reaction; (**f**) effect of time on reaction progress; (**g**) effect of ninhydrin concentration on colorimetric reaction; (**h**) effect of different amino acid concentrations on average grayscale intensity; (**i**) linear relationship between amino acid concentrations and average grayscale intensity; (**j**) effect of different amino acid concentrations on UV absorbance; (**k**) correlation between amino acid concentration and UV absorbance.

The pH, temperature, time, and concentration of ninhydrin on the pCBAA-μPAD were carefully optimized to achieve the optimal color reaction. Adjustments were made to the pH value, ranging from 3 to 7, in order to determine the ideal solution environment for achieving the best coloring effect. The results, as revealed in Figure 5d, indicated that the average relative intensity on the paper-based platform reached its peak at pH = 5. The impact of temperature on the reaction was also explored. As depicted in Figure 5e, temperatures below 40 °C led to minimal purple color formation and insignificant grayscale intensity. To minimize reaction time, the optimal reaction temperature was determined to be 60 °C, with the color of the system stabilizing after 5 min (Figure 5f). Ninhydrin, a key component of the color reaction, had its optimal concentration determined by analyzing grayscale intensity and UV absorbance, as illustrated in Figure 5g. The average relative intensity displayed an increasing trend with rising ninhydrin concentration, stabilizing at 1.5%. Consequently, all the aforementioned optimized conditions were utilized in pretreating the detection area of the pCBAA-µPAD, as demonstrated in Figure 5h,i. The relationship between amino acid concentration and average grayscale intensity exhibited good linearity within the range of 3–30 mmol/L, with a correlation equation of y = 2.538x + 58.023 and a correlation coefficient of R^2 = 0.992. The low detection limit was determined to be 0.236 mmol/L, which was calculated by Equation (1). Furthermore, the method was validated using UV spectrophotometry. It is revealed in Figure 5j that there is a positive linear relationship between absorbance at 570 nm and amino acid concentration (y = 0.059x − 0.223, R^2 = 0.995). The consistency between the outcomes obtained from the pCBAA-µPAD and spectrophotometry underscores the vast potential value of paper-based microfluidics in the analysis and detection of fruit nutrients.

3.5. Detection of Vitamin C

The detection principle of vitamin C was based on its strong reducing ability. As shown in Figure 6a, the Fe^{3+} in $FeCl_3$ was reduced to Fe^{2+}, and a specific color reaction between potassium ferricyanide and divalent iron formed a coordination compound called Turnbull's blue [40]. The quantity of Turnbull's blue generated was reflected by the grayscale intensity, establishing the relationship between the concentration of vitamin C and the average grayscale intensity. From the ultraviolet absorbance values presented in Figure 6b, it was observed that the color of Turnbull's blue in the experimental group was more intense compared to the blank group without the addition of vitamin C, providing a solid foundation for colorimetric sensing on the pCBAA-µPAD. Given the diverse range of nutrients in fruits, it was essential to study the selectivity of this reaction. As observed in Figure 6c, in the presence of various interfering substances in fruits, only vitamin C exhibited strong coloration and the highest grayscale intensity. This specificity was an important requirement for analysis and detection using the pCBAA-µPAD.

In order to achieve a stable response of the pCBAA-µPAD, several important parameters were investigated in the experiment. Due to the rapid reaction between divalent iron and potassium ferricyanide, as shown in Figure 6d, the reaction was completed in almost one second and reached a steady state within 10 s. The colorimetric reagent was prepared from $FeCl_3$, $K_3[Fe(CN)_6]$, and HCl, so the concentrations of these three reagents were carefully optimized. $FeCl_3$ served as a key mediator in this reaction, and optimization was conducted based on both the average grayscale intensity and the ultraviolet absorbance, as shown in Figure 6e. From the results displayed on the pCBAA-µPAD, $FeCl_3$ revealed good grayscale values at concentrations above 60 mmol/L. Combining with UV spectrophotometry, 100 mmol/L $FeCl_3$ was chosen for the preparation of the colorimetric reagent mixture. Another important parameter, $K_3[Fe(CN)_6]$, when dissolved in deionized water, produced a yellow solution. If the concentration was too high, it could hinder the grayscale information processing of Turnbull's blue. Therefore, a concentration of 0.5% $K_3[Fe(CN)_6]$ was selected for the preparation of the colorimetric reagent mixture (Figure 6f). Furthermore, the results for HCl with concentrations ranging from 0 to 12 mol/L indicated that a strong acid environment promoted the progress of the reaction. Therefore, 12 mol/L HCl was chosen as the optimal condition (Figure 6g). Ultimately, the colorimetric reagent was prepared by combining 100 mmol/L $FeCl_3$, 0.5% $K_3[Fe(CN)_6]$ and 12 mol/L HCl in a 3:2:1 ratio. When vitamin C interacted with the colorimetric reagent on the pCBAA-µPAD, the grayscale intensity enhanced with the increase in vitamin C concentration, showing a good linear response within the range of 1–10 mmol/L. The correlation equation was y = 4.095x + 95.734, with an R^2 value of 0.990 (Figure 6h,i). According to Equation (1), the detection limit for vitamin C was determined to be 0.125 mmol/L. Similarly, by collecting absorbance data of Turnbull's blue at 500 nm, a good linear equation was established (y = 0.082x + 1.293, R^2 = 0.988) (Figure 6j,k). These results demonstrated that the pCBAA-μPAD exhibited excellent sensing performance and has good detection capabilities for vitamin C.

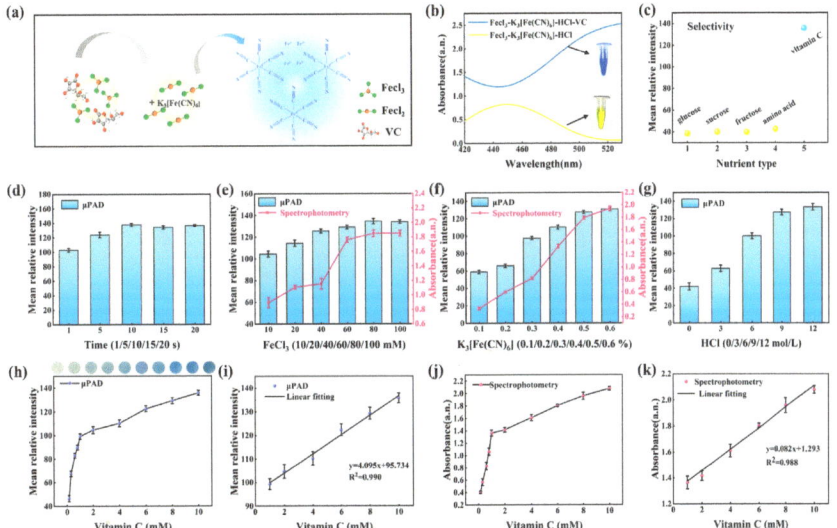

Figure 6. (**a**) Principle of vitamin C detection; (**b**) UV absorbance of experimental group and blank control group; (**c**) selective testing of detection method; (**d**) effect of time on reaction progress. (**e**) Effect of $FeCl_3$ concentration on detection method; (**f**) effect of $K_3[Fe(CN)_6]$ concentration on colorimetric reaction; (**g**) effect of HCl concentration on detection method; (**h**) effect of different vitamin C concentrations on average grayscale intensity; (**i**) linear relationship between vitamin C concentrations and average grayscale intensity; (**j**) effect of different vitamin C concentrations on UV absorbance; (**k**) correlation between vitamin C concentration and UV absorbance.

3.6. Real Samples Study

The detection performance of the pCBAA-μPAD was further evaluated by analyzing the nutrient content in 13 common real fruit samples (such as peach, grape, pear, and kiwi). The obtained results were compared with those obtained using UV spectrophotometry. From Tables 1–3, it was observed that there was no difference between the concentrations of glucose, free amino acids, and vitamin C obtained by the pCBAA-μPAD and those obtained by UV spectrophotometry. In order to verify the correlation between the two methods, regression analysis was conducted on each group of test results, as shown in Figure 7. UV spectrophotometry was positively correlated with the μPAD in glucose with r = 0.970, where the regression equation was significant with R^2 = 0.941 and $p < 0.001$. Meanwhile, UV spectrophotometry was positively correlated with the μPAD in amino acids and vitamin C, and the regression equation was significant with R^2 values of 0.989 and 0.986, respectively. Similarly, the p value of both analyses was less than 0.001. Thus, the feasibility of the paper-based sensing device for rapid detection of fruit nutrients was confirmed. Additionally, compared to UV spectrophotometry, the proposed pCBAA-μPAD required less analysis time and reagent consumption, which were also enabling efficient detection of three nutrients simultaneously. These results revealed that the sensors had excellent advantages and held great value for routine testing of fruit products.

Table 1. The detection of glucose in fruit samples with pCBAA-μPAD and UV–visible spectrophotometry.

Samples	Glucose (mg/100 g)		RSD (%) (n = 3)	
	μPAD	UV–vis Spectrum	μPAD	UV–vis Spectrum
Yellow peach	762.284	756.668	2.6	2.5
Xinyi peach	646.173	640.763	2.1	3.0
Longquan peach	928.909	950.763	3.1	3.1
Yangshan peach	752.196	719.653	3.9	2.5
Jasmine grapes	1053.776	996.987	2.6	2.9
Sunshine grapes	1383.334	1406.763	2.9	2.5
Kiwi fruit	1574.912	1489.866	3.5	2.9
Litchi	1077.419	1165.763	3.1	2.3
Sugar pear	1083.361	1134.862	2.9	2.5
Crown pear	1115.616	1198.765	2.9	3.1
Su Crisp pear	1016.362	1068.762	2.2	2.9
Mangosteen	1094.019	999.605	2.1	2.4
Longan	1051.619	1110.384	3.2	3.1

Table 2. The detection of free amino acids in fruit samples with pCBAA-μPAD and UV–visible spectrophotometry.

Samples	Amino Acids (mg/100 g)		RSD (%) (n = 3)	
	μPAD	UV–vis Spectrum	μPAD	UV–vis Spectrum
Yellow peach	165.217	178.753	2.9	3.4
Xinyi peach	41.425	32.7527	4.5	2.9
Longquan peach	133.072	148.763	3.9	4.2
Yangshan peach	120.271	110.763	2.9	2.7
Jasmine grapes	144.995	136.7652	2.4	3.8
Sunshine grapes	147.597	156.7573	3.5	2.7
Kiwi fruit	51.811	67.753	4.3	3.5
Litchi	357.854	329.7653	3.8	2.3
Sugar pear	82.969	73.762	2.5	2.6
Crown pear	82.969	72.656	3.4	4.1
Su Crisp pear	93.355	86.767	4.2	4.3
Mangosteen	117.103	126.763	3.7	2.9
Longan	530.310	553.763	3.9	3.7

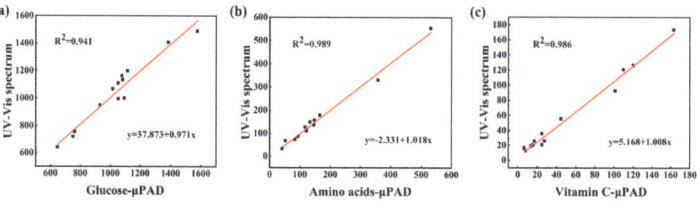

Figure 7. Linear regression analysis of μPAD and UV spectrophotometry. (**a**) Glucose; (**b**) amino acids; (**c**) vitamin C.

Table 3. The detection of vitamin C in fruit samples with pCBAA-µPAD and UV–visible spectrophotometry.

Samples	Vitamin C (mg/100 g)		RSD (%) (n = 3)	
	µPAD	UV–vis Spectrum	µPAD	UV–vis Spectrum
Yellow peach	24.810	20.876	3.5	2.7
Xinyi peach	101.035	92.878	4.2	3.8
Longquan peach	15.451	20.863	2.6	3.5
Yangshan peach	44.363	55.864	3.8	3.2
Jasmine grapes	24.6115	35.763	4.1	3.1
Sunshine grapes	7.608	12.733	2.6	2.3
Kiwi fruit	162.976	173.763	4.3	2.6
Litchi	119.783	126.733	2.5	2.4
Sugar pear	6.386	16.733	4.2	2.8
Crown pear	13.481	18.873	3.5	3.1
Su Crisp pear	17.147	25.763	3.6	4.0
Mangosteen	27.532	25.863	2.9	2.5
Longan	109.637	120.863	3.8	2.8

4. Discussion

In recent years, numerous research groups have been actively searching for new breakthroughs in the detection methods of glucose, amino acids, and vitamin C. As represented in Table 4, there are comparative analyses of these methods. Filiz et al. [41] designed a stable electrospun nanofiber composed of chitosan (CS) and polyvinyl alcohol (PVA). They utilized UV–visible spectrophotometry for colorimetric detection of glucose in aqueous media. This method exhibited strong stability, but its accuracy needed further improvement. Georgelis et al. [42] combined modern HPLC systems with high-precision mass spectrometers (HPLC-MS) to rapidly determine multiple sugars in mature potato tubers and strawberry fruits. This approach offered high accuracy and sensitivity. However, the equipment cost was high, and the operation was complex. Su et al. [43] developed three methods for the determination of small molecule carbohydrates in jujube extracts: high-performance liquid chromatography–evaporative light scattering detection (HPLC-ELSD), liquid chromatography–electrospray ionization tandem mass spectrometry (LC-ESI-MS/MS), and gas chromatography–mass spectrometry (GC-MS). HPLC-ELSD and LC-ESI-MS/MS presented high accuracy, which was suitable for quantitative analysis, but they required a skilled operator for assessment. GC-MS is more suitable for qualitative analysis. The detection of amino acids mainly relies on UV–vis spectrophotometry [44], near-infrared (NIR) [45], and HPLC-MS [46] methods. According to these studies, UV–vis spectrophotometry was simple to operate but has moderate accuracy. NIR was fast but not suitable for dispersed sample systems. The HPLC-MS combination offered high sensitivity but was not operationally convenient. Shrivas [47] developed reverse-phase high-performance liquid chromatography (RP-HPLC) with a diode array detector (DAD) for vitamin detection. Paper spray mass spectrometry (PS-MS) [48] and electroanalysis [49] were effective methods for detecting vitamin C content.

Table 4. Comparison of detection methods for glucose, amino acids, and vitamin C.

Method	Target	LOD	Characteristic	Path	Ref.
UV spectrophotometry	Glucose	2.70 mM	Strong stability, insufficient accuracy	CS/PVA	[41]
HPLC/MS		0.10 ng	High accuracy, complex operation	-	[42]
HPLC-ELSD		1.03 µg/mL	Good repeatability, moderate sensitivity		
LC–ESI–MS/MS		0.01 µg/mL	High sensitivity, high accuracy	-	[43]
GC–MS		0.65 µg/mL	Moderate accuracy, qualitative analysis		
UV spectrophotometry	Amino acids	0.15 µM	Strong stability, insufficient accuracy	AgNPs	[44]
NIR		52 nM	High sensitivity, simple operation.	-	[45]
HPLC-FLD-MS/MS		0.13–1.13 nM	High sensitivity, expensive instrument	-	[46]
RP-HPLC	Vitamin C	0.1 µg/mL	High accuracy, complex operation	DAD	[47]
PS-MS		0.3 µg/mL	Moderate accuracy, short duration.	-	[48]
Electroanalysis		0.067 µM	High accuracy, high sensitivity	SO_2NPs	[49]
pCBAA-µPAD	Three analytes	0.049/0.236/0.125 mM	High accuracy, portable, simple operation	Paper sensor	This work

In comparison to the various detection methods mentioned above, the as-prepared pCBAA-µPAD in this study enabled parallel detection of glucose, amino acids, and vitamin C. The advantages of this method included high efficiency, good selectivity, extremely low cost, strong portability, simple operation, and short duration. Thus, it is suitable for real-time on-site testing, which can complement the inconveniences of large analytical instruments.

5. Conclusions

In this study, a simple and rapid paper-based sensing strategy was developed for the simultaneous determination of glucose, free amino acids, and vitamin C in fruits. The fabrication of the pCBAA-µPAD was cost-effective and easily assembled without external devices. Detection of the three analytes was achieved through colorimetric reactions using corresponding selective substrates on the paper-based platform. After optimizing the experimental conditions, the detection limits for glucose, free amino acids, and vitamin C on the paper-based sensing platform were determined to be 0.049 mmol/L, 0.236 mmol/L, and 0.125 mmol/L, respectively. Through comparison between the evaluation of the nutritional composition from actual fruit samples by the paper-based platform and UV spectrophotometry, the feasibility of the method was validated. These findings reveal the outstanding application potential of the developed paper-based sensor for rapid and routine analysis of nutrients in fruit products, which is closely related to fruit quality control.

Supplementary Materials: The following supporting information can be downloaded at: https://www.mdpi.com/article/10.3390/bios13121001/s1. Figure S1: Reaction scheme for grafting pCBAA onto CF [50,51]; Figure S2: The XPS spectra of bare-CF and pCBAA-CF; Figure S3: Representative XPS high-resolution C1s spectra of bare-CF; Figure S4: TEM image of Ch-Ag NPs; Equation (S1): $\Delta R = \overline{R}_{after} - \overline{R}_{before}$; Equation (S2): $\Delta G = \overline{G}_{after} - \overline{G}_{before}$; Equation (S3): $\Delta B = \overline{B}_{after} - \overline{B}_{before}$; Equation (S4): $\Delta Gray = 0.30\Delta R + 0.59\Delta G + 0.11\Delta B$.

Author Contributions: Conceptualization, Y.Z. (Yong Zhao) and Y.Z. (Yongheng Zhu); methodology, C.Z. and X.Y.; software, X.Y.; validation, X.Y.; formal analysis, C.Z.; investigation, X.Y. and C.Z.; resources, Y.Z. (Yong Zhao) and Y.Z. (Yongheng Zhu); data curation, X.Y. and C.Z.; writing—original draft preparation, X.Y. and C.Z.; writing—review and editing, Y.Z. (Yongheng Zhu), X.Y. and C.Z.; supervision, Y.Z. (Yong Zhao); project administration, Y.Z. (Yongheng Zhu); funding acquisition, Y.Z. (Yong Zhao) and Y.Z. (Yongheng Zhu). All authors have read and agreed to the published version of the manuscript.

Funding: This research was funded by the Key Research Projects of Science and Technology for Agriculture of Shanghai (2021-02-08-00-12-F00763) from Shanghai Agriculture and Rural Committee.

Institutional Review Board Statement: Not applicable.

Informed Consent Statement: Not applicable.

Data Availability Statement: The data presented in this study are available on request from the corresponding author.

Acknowledgments: Thanks to Zhaohuan Zhang and Yingjie Pan of Shanghai Ocean University and Weiyi Zhang of Shanghai Agricultural Product Quality and Safety Center for their contributions to this article.

Conflicts of Interest: The authors declare no conflict of interest.

References

1. Sui Kiat, C.; Alasalvar, C.; Shahidi, F. Superfruits: Phytochemicals, antioxidant efficacies, and health effects-a comprehensive review. *Crit. Rev. Food Sci. Nutr.* **2019**, *59*, 1580–1604.
2. Ho, K.; Ferruzzi, M.G.; Wightman, J.D. Potential health benefits of (poly)phenols derived from fruit and 100% fruit juice. *Nutr. Rev.* **2020**, *78*, 145–174. [CrossRef] [PubMed]
3. Mason-D'Croz, D.; Bogard, J.R.; Sulser, T.B.; Cenacchi, N.; Dunston, S.; Herrero, M.; Wiebe, K. Gaps between fruit and vegetable production, demand, and recommended consumption at global and national levels: An integrated modelling study. *Lancet Planet. Health* **2019**, *3*, E318–E329. [CrossRef]
4. Gu, Y.X.; He, Y.S.; Ali, S.H.; Harper, K.; Dong, H.J.; Gittelsohn, J. Fruit and vegetable intake and all-cause mortality in a Chinese population: The China health and nutrition survey. *Int. J. Environ. Res. Public Health* **2021**, *18*, 342. [CrossRef] [PubMed]
5. Yuan, S.; Yu, H.J.; Liu, M.W.; Tang, B.W.; Zhang, J.; Gasevic, D.; Larsson, S.C.; He, Q.Q. Fat intake and hypertension among adults in China: The modifying effects of fruit and vegetable intake. *Am. J. Prev. Med.* **2020**, *58*, 294–301. [CrossRef] [PubMed]
6. Lin, P.H.; Sheu, S.C.; Chen, C.W.; Huang, S.C.; Li, B.R. Wearable hydrogel patch with noninvasive, electrochemical glucose sensor for natural sweat detection. *Talanta* **2022**, *241*, 11. [CrossRef]
7. Lutt, N.; Brunkard, J.O. Amino acid signaling for TOR in eukaryotes: Sensors, transducers, and a sustainable agricultural fuTORe. *Biomolecules* **2022**, *12*, 387. [CrossRef]
8. Shen, S.; Du, L.N.; Hu, X.B. Determiination of vitamin C in hordeum vulgare L. Seedling powder by HPLC. *Fresenius Environ. Bull.* **2020**, *29*, 7832–7839.
9. Serafim, J.A.; Silveira, R.F.; Vicente, E.F. Fast determination of short-chain fatty acids and glucose simultaneously by ultraviolet/visible and refraction index detectors via high-performance liquid chromatography. *Food Anal. Meth.* **2021**, *14*, 1387–1393. [CrossRef]
10. Guelpa, A.; Marini, F.; du Plessis, A.; Slabbert, R.; Manley, M. Verification of authenticity and fraud detection in South African honey using NIR spectroscopy. *Food Control* **2017**, *73*, 1388–1396. [CrossRef]
11. Li, L.; Hu, D.Y.; Tang, T.Y.; Tang, Y.L. Non-destructive detection of the quality attributes of fruits by visible-near infrared spectroscopy. *J. Food Meas. Charact.* **2023**, *17*, 1526–1534. [CrossRef]
12. del Barrio, M.; Cases, R.; Cebolla, V.; Hirsch, T.; de Marcos, S.; Wilhelm, S.; Galban, J. A reagentless enzymatic fluorescent biosensor for glucose based on upconverting glasses, as excitation source, and chemically modified glucose oxidase. *Talanta* **2016**, *160*, 586–591. [CrossRef]
13. Du, R.R.; Yang, D.T.; Jiang, G.J.; Song, Y.R.; Yin, X.Q. An approach for in situ rapid detection of deep-sea aromatic amino acids using laser-induced fluorescence. *Sensors* **2020**, *20*, 1330. [CrossRef] [PubMed]
14. Lu, Q.J.; Chen, X.G.; Liu, D.; Wu, C.Y.; Liu, M.L.; Li, H.T.; Zhang, Y.Y.; Yao, S.Z. A turn-on fluorescent probe for vitamin C based on the use of a silicon/CoOOH nanoparticle system. *Microchim. Acta* **2019**, *186*, 8. [CrossRef] [PubMed]
15. Asanica, A.C.; Catana, L.; Catana, M.; Burnete, A.G.; Lazar, M.A.; Belc, N.; Sanmartin, A.M. Internal validation of the methods for determination of water-soluble vitamins from frozen fruits by HPLC-HRMS. *Rom. Biotech. Lett.* **2019**, *24*, 1000–1007. [CrossRef]
16. Harada, M.; Karakawa, Y.; Yamada, N.; Miyano, H.; Shimbo, K. Biaryl axially chiral derivatizing agent for simultaneous separation and sensitive detection of proteinogenic amino acid enantiomers using liquid chromatography-tandem mass spectrometry. *J. Chromatogr. A* **2019**, *1593*, 91–101. [CrossRef] [PubMed]
17. Tao, Y.Z.; Shen, H.C.; Deng, K.Y.; Zhang, H.M.; Yang, C.Y. Microfluidic devices with simplified signal readout. *Sens. Actuator B-Chem.* **2021**, *339*, 14. [CrossRef]
18. Wu, K.M.; He, X.L.; Wang, J.L.; Pan, T.; He, R.; Kong, F.Z.; Cao, Z.M.; Ju, F.Y.; Huang, Z.; Nie, L.B. Recent progress of microfluidic chips in immunoassay. *Front. Bioeng. Biotechnol.* **2022**, *10*, 16. [CrossRef]
19. Cui, P.; Wang, S.C. Application of microfluidic chip technology in pharmaceutical analysis: A review. *J. Pharm. Anal.* **2019**, *9*, 238–247. [CrossRef]
20. Martinez, A.W.; Phillips, S.T.; Butte, M.J.; Whitesides, G.M. Patterned paper as a platform for inexpensive, low-volume, portable bioassays. *Angew. Chem. Int. Ed.* **2007**, *46*, 1318–1320. [CrossRef]
21. Wen, G.; Guo, Z.G. A paper-making transformation: From cellulose-based superwetting paper to biomimetic multifunctional inorganic paper. *J. Mater. Chem. A* **2020**, *8*, 20238–20259. [CrossRef]

22. Andini; Andayani, U.; Anneke; Sari, M.I.; Sabarudin, A. IOP Printed Low-Cost Microfluidic Paper-based Analytical Devices for Quantitative Detection of Vitamin C in Fruits. In Proceedings of the 9th Annual Basic Science International Conference (BaSIC)—Recent Advances in Basic Sciences Toward 4.0 Industrial Revolution, Malang, Indonesia, 20–21 March 2019.
23. Aksorn, J.; Teepoo, S. Development of the simultaneous colorimetric enzymatic detection of sucrose, fructose and glucose using a microfluidic paper-based analytical device. *Talanta* **2020**, *207*, 8. [CrossRef] [PubMed]
24. Kaewchuay, N.; Jantra, J.; Khettalat, C.; Ketnok, S.; Peungpra, N.; Teepoo, S. On-site microfluidic paper- based titration device for rapid semi-quantitative vitamin C content in beverages. *Microchem. J.* **2021**, *164*, 8. [CrossRef]
25. Prasad, A.; Tran, T.; Gartia, M.R. Multiplexed paper microfluidics for titration and detection of ingredients in beverages. *Sensors* **2019**, *19*, 1286. [CrossRef] [PubMed]
26. Laschewsky, A.; Rosenhahn, A. Molecular design of zwitterionic polymer interfaces: Searching for the difference. *Langmuir* **2019**, *35*, 1056–1071. [CrossRef] [PubMed]
27. Li, D.X.; Wei, Q.L.; Wu, C.X.; Zhang, X.F.; Xue, Q.H.; Zheng, T.R.; Cao, M.W. Superhydrophilicity and strong salt-affinity: Zwitterionic polymer grafted surfaces with significant potentials particularly in biological systems. *Adv. Colloid Interface Sci.* **2020**, *278*, 18. [CrossRef]
28. Wang, S.Y.; Fang, L.F.; Matsuyama, H. Construction of a stable zwitterionic layer on negatively-charged membrane via surface adsorption and cross-linking. *J. Membr. Sci.* **2020**, *597*, 9. [CrossRef]
29. Fu, F.F.; Wang, J.L.; Tan, Y.R.; Yu, J. Super-hydrophilic zwitterionic polymer surface modification facilitates liquid transportation of microfluidic sweat sensors. *Macromol. Rapid Commun.* **2022**, *43*, 8. [CrossRef]
30. Huang, H.Z.; Yuan, Q.; Yang, X.R. Preparation and characterization of metal-chitosan nanocomposites. *Colloid Surf. B Biointerfaces* **2004**, *39*, 31–37. [CrossRef]
31. Liu, P.S.; Chen, Q.; Liu, X.; Yuan, B.; Wu, S.S.; Shen, J.; Lin, S.C. Grafting of zwitterion from cellulose membranes via ATRP for improving blood compatibility. *Biomacromolecules* **2009**, *10*, 2809–2816. [CrossRef]
32. Xu, M.; Ji, F.; Qin, Z.H.; Dong, D.Y.; Tian, X.L.; Niu, R.; Sun, D.; Yao, F.L.; Li, J.J. Biomimetic mineralization of a hydroxyapatite crystal in the presence of a zwitterionic polymer. *Crystengcomm* **2018**, *20*, 2374–2383. [CrossRef]
33. Wu, R.J.; Li, L.; Dong, G.L.; Qin, Y.Q.; Li, M.W.; Hao, H. Fabrication and characterization of zwitterionic coatings with anti-oil and anti-biofouling activities. *J. Macromol. Sci. Part B Phys.* **2022**, *61*, 825–843. [CrossRef]
34. Shen, X.; Liu, T.; Xia, S.B.; Liu, J.J.; Liu, P.; Cheng, F.X.; He, C.X. Polyzwitterions grafted onto polyacrylonitrile membranes by Thiol- Ene click chemistry for oil/water separation. *Ind. Eng. Chem. Res.* **2020**, *59*, 20382–20393. [CrossRef]
35. Elgamouz, A.; Kawde, A.; Alharthi, S.; Laghoub, M.; Miqlid, D.; Nassab, C.; Bajou, K.; Patole, S.P. Cinnamon extract's phytochemicals stabilized Ag nanoclusters as nanozymes "peroxidase and xanthine oxidase mimetic" for simultaneous colorimetric sensing of H_2O_2 and xanthine. *Colloid Surf. A Physicochem. Eng. Asp.* **2022**, *647*, 10. [CrossRef]
36. Wang, Y.; Cheng, C.; Ma, R.F.; Xu, Z.R.; Ozaki, Y. In situ SERS monitoring of intracellular H_2O_2 in single living cells based on label-free bifunctional Fe_3O_4@Ag nanoparticles. *Analyst* **2022**, *147*, 1815–1823. [CrossRef]
37. Laudenslager, M.J.; Schiffman, J.D.; Schauer, C.L. Carboxymethyl chitosan as a matrix material for platinum, gold, and silver nanoparticles. *Biomacromolecules* **2008**, *9*, 2682–2685. [CrossRef]
38. Shen, X.L.; Wu, J.M.; Chen, Y.H.; Zhao, G.H. Antimicrobial and physical properties of sweet potato starch films incorporated with potassium sorbate or chitosan. *Food Hydrocoll.* **2010**, *24*, 285–290. [CrossRef]
39. Pilicer, S.L.; Wolf, C. Ninhydrin revisited: Quantitative chirality recognition of amines and amino alcohols based on nondestructive dynamic covalent chemistry. *J. Org. Chem.* **2020**, *85*, 11560–11565. [CrossRef]
40. Rukmini, N.; Kavitha, V.S.; Devendra Vijaya, K. Determination of ascorbic acid with ferricyanide. *Talanta* **1981**, *28*, 332–333. [CrossRef]
41. Filiz, B.C.; Elalmis, Y.B.; Bektas, I.S.; Figen, A.K. Fabrication of stable electrospun blended chitosan-poly(vinyl alcohol) nanofibers for designing naked-eye colorimetric glucose biosensor based on GOx/HRP. *Int. J. Biol. Macromol.* **2021**, *192*, 999–1012. [CrossRef]
42. Georgelis, N.; Fencil, K.; Richael, C.M. Validation of a rapid and sensitive HPLC/MS method for measuring sucrose, fructose and glucose in plant tissues. *Food Chem.* **2018**, *262*, 191–198. [CrossRef] [PubMed]
43. Sun, S.H.; Wang, H.; Xie, J.P.; Su, Y. Simultaneous determination of rhamnose, xylitol, arabitol, fructose, glucose, inositol, sucrose, maltose in jujube (*Zizyphus jujube* Mill.) extract: Comparison of HPLC-ELSD, LC-ESI-MS/MS and GC-MS. *Chem. Cent. J.* **2016**, *10*, 9. [CrossRef] [PubMed]
44. Shrivas, K.; Naik, W.; Kumar, D.; Singh, D.; Dewangan, K.; Kant, T.; Yadav, S.; Tikeshwari; Jaiswal, N. Experimental and theoretical investigations for selective colorimetric recognition and determination of arginine and histidine in vegetable and fruit samples using bare-AgNPs. *Microchem. J.* **2021**, *160*, 9. [CrossRef]
45. Yang, Q.M.; Xie, C.; Luo, K.; Tan, L.B.; Peng, L.P.; Zhou, L.Y. Rational construction of a new water soluble turn-on colorimetric and NIR fluorescent sensor for high selective Sec detection in Se-enriched foods and biosystems. *Food Chem.* **2022**, *394*, 8. [CrossRef]
46. Zhou, W.; Wang, Y.W.; Yang, F.; Dong, Q.; Wang, H.L.; Hu, N. Rapid determination of amino acids of nitraria tangutorum Bobr. from the Qinghai-tibet Plateau using HPLC-FLD-MS/MS and a highly selective and sensitive pre-column derivatization method. *Molecules* **2019**, *24*, 1665. [CrossRef] [PubMed]
47. Patle, T.K.; Shrivas, K.; Patle, A.; Patel, S.; Harmukh, N.; Kumar, A. Simultaneous determination of B_1, B_3, B_6 and C vitamins in green leafy vegetables using reverse phase-high performance liquid chromatography. *Microchem. J.* **2022**, *176*, 7.

48. Yu, M.Q.; Wen, R.Z.; Jiang, L.; Huang, S.; Fang, Z.F.; Chen, B.; Wang, L.P. Rapid analysis of benzoic acid and vitamin C in beverages by paper spray mass spectrometry. *Food Chem.* **2018**, *268*, 411–415. [CrossRef]
49. Bakhsh, H.; Palabiyik, I.M.; Oad, R.K.; Qambrani, N.; Buledi, J.A.; Solangi, A.R.; Sherazi, S.T.H. SnO_2 nanostructure based electroanalytical approach for simultaneous monitoring of vitamin C and vitamin B_6 in pharmaceuticals. *J. Electroanal. Chem.* **2022**, *910*, 8. [CrossRef]
50. Rodriguez-Emmenegger, C.; Houska, M.; Alles, A.B.; Brynda, E. Surfaces resistant to fouling from biological fluids: Towards bioactive surfaces for real applications. *Macromol. Biosci.* **2012**, *12*, 1413–1422. [CrossRef]
51. Rodriguez-Emmenegger, C.; Schmidt, B.; Sedlakova, Z.; Subr, V.; Alles, A.B.; Brynda, E.; Barner-Kowollik, C. Low temperature aqueous living/controlled (RAFT) polymerization of carboxybetaine methacrylamide up to high molecular weights. *Macromol. Rapid Commun.* **2011**, *32*, 958–965. [CrossRef]

Disclaimer/Publisher's Note: The statements, opinions and data contained in all publications are solely those of the individual author(s) and contributor(s) and not of MDPI and/or the editor(s). MDPI and/or the editor(s) disclaim responsibility for any injury to people or property resulting from any ideas, methods, instructions or products referred to in the content.

Article

Evaluation of Transducer Elements Based on Different Material Configurations for Aptamer-Based Electrochemical Biosensors

Ivan Lopez Carrasco [1], Gianaurelio Cuniberti [2], Jörg Opitz [1] and Natalia Beshchasna [1,*]

[1] Fraunhofer Institute for Ceramic Technologies and Systems IKTS, Maria-Reiche-Strasse 2, 01109 Dresden, Germany; ivan.lopez.carrasco@ikts.fraunhofer.com (I.L.C.); joerg.opitz@ikts.fraunhofer.de (J.O.)
[2] Faculty of Mechanical Science and Engineering, Institute of Materials Science and Max Bergmann Center of Biomaterials, Technische Universität Dresden, 01062 Dresden, Germany; office.nano@tu-dresden.de
* Correspondence: natalia.beshchasna@ikts.fraunhofer.de

Abstract: The selection of an appropriate transducer is a key element in biosensor development. Currently, a wide variety of substrates and working electrode materials utilizing different fabrication techniques are used in the field of biosensors. In the frame of this study, the following three specific material configurations with gold-finish layers were investigated regarding their efficacy to be used as electrochemical (EC) biosensors: (I) a silicone-based sensor substrate with a layer configuration of 50 nm SiO/50 nm SiN/100 nm Au/30–50 nm WTi/140 nm SiO/bulk Si); (II) polyethylene naphthalate (PEN) with a gold inkjet-printed layer; and (III) polyethylene terephthalate (PET) with a screen-printed gold layer. Electrodes were characterized using electrochemical impedance spectroscopy (EIS) and cyclic voltammetry (CV) to evaluate their performance as electrochemical transducers in an aptamer-based biosensor for the detection of cardiac troponin I using the redox molecule hexacyanoferrade/hexacyaniferrade ($K_3[Fe(CN)_6]/K_4[Fe(CN)_6]$). Baseline signals were obtained from clean electrodes after a specific cleaning procedure and after functionalization with the thiolate cardiac troponin I aptamers "Tro4" and "Tro6". With the goal of improving the PEN-based and PET-based performance, sintered PEN-based samples and PET-based samples with a carbon or silver layer under the gold were studied. The effect of a high number of immobilized aptamers will be tested in further work using the PEN-based sample. In this study, the charge-transfer resistance (R_{ct}), anodic peak height (I_{pa}), cathodic peak height (I_{pc}) and peak separation (ΔE) were determined. The PEN based electrodes demonstrated better biosensor properties such as lower initial R_{ct} values, a greater change in Rct after the immobilization of the Tro4 aptamer on its surface, higher I_{pc} and I_{pa} values and lower ΔE, which correlated with a higher number of immobilized aptamers compared with the other two types of samples functionalized using the same procedure.

Keywords: electrochemical biosensor; gold electrodes; aptamer; cardiac troponin; EIS; CV

Citation: Lopez Carrasco, I.; Cuniberti, G.; Opitz, J.; Beshchasna, N. Evaluation of Transducer Elements Based on Different Material Configurations for Aptamer-Based Electrochemical Biosensors. *Biosensors* 2024, *14*, 341. https://doi.org/10.3390/bios14070341

Received: 3 June 2024
Revised: 10 July 2024
Accepted: 11 July 2024
Published: 13 July 2024

Copyright: © 2024 by the authors. Licensee MDPI, Basel, Switzerland. This article is an open access article distributed under the terms and conditions of the Creative Commons Attribution (CC BY) license (https://creativecommons.org/licenses/by/4.0/).

1. Introduction

Biosensors are analytical devices that combine two main elements, a biorecognition element and a transducer. The device recognizes certain biological phenomena and translates them into measurable signals. Biosensors have a wide range of applications, including healthcare diagnostics, drug discovery, biomedicine, food processing and safety and environmental monitoring. These areas are made possible by the selection of appropriate biorecognition elements and transducers. Since the development of the first biosensors, scientists have worked closely with new types of bioreceptors, transducers, immobilization protocols and transducer manufacturing technologies to produce biosensors that are reliable and inexpensive, with low detection limits and high specificity [1–4].

There are several classifications and subclassifications that can be used for different biosensors. The classification can be based on the type of transducer or the biorecognition element used. If the detection mechanism is considered, we have a biocatalytic group, including enzymes; a bioaffinity group, including antibodies and nucleic acids; and a microbe-based group containing microorganisms. If the transducer type is considered, the main classifications of biosensors include mass sensitive, optical, electrochemical and thermal [5].

Currently, there are several commercially available biosensors; examples include enzyme-based and tissue-based as well as immunosensors, DNA biosensors, thermal and piezoelectric biosensors [6]. In general, the field of biosensors continues to grow as studies mention that the global biosensor market size was valued at USD 26.8 billion in 2022 and is expected to grow at a compound annual growth rate (CAGR) of 8.0% from 2023 to 2030 [7]. Some of the more specific drivers of this growth are the prevalence of chronic diseases such as cancer, HIV, antimicrobial resistance (AMR) and cardiovascular problems. Here, electrochemical biosensors have an opportunity because they can be precise, fast and non-invasive, with other advantages such as low production costs, robustness, miniaturization capabilities and ease of use for diagnostic and monitoring purposes [8,9]. The market for electrochemical biosensors is expected to have a global CAGR of 6.65% and reach 26.8 billion USD by 2030. The European region has the second highest market share [10].

The electrochemical category of biosensors uses an electrochemical transducer that converts the transducer signal to an electronic signal and amplifies it. A computer then converts the signal into a physical parameter that can be interpreted and presented to the user. The direct monitoring of the analyte or the biological activity associated with the analyte are the two measurement approaches. Based on the operating principle of the electrochemical biosensor, the electrochemical transducer or electrode are the key components used to generate detectable signals that are the product of the interaction between the target and immobilized molecules. We aimed to investigate the electrochemical properties of different gold transducers fabricated using different technologies and using different materials as substrates. The selected materials included silicon, which has been used in electronics for years due to its mechanical strength and resilience against harsh environments. Other advantages include uniform structural attributes, the number of processing approaches available and surfaces modifications with other materials when use as substrates as well as access to miniaturization, a lightweight nature, biocompatibility and control at a microscale level [11,12]. The other materials used as substrates were PET and PEN, which have the potential to upscale production due to their mechanically flexible properties with no measurable changes that are high in demand. These materials present a low-cost alternative, with low-temperature manufacturing, a light weight and the easy integration of a gold layer using printing technologies [13]. Such properties open the doors to the possibility of using these materials in wearable applications [14,15]. The selected electrodes enabled us to choose the most suitable option as a starting electrode for the development of an aptamer-based electrochemical sensor to detect cardiac biomarkers. The cardiac biomarker chosen was cardiac troponin I (cTnI); an increase in its normal concentration indicates cardiac muscle damage [16]. The selected electrode was used as a label-free approach for the electrochemical detection of cTnI in biological samples; for sensors with a strong and reliable reference signal, further modifications of the sensor surface with a target solution can be identified using changes in the reference signal. The general approach was that we generated baseline signals with techniques such as DPV or SWV using a blank aptamer-modified sensor and [Fe (CN)6]$^{3-/4-}$ freely in the solution (see Figure 1A). Then, the target was incubated on the sensor surface. After the incubation, a new electrochemical characterization was performed to determine the change in signal (see Figure 1B). The same approach was applied to the other tests.

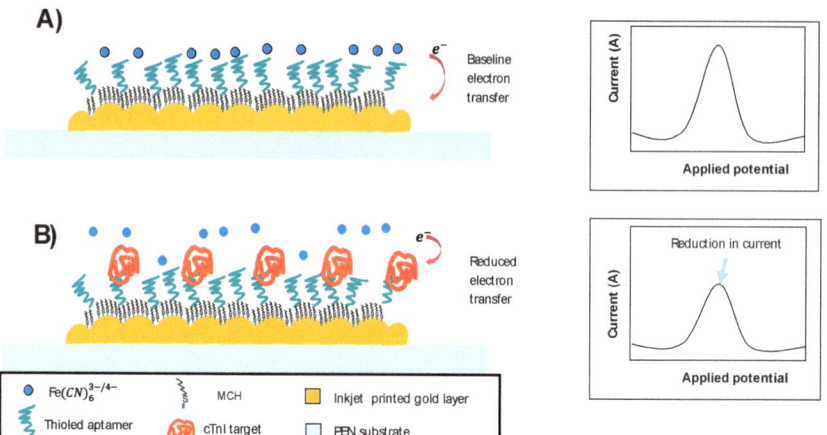

Figure 1. General scheme for the label-free aptamer-based sensors used to detect cardiac troponin I. (**A**) Graphical representation of aptamer-based sensor electron transfer process without target and recoding of signal without target binding to the aptamer-modified surface using [Fe (CN)6]$^{3-/4-}$ redox probe freely in a solution (baseline signal). (**B**) Graphical representation of aptamer-based sensor electron transfer process with target and recoding of signal with target bound to the aptamer-modified surface using [Fe (CN)6]$^{3-/4-}$ redox probe freely in a solution (reduced signal).

2. Materials and Methods

2.1. Materials

Potassium hydroxide (KOH), hydrogen peroxide (30% H_2O_2), phosphate-buffered saline (PBS), magnesium chloride ($MgCl_2$), potassium ferricyanide K3[Fe (CN)6] and potassium ferrocyanide K4[Fe (CN)6] were purchased from Carl Roth (Karlsruhe, Germany). Aptamers Tro4 and Tro6 were purchased from Eurofins Genomic (https://www.eurofins.de/, accessed on 3 July 2024, Ebersberg, Germany) and 6-mercapto-1-hexanol (MCH) was purchased from Sigma-Aldrich. All solutions were prepared in ultrapure water (18.2 MΩ.cm at 25 °C) produced using a Direct-UV Water Purification System purchased from Merck (Darmstadt, Germany).

2.2. Working Electrodes

Three types of electrodes manufactured using different technologies were tested as working electrodes. All electrode substrates had dimensions of 34 mm × 10 mm. The gold sensing layer had a geometrical area of 0.28 cm^2 in a circular shape. The first type of electrode used PEN as a substrate and had an inkjet-printed gold layer 125 μm in thickness and a resistance of 20 to 30 ohms. Three batches were tested. Batch one had a non-sintered gold layer (the printer had a dpi of 846), batch two had a sintering process at 220 °C for 30 min and the dpi was 846 and batch three had a sintering process at 220 °C for 30 min and the dpi was 1016. The PEN-based electrodes were fabricated at the Fraunhofer Institute for Ceramic Technologies and Systems IKTS (Dresden, Germany). The second type used PET as the substrate and used paste for the gold layer with a gold thickness of 1 μm. Three batches were considered. The first batch comprised the PET-based electrode with gold paste only. The second batch comprised PET-based electrodes with silver paste as the first layer and a gold layer dropcasted onto the silver layer. The third batch comprised PET-based electrodes with carbon paste as the first layer and a gold layer dropcasted onto the carbon layer. The PET-based electrodes were fabricated at Innome GmBH (Dresden, Germany).

The last type of electrodes comprised silicon-based electrodes with a layer configuration of 50 nm SiO/50 nm SiN/100 nm Au/30–50 nm WTi/140 nm SiO/bulk that were fabricated by TU Dresden (https://tu-dresden.de/, Dresden, Germany).

A POSTAT204 potentiostat from Metrohm Autolab (Stuttgart, Germany) was used for the electrochemical characterization and electrochemical cleaning, with the following three-electrode configuration: the working electrode (WE) comprised the gold electrode samples, a platinum rod was the counter electrode (CE) and the Ag/AgCl electrode was the reference electrode. The electrochemical cell used in this study was designed and fabricated at the Fraunhofer Institute for Ceramic Technologies and Systems IKTS (Dresden, Germany).

2.3. Methods

2.3.1. Cleaning Procedure

The electrodes were immersed in a beaker of ethanol and sonicated for 2 min, then rinsed with distilled water (DW) and dried with nitrogen. Once dry, the electrodes were placed in a UV/ozone cleaner for 30 min to remove organic compounds (e.g., water, carbon dioxide and nitrogen). After the UV/ozone cleaner, the electrodes were chemically cleaned in a beaker using a 3:1 solution of 0.05 M KOH and 0.05 M H_2O_2 for 10 min, shaking the beaker gently during the 10 min. The electrodes were then removed, rinsed with DW and dried with nitrogen. The final step was an electrochemical cleaning procedure consisting of a single linear potential sweep from −200 mV to 1200 mV [vs. Ag/AgCl (sat. 4 M KCl)] in 50 mM KOH at a sweep rate of 50 mV/s. They were thoroughly rinsed with ultrapure water (double-distilled water; Milli-Q water) and dried with N_2. After this step, they were ready for the electrochemical characterization.

2.3.2. Electrochemical Characterization

Cyclic voltammetry and electrochemical impedance spectroscopy were the two techniques used in this study for the characterization of the bare and functionalized samples. All CV and EIS experiments were performed using a standard PBS buffer The EIS measurements were obtained using at open circuit potential (OCP) and the frequency range from 100 kHz to 0.1 Hz, E_{ac} set at 0.01 Vrms; in total 61 frequencies were measured. Cyclic voltammetry (CV) measurements were obtained by sweeping a potential range from −0.1 V to +0.5 V at 0.02 V/s with a step of 0.003 V.

2.3.3. Functionalization Procedure

The functionalization of the gold surfaces with Tro4 and Tro6 was carried out using a thiol anchor via a dropcasting approach. First, the stock solution of the aptamer (100 uM) was diluted to the final working concentration of 25 μM in PBS (1 mM $MgCl_2$). A 10 μL drop of the aptamer solution was deposited onto a dried sample. The sample was stored in a sealed container at 4 °C for 20 h. After this time, the sample was rinsed with ultrapure water (MilliQ-water) to remove non-immobilized aptamers. A drop of 20 μL of 1 mM MCH was then applied to the aptamer-modified surface to block the exposed areas and prevent unwanted interactions with the surface. The passivation with MCH took 2 h at room temperature. The remaining MCH solution was removed after rinsing the sample with MilliQ-water. After this final step, the sample was ready to be characterized.

3. Results

The characterization and comparison were carried out at different stages in the development of the aptamer-based sensor. The stages considered included bare electrodes after the cleaning stage and functionalized samples using the aptamers Tro4 and Tro6. The electrodes used in this study included PEN-based electrodes, silicon-based electrodes and PET-based electrodes. Each type of electrode test was performed using three samples. The selected manufacturing technologies had been used in other electrochemical biosensor applications [17–19].

3.1. Cyclic Voltammetry Tests

The three conditions (bare samples (non-functionalized), Tro6-functionalized samples and Tro4Tro4-functionalized samples) were tested according to the parameters of the cyclic voltammetry (CV) experiments described in the Section 2.3. From the data generated by each test, the peak potential of both parts of the redox process as well as the peak heights of the redox process were collected as these are commonly used parameters from CV data [20,21]. The datum values of the anodic peak current (I_{pa}) and cathodic peak current (I_{pc}) for the PEN-based electrodes, silicon-based electrodes and PET-based electrodes of the samples are shown in Figures 2 and 3, respectively. The I_{pa} and I_{pc} values obtained for each type of electrode were different due to the different morphologies, which were different from the measured geometric area [22]. All results were consistent with the literature for the solution-phase redox reporters hexacyanoferrade/hexacyaniferrade (K3 [Fe(CN)6]/K4[Fe (CN)6) and had the typical "duck" shape, where the peak shrinks due to the surface-modification blocking of the surface and the decrease in the current of the voltammogram [23,24]. The samples with the highest I_{pa} and I_{pc} for non-functionalized samples were the PEN-based samples. In second place were the silicon-based samples and those with a lower I_{pa} were the PET-based samples. The peak current ratios (I_{pa}/I_{pc}) were taken into account. For the reversible process used, it should ideally be equal to 1 for all samples at an equilibrium [25,26]. Taking into account the I_{pa} and I_{pc} values, we obtained I_{pa}/I_{pc} values of 0.923171621, 0.655710401 and 0.982881907 for the PEN-based, PET-based and silicon-based samples, respectively. The PEN-based and silicon-based samples were closer to the ideal value, but the PET-based samples were far from it, which may have been due to the adhesion problems observed during the study. Based on the redox process considered in this study, the peak-to-peak separation (ΔE) of the samples was analyzed against the theoretical value of 59.2/n mV (n = 1 at all scan rates, at 25 °C) [27]. Considering the values observed in Figure 4, the clean and non-functionalized PEN-based samples had a lower ΔE and were closer to the theoretical value. The sputtered gold layer on the silicon-based electrodes presented the highest peak separation ΔE of the sample types investigated. Based on this result, PEN-based samples might be more suitable for implementation in biosensors due to the highest basic output current signal and lower peak-to-peak separation, ensuring that the redox process of the reporter is not disturbed.

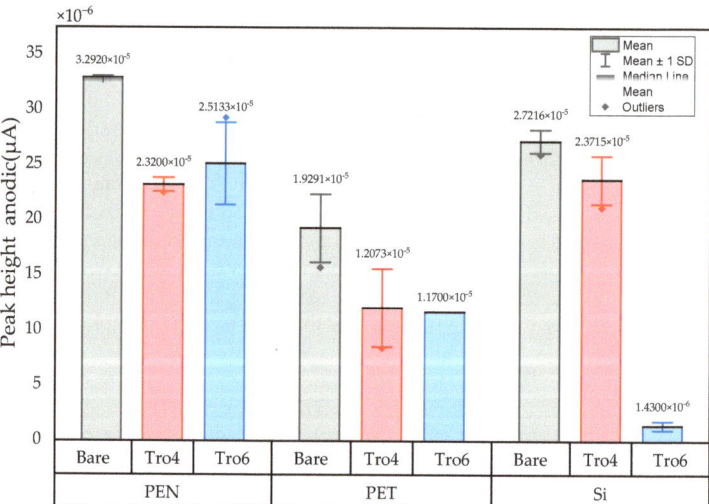

Figure 2. Mean and standard deviation of I_{pa} values of bare samples (gray), Tro4-functionalized samples (red) and Tro6-functionalized samples (blue) for PEN-, PET- and silicon-based substrates.

After obtaining the baselines for each type of sensor, they were functionalized using two cardiac troponin I-specific thiolated aptamers and 6-mercapto-1-heaxanol as a passivation agent, based on the functionalization protocol listed in the Section 2. CV techniques were performed for characterization. A summary of the results of the I_{pa}, I_{pc} and ΔE are shown in Figures 2–4, respectively. The CV plots used to obtain the I_{pa}, I_{pc} and ΔE can be found in the Supplementary Materials. As reported in the literature, the aptamer and MCH acted as a blocking biolayer of the redox process, reducing the diffusion from the bulk solution to the surface [28–30]. The blocking of the transport process could be seen in the reduction in the I_{pa} and I_{pc}, as shown in Figures 2 and 3, and the increase in the ΔE was due to the drift in peak positions caused by the reduction in the electron-transfer process (see Figure 4) [31].

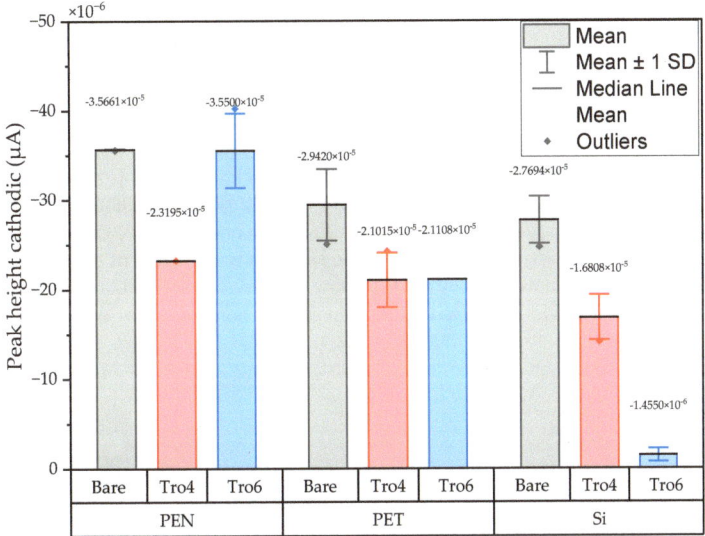

Figure 3. Mean and standard deviation of I_{pc} values of bare samples (gray), Tro4-functionalized samples (red) and Tro6-functionalized samples (blue) for PEN-, PET- and silicon-based substrates.

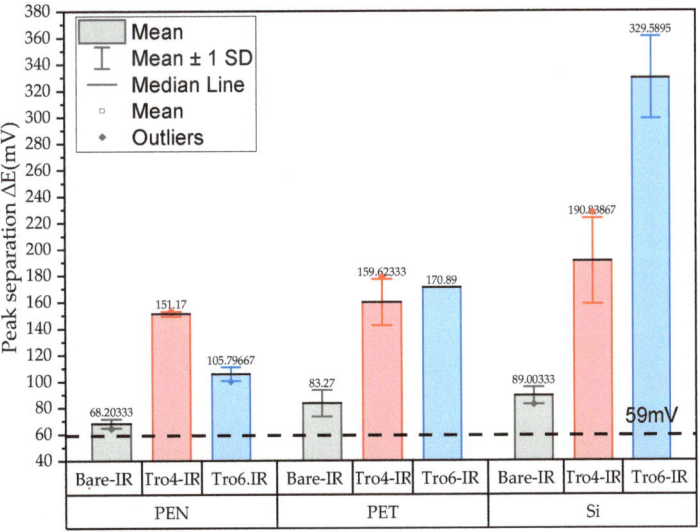

Figure 4. Mean and standard deviation of the peak separation (ΔE) of bare samples (gray), Tro4-functionalized samples (red) and Tro6-functionalized samples (blue) for PEN-, PET- and silicon-based substrates.

In general, the standard deviation was small for the whole set of cleaned samples. Once the samples had been functionalized using Tro4Tro4 and Tro6, the oxidation and reduction peaks decreased due to the blockade created by the aptamers, with their negatively charged backbone preventing the charge transfer from the bulk solution to the surface.

Cyclic Voltammetry of Modified PEN- and PET-Based Samples

Based on the results of the first batch of samples used, we made some modifications to the PET- and PEN-based samples. For the PEN-based samples, the change was to modify the surface area resulting from the standard manufacturing process that is used to give more uniformity and conductivity to a gold electrode [32–35]. In the case of the PET-based samples, the aim was to improve the adhesion of the gold layer to the PET substrate by adding an additional carbon or silver layer under the gold layer. Silver and carbon layers have been used on PET and the formation of gold on top has been possible for other applications such as transducers [13,36–38]. The surface morphology and adhesion current state of the gold surfaces for PEN and PET, respectively, are shown in Figure 4.

As shown in Figure 5A, the poor adhesion of the gold to the PET substrate was significant. In the sample shown, only the application of water released from the 1000 µL pipette used to rinse the samples was sufficient to remove part of the gold layer in some samples. Figure 5A,B show the lines on the gold-layer product of the inkjet-printing process, which affected the standard deviation values for the Tro4- and Tro6-functionalized samples.

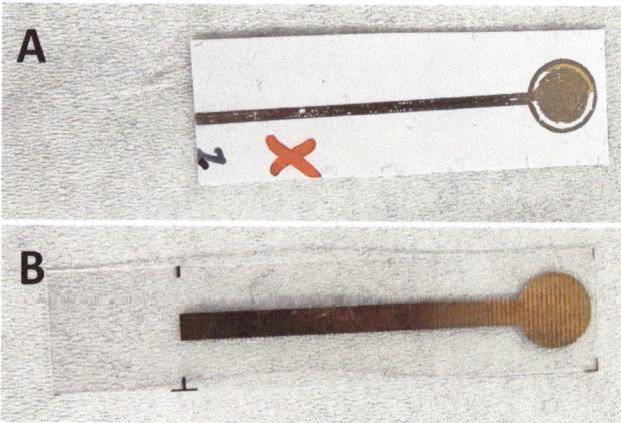

Figure 5. Examples of the initial batch of PET-based (**A**) and PEN-based (**B**) samples. Red "x" indicated samples that delaminated sections.

The morphology of the PEN-based samples was modified using a sintering process applied to two batches, one at 846 dpi and the other at 1016 dpi. To improve the adhesion of the gold to the PET-based samples, the new batches contained a silver layer under the gold layer for the first batch and a carbon layer for the second batch. All samples were electrochemically characterized. A summary of the CV results showing the I_{pc}, I_{pc} and ΔE obtained from the new batches of sintered PEN, including the unsintered samples, are shown in Figures 6–8 respectively.

The results showed that the I_{pa} decreased and the I_{pc} increased for both the 846 dpi and 1016 dpi samples compared with the non-sintered samples. The SD also increased. In the case of the peak current ratio (I_{pa}/I_{pc}), the values were 0.921, 0.765 and 0.758 for the non-sintered, 846 and 1016 samples, respectively, which were not desirable results. Based on these results, modifications using a sintering process affected the reversibility of the redox process, based on the I_{pa}/I_{pc} values of the samples. Considering the peak separation, we observed that the reversibility of the redox process used for characterization was better for the non-sintered samples as we observed a lower ΔE.

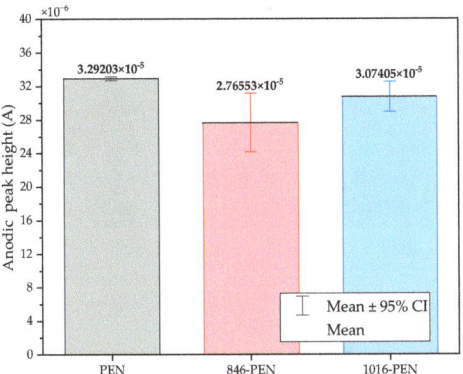

Figure 6. Mean and standard deviation of I_{pa} values of bare samples of unmodified PEN-based samples (gray), 846-PEN-based samples (red) and 1016-PEN-based samples (blue).

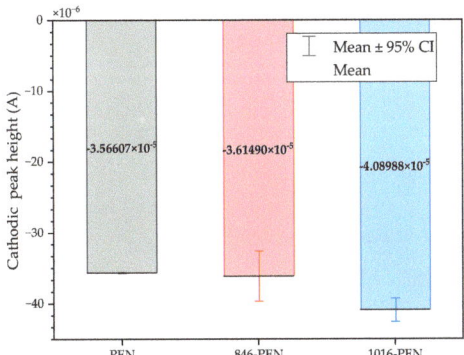

Figure 7. Mean and standard deviation of I_{pc} values of bare samples of unmodified PEN-based samples (gray), 846-PEN-based samples (red) and 1016-PEN-based samples (blue).

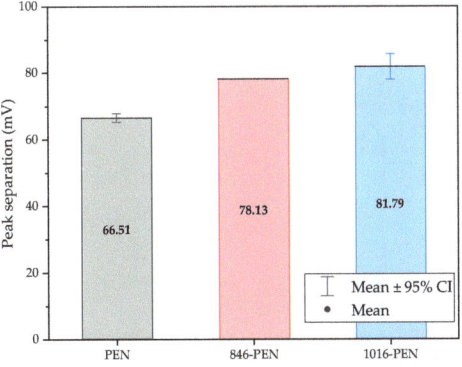

Figure 8. Mean and standard deviation of peak separation (ΔE) values of bare samples of unmodified PEN-based samples, 846-PEN-based samples and 1016-PEN-based samples.

The CV results of the modified samples with carbon are shown in Figures 9 and 10 and the CV results of the modified samples with silver are shown in Figure 11. Based on the data in Figures 8 and 10, the I_{pa} and I_{pc} values for the non-functionalized and functionalized samples were higher for the unmodified samples than for the samples with a carbon layer. For the Tro4Tro4-functionalized samples with a carbon layer, a peak search to determine the I_{pa} and I_{pc} was not achievable for the software, hence the columns are not visible in the figures.

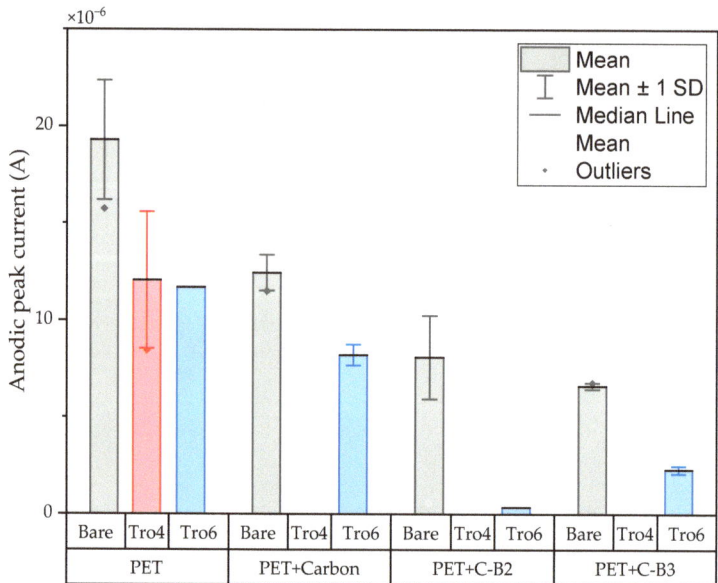

Figure 9. Mean and standard deviation of I_{pa} values of bare samples (gray), Tro4-functionalized samples (red) and Tro6-functionalized samples (blue) for carbon-modified PET-based samples from the initial batch and 3 other batches.

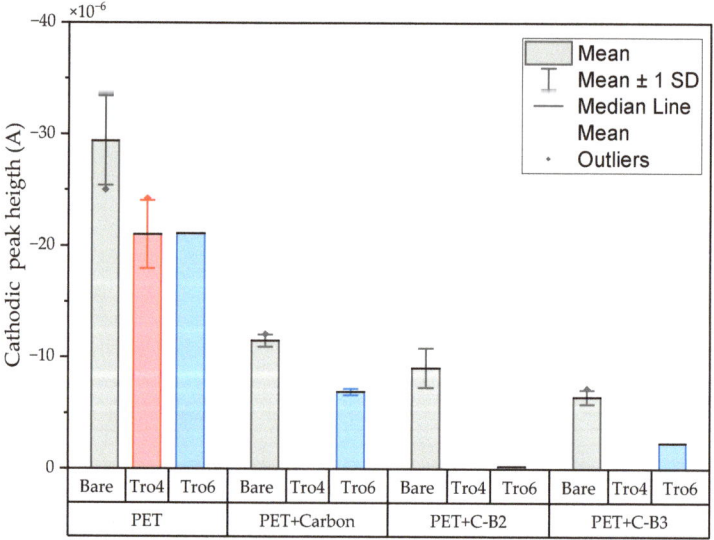

Figure 10. Mean and standard deviation of I_{pc} values of bare samples (gray), Tro4-functionalized samples (red) and Tro6-functionalized samples (blue) for carbon-modified PET-based samples from the initial batch and 3 other batches.

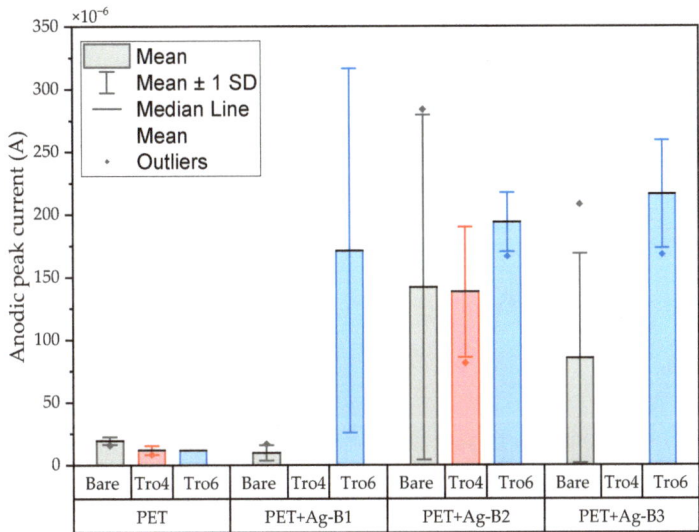

Figure 11. Mean and standard deviation of I_{pa} values of bare samples (gray), Tro4-functionalized samples (red) and Tro6-functionalized samples (blue) for silver-modified PET-based samples from the initial batch and 3 other batches.

The PET-based samples modified with a silver layer presented greater limitations in the obtention of CV parameters due to the recorded current values from the CV experiment. The I_{pa} and SD values of the clean samples in batch two increased and this trend was also observed in the functionalized samples for the three batches with a silver layer. It was not possible to estimate the I_{pc} and ΔE from the available datasets of the samples tested with a silver layer. An example cyclic voltammogram for these samples is shown in Figure 12.

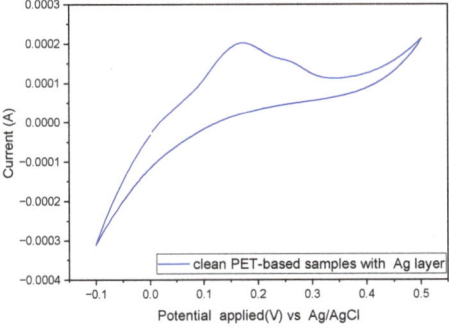

Figure 12. Cyclic voltammogram of PET-based sample with a silver layer under a gold layer in a standard PBS buffer containing 1 mM $[Fe(CN)6]^{3-/4-}$.

3.2. Electrochemical Impedance Spectroscopy (EIS)

EIS was carried out to complement and corroborate the CV results. The tests were carried out using the same samples as in the CV section and the parameter we focused on was the charge-transfer resistance, which was obtained by fitting the impedance spectra generated in the EIS experiment to a modified Randle's equivalent circuit and then presenting the results in a Nyquist plot. An example of a modified Randle's equivalent circuit is shown in Figure 13 [27], where Rs is the resistance of the solution, Rct is the charge-transfer resistance, CPE is the constant phase element and W is the Warburg impedance. The fitting of the impedance spectra to such a model was carried out using NOVA software (version 2.1.5) from Metrohm Autolab. The fitted charge-transfer resistance (Rct) value was used for the analysis as it is commonly used in the literature [39]. The first set of samples comprised unmodified PEN-based, PET-based and Si-based samples and a summary of the Rct values of the different samples is shown in Figure 14. The Nyquist plots used to obtain the Rct values can be found in the Supplementary Materials.

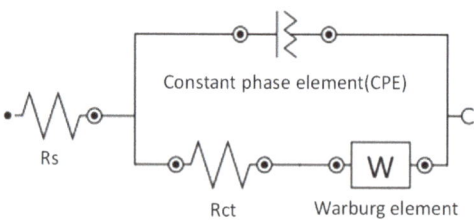

Figure 13. Modified Randle's equivalent circuit.

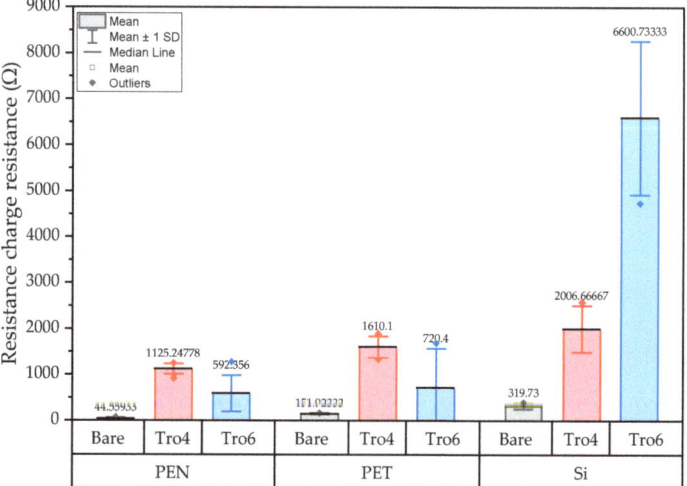

Figure 14. Mean and standard deviation of the Rct values of bare samples (gray), Tro4-functionalized samples (red) and Tro6-functionalized samples (blue) for PEN-, PET- and silicon-based substrates.

Based on the EIS results shown in Figure 14, the Rct value of the PEN-based samples showed a lower Rct for the bare samples. The bare samples with the highest Rct values were the silicon-based samples, with an average value of 319.73 Ω. For the PET-based samples, the Rct value was three times higher than the PEN samples. When fitting the impedance spectra for the PEN-based sample, the semicircle resulting from the parallel arrangement of the Rct and CPE values in the circuit could not be well-observed, as shown in Figure 15 (blue line and blue dotted line). The NOVA software required more time and better starting values of the electrical circuit in the simulation to obtain a better fit of the data [40]. In comparison, the Rct values of the PET-based and silicon-based samples (Figure 15 red curves and black curves respectively) were obtained faster and with a smaller modification of the initial values used. The impedance spectra of the bare PEN-based samples showed a Faradic process, mainly driven by the diffusion of ions to the surface.

The Tro4- and Tro6-functionalized surfaces showed an increased charge-transfer resistance for all samples, as shown in Figure 14. The values of the Tro4-functionalized sample showed a lower standard deviation (SD) compared with the values of the Tro6-functionalized sample.

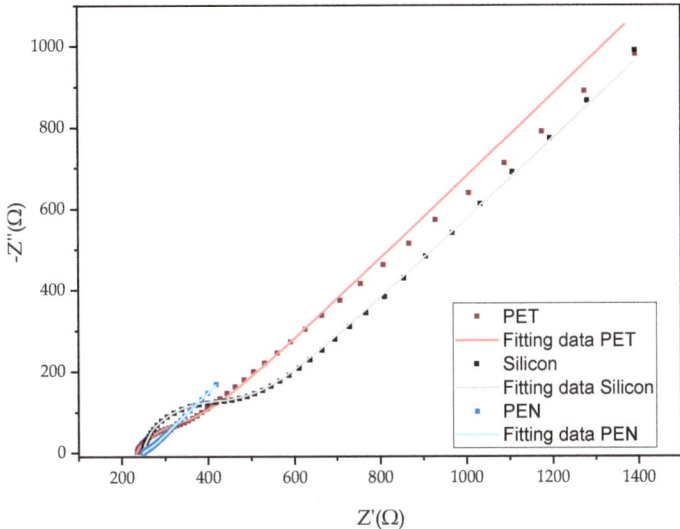

Figure 15. Nyquist plots of 3 individual samples after a cleaning step recorded in standard PBS buffer containing 1 mM [Fe (CN)6]$^{3-/4-}$ at OCP. Blue dotted line: experimental data of inkjet-printed gold layer on PEN and blue line fitted data. Red dotted line: experimental data of screen-printed gold layer on PET and red line fitted data. Black dotted line: experimental data of sputtered gold layer on silicon oxide wafer and black line fitted data.

We used the percentage of change of the charge-transfer resistance (ΔRct (%)) to better understand the changes. ΔRct (%) could be calculated using Equation (1), as follows:

$$\Delta\text{Rct }(\%) = \frac{R_{ct}final - R_{ct}initial}{R_{ct}initial} \times 100\% \tag{1}$$

where $R_{ct}final$ represents the Rct value after functionalization and $R_{ct}initial$, represents the Rct value of the bare samples. The obtained ΔRct (%) figures are presented in Table 1. The Tro4-functionalized samples had the greatest change for the PEN-based samples, followed by the PET-based samples and then, with the lowest change, the silicon-based samples, at 528% ΔRct (%). Tro6 functionalization with the highest ΔRct (%) was obtained with the silicon-based samples, followed by the PEN-based samples; the lowest ΔRct (%) values were from the PET-based samples. The silicon-based samples had a higher ΔRct (%) with Tro6 functionalization; however, the SD was high. The PET-based samples showed a high SD for both functionalizations.

Table 1. ΔRct (%) values for the different groups of samples tested after immobilization with the aptamers Tro4 and Tro6.

ΔRct (%)	PEN	PET	Silicon
Tro4	2425.28%	960%	528%
Tro6	1229%	374%	1964%

Electrochemical Impedance Spectroscopy of Modified PEN- and PET-Based Samples

The EIS characterization parameters for the modified PEN-based and PET-based samples were the same as for the first batch. The results for the modified PEN-based samples without functionalization are shown in Figure 16. The results showed that the initial Rct increased for the sintered samples, but the semicircle where the kinetic control had an influence remained the same as the example shown in Figure 15.

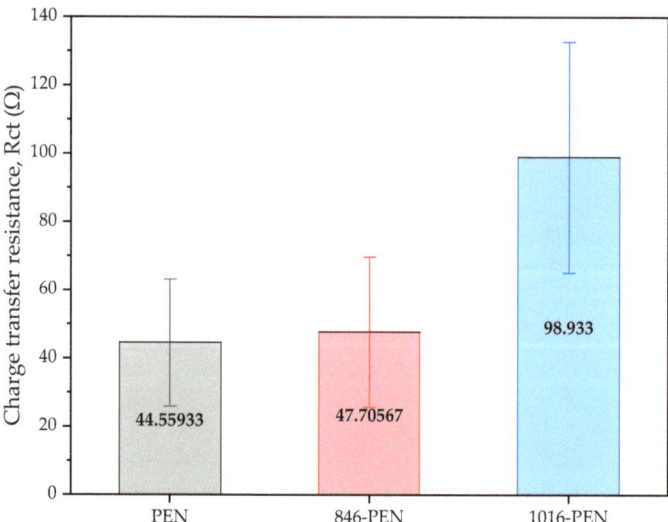

Figure 16. Mean and standard deviation of the Rct values of bare samples of unmodified PEN-based samples, 846-PEN-based samples and 1016-PEN-based samples.

The impedance spectra of the modified PET samples with either a carbon or a silver layer showed several changes compared with the unmodified samples, which interfered with the fitting of the data to the initial electrical model used previously. Two impedance spectrum datum examples from the characterized samples are shown in Figure 17. The impedance data suggested that the electrochemical system had a different layer configuration compared with the proposed Randle's equivalent circuit. The data showed two semicircles, which altered the electrical model required for fitting [40]. The modelling of a new electrical circuit, which deviated from the proposed idea that the new layer only improved the adhesion of the gold layer to the substrate without affecting the EC parameters, was not possible with the manufacturing process used. The Supplementary Materials presents the Nyquist plots of the experimental data for the modified PET-based samples with silver and carbon layers.

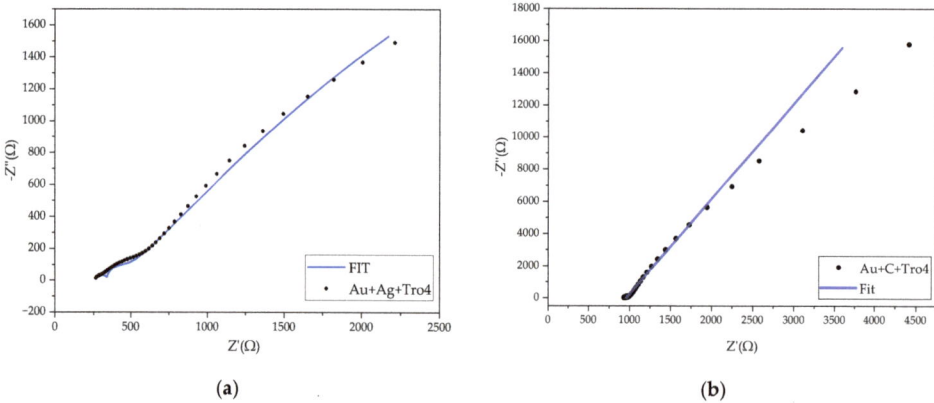

Figure 17. Nyquist plots of modified PET-based samples with Tro4 functionalization. (**a**) Nyquist plot of PET samples with an Ag layer under the gold and functionalized using a Tro4 aptamer and (**b**) Nyquist plot of PET samples with a C layer under the gold and functionalized using a Tro4 aptamer.

4. Discussion

The electrochemical characterization of the three types of samples before and after functionalization was achievable and the overall results for the three types of samples had comparable outputs in the CV and EIS experiments, where the inclusion of an aptamer–MCH layer affected the electron transfer of the surface due to the negatively charged aptamer [41].

The respective change in Rct for each sample type varied based on the aptamer, the aptamer concentration and the working electrode used. In the literature, the functionalization of Tro4 and Tro6 thiolated aptamers on electrodes was demonstrated to be possible on three types of samples at different levels due to the surface morphology [42,43]. The change in Rct in the literature was the result of reducing the electron-transfer process, which increased the Rct values, reduced the I_{pc}–I_{pa} values and increased the ΔE [44–46]. As described in the literature, each manufacturing technology has different baseline values depending on the functionalization protocol and the specific aptamer used, as observed in this study where the same functionalization protocol was used for two cardiac troponin I aptamers. The bare samples showed a peak-to-peak separation close to the theoretical value for a reversible process and these values were similar to those found in the literature for each type of manufacturing technology [44,47,48]. From the three types of samples, the peak separation of the PEN-based samples was the lower (68 mV) by at least 10 mV with respect to the other two types of samples, indicating better conditions of the reversibility of the process. Having lower starting ΔE values helps because the incorporation of aptamers is known to slow the electron transfer even further. The smaller shift of the current peaks could have a lower impact on the aptamer–MCH layer for the long-term use of electrodes with voltametric approaches by applying smaller potential windows with lower potentials. In a similar manner, having higher baseline I_{pc} and I_{pa} values, as was the case for the PEN samples, could help improve the resolution in the further developmental steps of electrochemical biosensors [49]. Once the aptamers were immobilized, the CV analysis indicated that the PEN-based samples had low variations between samples and the current values were higher for both functionalizations using PEN-based samples. Such baselines are preferable for the further development of sensors in POCT applications or wearable applications where the robustness of the readout system is lower than that of a laboratory setup. With respect to the EIS measurements, many examples have shown ΔRct (%) values from 200% to 600% for aptamer concentrations ranging from 0.5 µM to 15 µM [19,50], which are indeed smaller than the values of ΔRct (%) obtained for the 25 µM aptamer concentration used in this study, based on an increase in aptamer density at the surface. Other baseline Rct values that are different from those investigated here can be found in the literature for different gold electrodes [9,40,41,43]. Similar to the CV analysis, the EIS showed that the PEN-based samples had higher Rct changes while maintaining a lower sample-to-sample variation for the Tro4-functionalized samples and Tro6-functionalized samples, a quality that will be helpful with further surface modifications and detection experiments where possible target interactions cannot be distinguished because the standard deviation of the values used as a baseline overshadows these small changes.

For the first set of samples tested, the EIS response was smooth and had easy parametrization when compared with other working electrodes in the literature [51]. However, once the modified versions of the PEN and PET samples were studied, the sintering process for the PEN-based samples did not reduce the Rct values, as in most cases in the literature [52]. In the case of the PET-based samples, a different equivalent circuit was required due to the new active layer configuration that the system produced during the characterization. This may have occurred due to the aptamer and MCH layer, as reported in other literature, although this could not be the case as the other samples tested in this study did not have such results [53].

The reproducibility of the modified PET-based samples after functionalization suggested that the uncontrolled roughness hindered the adhesion of the thiol anchor, which affected the uniformity and density of the aptamer + MCH layer. This was noticeable in the increase in SD and low ΔRct (%) when the functionalized samples were characterized. It is also important to mention that the addition of a carbon or silver layer produced unexpected results when compared with examples in the literature where glassy carbon electrodes were presented with a gold layer on top, which showed good conductivity, and no modification models were used for the EIS analysis, something that was not the case in this study. In the case of the silver-modified electrodes, the electrochemical test results could be attributed to the oxidation of the silver layer. Such a layer should not be in contact with the electrochemical solution to avoid this interaction [54].

The improvement in the metal–polymer adhesion could be tackled using other strategies such as using an argon treatment instead of adding an extra layer [55,56]. With the current results, the use of carbon- or silver-modified samples was not a comparable option with other cases in the literature. In the case of the silicon-based option, the example that we found in the literature could be compared in the non-functionalized state; however, the introduction of the aptamer had a lower priority compared with the PEN-based option [47]. A low variability between samples for each electrode technology relative to the baseline measurements taken is something that has been obtained in other studies for biological fluids and at a more fundamental level [57]. The results presented in this study suggested that further testing and optimization are required to ensure a low variability so that transducers are suitable for aptamer sensor applications; however, the results showed a good starting point for the electrodes tested.

5. Conclusions

Based on an electrochemical characterization, PEN-based samples were the easiest to characterize compared with PET- and silicon-based samples due to the artefact presented in the electrochemical data where the redox process was not fully observed in the initial potential window used. The PEN-based samples also showed higher current values in the CV and lower peak separation. In the case of the PET samples, the variation in the EC parameter of the measured samples and the consistency of the prepared samples were not suitable for a better EC characterization, mainly due to the delamination of the gold layer, which changed the ECSA.

Once the aptamers Tro4 and Tro6 were introduced to the surface by immobilization with a thiol anchor, the PEN-based samples showed a greater change in ΔRct (%) values with respect to the non-functionalized samples for Tro4. In second place was the Tro6 aptamer, indicating that there was greater immobilization on the surface of these samples, which correlated with a greater number of aptamers on the surface blocking the electron-transfer process. The modifications presented in this study for the PEN- and PET-based samples did not improve the results with the immobilization of aptamers as we had proposed because the electrochemical characterization showed that the different EC parameter values obtained were deficient when compared with the unmodified samples for most of the parameters used in this study, which made them less suitable for aptamer immobilization and the detection of cardiac troponin I using a label-free approach compared with the unmodified samples.

Although the PEN-based samples gave better results, this technology should be improved to reduce the variation between samples and to propose surface modifications to improve the electrochemical parameters. Possible modifications include the addition of nanomaterials such as carbon nanotubes or other nanoparticles such as hematite to the surface, which will improve the electrochemical properties of the surface. Another modification that can be made to the substrate is a specific morphology and roughness so that the gold layer on top changes accordingly and the ECSA could be modified in a positive way. Other sintering protocols can also be considered to improve conductivity and reduce the sample-to-sample variation. These modifications should maintain the suitability of the surface to immobilize Tro4 and Tro6 aptamers for their use in detection experiments for cardiac troponin I.

Supplementary Materials: The following supporting information can be downloaded at: https://www.mdpi.com/article/10.3390/bios14070341/s1.

Author Contributions: Conceptualization, I.L.C.; methodology, I.L.C.; formal analysis I.L.C.; investigation, I.L.C.; resources, N.B.; writing—original draft preparation, I.L.C.; writing—review and editing, N.B. and J.O.; supervision, G.C.; project administration, N.B.; funding acquisition, N.B. and J.O. All authors have read and agreed to the published version of the manuscript.

Funding: This Research was funded by Bundesministerium für Wirtschaft und Klimaschutz (BMWK) in the frame of "Zentrales Innovationsprogramm Mittelstand (ZIM)", grant number KK5033910.

Institutional Review Board Statement: Not applicable.

Informed Consent Statement: Not applicable.

Data Availability Statement: The original contributions presented in the study are included in the article/Supplementary Materials; further inquiries can be directed to the corresponding author.

Conflicts of Interest: The authors declare no conflicts of interest. The funders had no role in the design of the study; in the collection, analyses or interpretation of data; in the writing of the manuscript; or in the decision to publish the results.

References

1. Naresh, V.; Lee, N. A Review on Biosensors and Recent Development of Nanostructured Materials-Enabled Biosensors. *Sensors* **2021**, *21*, 1109. [CrossRef]
2. Shanbhag, M.M.; Manasa, G.; Mascarenhas, R.J.; Mondal, K.; Shetti, N.P. Fundamentals of bio-electrochemical sensing. *Chem. Eng. J. Adv.* **2023**, *16*, 100516. [CrossRef]
3. Wu, J.; Liu, H.; Chen, W.; Ma, B.; Ju, H. Device integration of electrochemical biosensors. *Nat. Rev. Bioeng.* **2023**, *1*, 346–360. [CrossRef] [PubMed]
4. Sumitha, M.; Xavier, T. Recent advances in electrochemical biosensors—A brief review. *Hybrid Adv.* **2023**, *2*, 100023. [CrossRef]
5. Polat, E.O.; Cetin, M.M.; Tabak, A.F.; Bilget Güven, E.; Uysal, B.Ö.; Arsan, T.; Kabbani, A.; Hamed, H.; Gül, S.B. Transducer Technologies for Biosensors and Their Wearable Applications. *Biosensors* **2022**, *12*, 385. [CrossRef]
6. Mehrotra, P. Biosensors and their applications—A review. *J. Oral Biol. Craniofacial Res.* **2016**, *6*, 153–159. [CrossRef] [PubMed]
7. Global Market Insights Inc. Biosensors Market Size Analysis|Global Statistics Report—2032. 5 March 2024. Available online: https://www.gminsights.com/industry-analysis/biosensors-market?gclid=EAIaIQobChMIl6yklbnxhQMVOT0GAB3VbwFgEAAYASAAEgIhdPD_BwE (accessed on 5 March 2024).
8. Grieshaber, D.; MacKenzie, R.; Vörös, J.; Reimhult, E. Electrochemical Biosensors—Sensor Principles and Architectures. *Sensors* **2008**, *8*, 1400–1458. [CrossRef] [PubMed]
9. Thévenot, D.R.; Toth, K.; Durst, R.A.; Wilson, G.S. Electrochemical biosensors: Recommended definitions and classification. *Biosens. Bioelectron.* **2001**, *16*, 121–131. [CrossRef] [PubMed]
10. Future, M.R. Bioelectronic Sensors Market Report Size, Share and Trends 2032. 5 March 2024. Available online: https://www.marketresearchfuture.com/reports/bioelectronic-sensors-market-12348?utm_term=&utm_campaign=&utm_source=adwords&utm_medium=ppc&hsa_acc=2893753364&hsa_cam=20543884685&hsa_grp=153457592316&hsa_ad=673752668768&hsa_src=g&hsa_tgt=dsa-2246460572593&hsa_kw=&hsa_mt=&hsa_net=adwords&hsa_ver=3&gad_source=1 (accessed on 5 March 2024).
11. Xu, Y.; Hu, X.; Kundu, S.; Nag, A.; Afsarimanesh, N.; Sapra, S.; Mukhopadhyay, S.C.; Han, T. Silicon-Based Sensors for Biomedical Applications: A Review. *Sensors* **2019**, *19*, 2908. [CrossRef]
12. Yan, X.; Almajidi, Y.Q.; Uinarni, H.; Bokov, D.O.; Mansouri, S.; Fenjan, M.N.; Saxena, A.; Zabibah, R.S.; Hamzah, H.F.; Oudah, S.K. Bio(sensors) based on molecularly imprinted polymers and silica materials used for food safety and biomedical analysis: Recent trends and future prospects. *Talanta* **2024**, *276*, 126292. [CrossRef]
13. Reddy, A.S.G.; Narakathu, B.B.; Atashbar, M.Z.; Rebros, M.; Hrehorova, E.; Joyce, M. Printed electrochemical based biosensors on flexible substrates. In Proceedings of the IEEE Sensors, Waikoloa, HI, USA, 1–4 November 2010; pp. 1596–1600.
14. Noguchi, Y.; Sekitani, T.; Someya, T. Organic-transistor-based flexible pressure sensors using ink-jet-printed electrodes and gate dielectric layers. *Appl. Phys. Lett.* **2006**, *89*, 253507. [CrossRef]
15. Cai, J.; Cizek, K.; Long, B.; McAferty, K.; Campbell, C.G.; Allee, D.R.; Vogt, B.D.; La Belle, J.; Wang, J. Flexible thick-film electrochemical sensors: Impact of mechanical bending and stress on the electrochemical behavior. *Sens. Actuators B Chem.* **2009**, *137*, 379–385. [CrossRef] [PubMed]
16. Sharma, S.; Jackson, P.G.; Makan, J. Cardiac troponins. *J. Clin. Pathol.* **2004**, *57*, 1025–1026. [CrossRef] [PubMed]
17. García-Miranda Ferrari, A.; Foster, C.W.; Kelly, P.J.; Brownson, D.A.; Banks, C.E. Determination of the Electrochemical Area of Screen-Printed Electrochemical Sensing Platforms. *Biosensors* **2018**, *8*, 53. [CrossRef] [PubMed]
18. Movilli, J.; Kolkman, R.W.; Rozzi, A.; Corradini, R.; Segerink, L.I.; Huskens, J. Increasing the Sensitivity of Electrochemical DNA Detection by a Micropillar-Structured Biosensing Surface. *Langmuir* **2020**, *36*, 4272–4279. [CrossRef]

19. Diaz-Amaya, S.; Lin, L.-K.; DiNino, R.E.; Ostos, C.; Stanciu, L.A. Inkjet printed electrochemical aptasensor for detection of Hg^{2+} in organic solvents. *Electrochim. Acta* **2019**, *316*, 33–42. [CrossRef]
20. Faria, A.M.; Peixoto, E.B.; Adamo, C.B.; Flacker, A.; Longo, E.; Mazon, T. Controlling parameters and characteristics of electrochemi-cal biosensors for enhanced detection of 8-hydroxy-2'-deoxyguanosine. *Sci. Rep.* **2019**, *9*, 7411. [CrossRef]
21. Elgrishi, N.; Rountree, K.J.; McCarthy, B.D.; Rountree, E.S.; Eisenhart, T.T.; Dempsey, J.L. A Practical Beginner's Guide to Cyclic Voltammetry. *J. Chem. Educ.* **2018**, *95*, 197–206. [CrossRef]
22. Martínez-Hincapié, R.; Wegner, J.; Anwar, M.U.; Raza-Khan, A.; Franzka, S.; Kleszczynski, S.; Čolić, V. The determination of the electrochemically active surface area and its effects on the electrocatalytic properties of structured nickel electrodes produced by additive manufacturing. *Electrochim. Acta* **2024**, *476*, 143663. [CrossRef]
23. Krishnaveni, P.; Ganesh, V. Electron transfer studies of a conventional redox probe in human sweat and saliva bio-mimicking conditions. *Sci. Rep.* **2021**, *11*, 7663. [CrossRef]
24. Koç, Y.; Morali, U.; Erol, S.; Avci, H. Investigation of electrochemical behavior of potassium ferricyanide/ferrocyanide redox probes on screen printed carbon electrode through cyclic voltammetry and electrochemical impedance spectroscopy. *Turk. J. Chem.* **2021**, *45*, 1895–1915.
25. Munteanu, I.G.; Apetrei, C. Tyrosinase-Based Biosensor—A New Tool for Chlorogenic Acid Detection in Nutraceutical Formulations. *Materials* **2022**, *15*, 3221. [CrossRef] [PubMed]
26. Ferrario, A.; Scaramuzza, M.; Pasqualotto, E.; de Toni, A.; Paccagnella, A. Development of a Disposable Gold Electrodes-Based Sensor for Electrochemical Measurements of cDNA Hybridization. *Procedia Chem.* **2012**, *6*, 36–45. [CrossRef]
27. Bard, A.J.; Faulkner, L.R.; White, H.S. *Electrochemical Methods: Fundamentals and Applications*; Wiley: Hoboken, NJ, USA; Wiley: Chichester, UK, 2022.
28. He, L.; Huang, R.; Xiao, P.; Liu, Y.; Jin, L.; Liu, H.; Li, S.; Deng, Y.; Chen, Z.; Li, Z.; et al. Current signal amplification strategies in aptamer-based electrochemical biosensor: A review. *Chin. Chem. Lett.* **2021**, *32*, 1593–1602. [CrossRef]
29. Li, L.; Liu, X.; Yang, L.; Zhang, S.; Zheng, H.; Tang, Y.; Wong, D.K. Amplified oxygen reduction signal at a Pt-Sn-modified TiO2 nanocomposite on an electrochemical aptasensor. *Biosens. Bioelectron.* **2019**, *142*, 111525. [CrossRef] [PubMed]
30. Sharma, A.; Bhardwaj, J.; Jang, J. Label-Free, Highly Sensitive Electrochemical Aptasensors Using Polymer-Modified Reduced Graphene Oxide for Cardiac Biomarker Detection. *ACS Omega* **2020**, *5*, 3924–3931. [CrossRef]
31. Leung, K.K.; Downs, A.M.; Ortega, G.; Kurnik, M.; Plaxco, K.W. Elucidating the Mechanisms Underlying the Signal Drift of Electrochemical Aptamer-Based Sensors in Whole Blood. *ACS Sens.* **2021**, *6*, 3340–3347. [CrossRef] [PubMed]
32. Zamani, M.; Klapperich, C.M.; Furst, A.L. Recent advances in gold electrode fabrication for low-resource setting biosensing. *Lab A Chip* **2023**, *23*, 1410–1419. [CrossRef]
33. Petrila, I.; Tudorache, F. Effects of sintering temperature on the microstructure, electrical and magnetic characteristics of copper-zinc spinel ferrite with possibility use as humidity sensors. *Sens. Actuators A Phys.* **2021**, *332*, 113060. [CrossRef]
34. Tortorich, R.P.; Shamkhalichenar, H.; Choi, J.-W. Inkjet-Printed and Paper-Based Electrochemical Sensors. *Appl. Sci.* **2018**, *8*, 288. [CrossRef]
35. Im, J.; Trindade, G.F.; Quach, T.T.; Sohaib, A.; Wang, F.; Austin, J.; Turyanska, L.; Roberts, C.J.; Wildman, R.; Hague, R.; et al. Functionalized Gold Nanoparticles with a Cohesion Enhancer for Robust Flexible Electrodes. *ACS Appl. Nano Mater.* **2022**, *5*, 6708–6716. [CrossRef] [PubMed]
36. Wang, Z.; Wang, Y.; Wang, M.; Zheng, Q. Analysis of Adhesion Strength between Silver Film and Substrate in Plain Silver Surface Plasmon Resonance Imaging Sensor. *Sens. Mater.* **2022**, *34*, 1629. [CrossRef]
37. Yoon, J.; Shin, M.; Lee, T.; Choi, J.W. Highly Sensitive Biosensors Based on Biomolecules and Functional Nanomaterials Depending on the Types of Nanomaterials: A Perspective Review. *Materials* **2020**, *13*, 299. [CrossRef] [PubMed]
38. Liang, S.; Schwartzkopf, M.; Roth, S.V.; Müller-Buschbaum, P. State of the art of ultra-thin gold layers: Formation fundamentals and applications. *Nanoscale Adv.* **2022**, *4*, 2533–2560. [CrossRef] [PubMed]
39. Rohrbach, F.; Karadeniz, H.; Erdem, A.; Famulok, M.; Mayer, G. Label-free impedimetric aptasensor for lysozyme detection based on carbon nanotube-modified screen-printed electrodes. *Anal. Biochem.* **2012**, *421*, 454–459. [CrossRef]
40. Lazanas, A.C.; Prodromidis, M.I. Electrochemical Impedance Spectroscopy-A Tutorial. *ACS Meas. Sci. Au* **2023**, *3*, 162–193. [CrossRef]
41. Tatarko, M.; Spagnolo, S.; Csiba, M.; Šubjaková, V.; Hianik, T. Analysis of the Interaction between DNA Aptamers and Cytochrome C on the Surface of Lipid Films and on the MUA Monolayer: A QCM-D Study. *Biosensors* **2023**, *13*, 251. [CrossRef]
42. Dutta, G.; Fernandes, F.C.; Estrela, P.; Moschou, D.; Bueno, P.R. Impact of surface roughness on the self-assembling of molecular films onto gold electrodes for label-free biosensing applications. *Electrochim. Acta* **2021**, *378*, 138137. [CrossRef]
43. Oberhaus, F.V.; Frense, D.; Beckmann, D. Immobilization Techniques for Aptamers on Gold Electrodes for the Electrochemical Detection of Proteins: A Review. *Biosensors* **2020**, *10*, 45. [CrossRef]
44. Määttänen, A.; Vanamo, U.; Ihalainen, P.; Pulkkinen, P.; Tenhu, H.; Bobacka, J.; Peltonen, J. A low-cost paper-based inkjet-printed platform for electrochemical analyses. *Sens. Actuators B Chem.* **2013**, *177*, 153–162. [CrossRef]
45. Xu, X.; Makaraviciute, A.; Kumar, S.; Wen, C.; Sjödin, M.; Abdurakhmanov, E.; Danielson, U.H.; Nyholm, L.; Zhang, Z. Structural Changes of Mercaptohexanol Self-Assembled Monolayers on Gold and Their Influence on Impedimetric Aptamer Sensors. *Anal. Chem.* **2019**, *91*, 14697–14704. [CrossRef] [PubMed]

46. Pandhi, T.; Cornwell, C.; Fujimoto, K.; Barnes, P.; Cox, J.; Xiong, H.; Davis, P.H.; Subbaraman, H.; Koehne, J.E.; Estrada, D. Fully inkjet-printed multilayered graphene-based flexible electrodes for repeatable electrochemical response. *RSC Adv.* **2020**, *10*, 38205–38219. [CrossRef] [PubMed]
47. Libansky, M.; Zima, J.; Barek, J.; Reznickova, A.; Svorcik, V.; Dejmkova, H. Basic electrochemical properties of sputtered gold film electrodes. *Electrochim. Acta* **2017**, *251*, 452–460. [CrossRef]
48. Wang, Y.; Wang, X.; Lu, W.; Yuan, Q.; Zheng, Y.; Yao, B. A thin film polyethylene terephthalate (PET) electrochemical sensor for detection of glucose in sweat. *Talanta* **2019**, *198*, 86–92. [CrossRef] [PubMed]
49. Watkins, Z.; Karajic, A.; Young, T.; White, R.; Heikenfeld, J. Week-Long Operation of Electrochemical Aptamer Sensors: New Insights into Self-Assembled Monolayer Degradation Mechanisms and Solutions for Stability in Serum at Body Temperature. *ACS Sens.* **2023**, *8*, 1119–1131. [CrossRef]
50. Qiao, X.; Li, K.; Xu, J.; Cheng, N.; Sheng, Q.; Cao, W.; Yue, T.; Zheng, J. Novel electrochemical sensing platform for ultrasensitive detection of cardiac troponin I based on aptamer-MoS2 nanoconjugates. *Biosens. Bioelectron.* **2018**, *113*, 142–147. [CrossRef] [PubMed]
51. Wang, B.; Jing, R.; Qi, H.; Gao, Q.; Zhang, C. Label-free electrochemical impedance peptide-based biosensor for the detection of cardiac troponin I incorporating gold nanoparticles modified carbon electrode. *J. Electroanal. Chem.* **2016**, *781*, 212–217. [CrossRef]
52. Meng, F.; Huang, J.; Zhang, H.; Zhao, P.; Li, P.; Wang, C. Metal Coating Synthesized by Inkjet Printing and Intense Pulsed-Light Sintering. *Materials* **2019**, *12*, 1289. [CrossRef]
53. Serafín, V.; Torrente-Rodríguez, R.M.; González-Cortés, A.; García de Frutos, P.; Sabaté, M.; Campuzano, S.; Yáñez-Sedeño, P.; Pingarrón, J.M. An electrochemical immunosensor for brain natriuretic peptide prepared with screen-printed carbon electrodes nanostructured with gold nanoparticles grafted through aryl diazonium salt chemistry. *Talanta* **2018**, *179*, 131–138.
54. Shafkat, A.; Rashed, A.N.Z.; El-Hageen, H.M.; Alatwi, A.M. The Effects of Adding Different Adhesive Layers with a Microstructure Fiber Sensor Based on Surface Plasmon Resonance: A Numerical Study. *Plasmonics* **2021**, *16*, 819–832. [CrossRef]
55. Kotál, V.; Švorčík, V.; Slepička, P.; Sajdl, P.; Bláhová, O.; Šutta, P.; Hnatowicz, V. Gold Coating of Poly(ethylene terephthalate) Modified by Argon Plasma. *Plasma Process. Polym.* **2007**, *4*, 69–76. [CrossRef]
56. Drobota, M.; Butnaru, M.; Vornicu, N.; Plopa, O.; Aflori, M. Facile Method for Obtaining Gold-Coated Polyester Surfaces with Antimicrobial Properties. *Adv. Polym. Technol.* **2020**, *2020*, 4504062. [CrossRef]
57. Pellitero, M.A.; Curtis, S.D.; Arroyo-Currás, N. Interrogation of Electrochemical Aptamer-Based Sensors via Peak-to-Peak Separation in Cyclic Voltammetry Improves the Temporal Stability and Batch-to-Batch Variability in Biological Fluids. *ACS Sens.* **2021**, *6*, 1199–1207. [CrossRef] [PubMed]

Disclaimer/Publisher's Note: The statements, opinions and data contained in all publications are solely those of the individual author(s) and contributor(s) and not of MDPI and/or the editor(s). MDPI and/or the editor(s) disclaim responsibility for any injury to people or property resulting from any ideas, methods, instructions or products referred to in the content.

Article

Using AuNPs-DNA Walker with Fluorophores Detects the Hepatitis Virus Rapidly

Baining Sun [1,†], Chenxiang Zheng [1,†], Dun Pan [1], Leer Shen [2], Wan Zhang [1], Xiaohua Chen [2,*], Yanqin Wen [1,*] and Yongyong Shi [1]

1. Bio-X Institutes, Key Laboratory for the Genetics of Developmental and Neuropsychiatric Disorders (Ministry of Education), Shanghai Jiao Tong University, Shanghai 200030, China; hibatsuna1827@sjtu.edu.cn (B.S.); cxzheng@sjtu.edu.cn (C.Z.); pandun@sjtu.edu.cn (D.P.); wanzh318@sjtu.edu.cn (W.Z.); shiyongyong@sjtu.edu.cn (Y.S.)
2. Department of Infectious Diseases, Shanghai Sixth People's Hospital Affiliated to Shanghai Jiao Tong University School of Medicine, Shanghai 200233, China; rachel_slr@sjtu.edu.cn
* Correspondence: chenxiaohua2000@163.com (X.C.); yqwen@sjtu.edu.cn (Y.W.)
† These authors contribute equally to this work.

Abstract: Viral hepatitis is a systemic infectious diseases caused by various hepatitis viruses, primarily leading to liver damage. It is widely prevalent worldwide, with hepatitis viruses categorized into five types: hepatitis A, B, C, D, and E, based on their etiology. Currently, the detection of hepatitis viruses relies on methods such as enzyme-linked immunosorbent assay (ELISA), immunoelectron microscopy to observe and identify viral particles, and in situ hybridization to detect viral DNA in tissues. However, these methods have limitations, including low sensitivity, high error rates in results, and potential false negative reactions due to occult serum infection conditions. To address these challenges, we have designed an AuNPs-DNA walker method that uses gold nanoparticles (AuNPs) and complementary DNA strands for detecting viral DNA fragments through a colorimetric assay and fluorescence detection. The DNA walker, attached to gold nanoparticles, comprises a long walking strand with a probe sequence bound and stem-loop structural strands featuring a modified fluorescent molecule at the 3′ end, which contains the DNAzyme structural domain. Upon the addition of virus fragments, the target sequence binds to the probe chains. Subsequently, the long walking strand is released and continuously hybridizes with the stem-loop structural strand. The DNAzyme undergoes hydrolytical cleavage by Mg^{2+}, breaking the stem-loop structural strand into linear single strands. As a result of these structural changes, the negative charge density in the solution decreases, weakening spatial repulsion and rapidly reducing the stability of the DNA walker. This leads to aggregation upon the addition of a high-salt solution, accompanied by a color change. Virus typing can be performed through fluorescence detection. The innovative method can detect DNA/RNA fragments with high specificity for the target sequence, reaching concentrations as low as 1 nM. Overall, our approach offers a more convenient and reliable method for the detection of hepatitis viruses.

Keywords: hepatitis viruses; gold nanoparticles; DNA walker

Citation: Sun, B.; Zheng, C.; Pan, D.; Shen, L.; Zhang, W.; Chen, X.; Wen, Y.; Shi, Y. Using AuNPs-DNA Walker with Fluorophores Detects the Hepatitis Virus Rapidly. *Biosensors* 2024, 14, 370. https://doi.org/10.3390/bios14080370

Received: 5 May 2024
Revised: 19 July 2024
Accepted: 23 July 2024
Published: 29 July 2024

Copyright: © 2024 by the authors. Licensee MDPI, Basel, Switzerland. This article is an open access article distributed under the terms and conditions of the Creative Commons Attribution (CC BY) license (https://creativecommons.org/licenses/by/4.0/).

1. Introduction

Viral hepatitis is widely spread worldwide, and there are five types of hepatitis viruses classified according to etiology [1,2], namely hepatitis A, B, C, D, and E. China is a high incidence area of viral hepatitis. Hepatitis A was reported to exist in China for over 5000 years [3]. The population prevalence of hepatitis A (anti-HAV positive) is about 80%. There were more than 292 million HBsAg carriers worldwide in 2016 [4], including 120 million in China, while the number increased to 296 million in 2019 and 820,000 deaths worldwide [5]. About 1% of the world's population is infected with HCV, and in some regions, such as West Africa or Central Africa, hepatitis C infections account for about 5% to 8% of the infected population [6]. The clinical manifestations of all viral hepatitis are similar, including mainly fatigue, loss of appetite, oil aversion, abnormal liver function, and jaundice in some cases [7]. In severe cases, acute liver failure and hepatitis cirrhosis may occur, as well as multiple complications such as hepatic encephalopathy, upper gastrointestinal bleeding, and hepatorenal syndrome, and some patients may even need liver transplantation to keep them alive later. About 887,000 people die every year from hepatitis B and related diseases, mainly related to advanced liver fibrosis and cirrhosis [4,5,8].

The hepatitis virus was identified in infected people and led to the development of diagnostic tests, molecular characterization, and propagation in cell culture [9]. Currently, the method to detect and diagnose hepatitis viruses relies on enzyme-linked immunosorbent assay (ELISA), radioimmunoassay, immunoelectron microscopy to observe and identify viral particles, and in situ hybridization to detect viral DNA in tissues [2,6,9–13]. Although enzyme-linked immunosorbent assay (ELISA) is relatively convenient, the detection sensitivity of this method is low and causes considerable errors. It may also lead to false negative results and missing detection due to the latent serum infection. Compared with serology, which relies on the concentration of the protein immune response, developed nucleic acid molecular detection technology is considered a more sensitive and accurate diagnostic method.

To detect diseases more quickly and accurately (such as the influenza virus genome and cancer markers), some new detection technologies are developed, such as DNA nanotechnology [14,15], the CRISPR/Cas system [16], electrochemical biosensors [17], etc. Nucleic acid detection has been widely used to detect hepatitis virus in samples of different origins, like blood, saliva, and other clinical specimens [7]. Hepatitis virus has been detected by techniques such as restriction fragment length polymorphism [18], Southern blotting [19], amplification based on nucleic acid sequencing [20], reverse transcription-PCR (RT-PCR) [21], antigen capture RT-PCR [22], etc. There are also many ways to test for viral load, such as ultraviolet (UV) spectrophotometry, polymerase chain reaction (PCR), real-time PCR (rt-PCR), digital PCR, loop-mediated isothermal amplification (LAMP), transcription-mediated amplification (TMA), nucleic acid sequence-based amplification (NASBA), rolling circle amplification (RCA) as well as electrochemical, quartz crystal microbalance, microcantilever, and surface plasmon resonance biosensors [4,5,23–25]. However, these methods have certain limitations. For example, PCR is not fit for short-length oligonucleotides; the cost of fluorescence assay analysis can be high as expensive reagents and instruments may be required. Therefore, developing convenient and sensitive virus DNA/RNA detection methods is urgently needed.

The gold nanoparticles (AuNPs) method is one of the most studied nanomaterials for biomedical applications. AuNPs can be functionalized with various biomolecules, such as nucleic acids or antibodies, to recognize and bind to specific targets [26]. AuNPs have been tested with impressive results as a biosensor, contrast agent, and therapeutic agent [27]. It has been reported that small-size AuNPs (about 3 nm) exhibit antitumor activity against breast cancer (MCF-7) cell lines and colon carcinoma (HCT-116) cell lines, but no cytotoxicity to the human embryonic normal kidney cell line (HEK 293) [28]. In addition, AuNPs can be connected to anti-cancer drugs through surface functionalization, thus playing a role in drug delivery. Small-size AuNPs were used to pair with the anti-cancer drug methotrexate (MTX) and evaluate the stability and specificity of its drug delivery [29]. In addition to drugs and proteins, DNA can also be linked to AuNP by forming covalent bonds with functional groups (such as amine or mercaptan groups) on the DNA molecule, or by hybridizing single-stranded DNA to complementary sequences attached to the AuNP surface [30]. AuNPs-based biosensors are mainly used to detect small molecules, DNA and proteins, and use AuNPs' surface plasmonic resonance (SPR) characteristics to detect target molecules with high sensitivity and selectivity [31]. DNA nanotechnology has made great strides in the past year. The DNA walker mimics natural molecular motors by biasing chemical energy against Brownian motion [32–34]. The DNA walker can automatically move in one direction along the track without intervention. According to this characteristic, the synthetic DNA walker has been widely used in nucleic acid amplification detection in recent years [35–38].

In this work, we designed a DNA walker using gold nanoparticles (AuNPs) and DNA complementary strands to detect hepatitis viral DNA fragments using colorimetric assay and fluorescence detection. Our study shows that the DNA walker operates at room temperature without the requirement of protein enzymes and temperature controllers after adding the target sequences and obtains the color change of the solution and fluorescence intensity change as an output signal. The method detects DNA fragments down to 1 nM with high specificity for the target sequence. This method provides a more convenient method for the reliable detection of hepatitis viruses.

2. Materials and Methods

2.1. Apparatus

UV-vis absorption spectra and fluorescence emission spectra were measured with a UV Spectrometer (Thermo Fisher Technology (China) Co., Ltd., Shanghai, China) and Steady State and Transient State Fluorescence Spectrometer (Edinburgh Instruments, Livingston, Scotland, UK), respectively, at the Instrumental Analysis Center of Shanghai Jiao Tong University. The centrifuge was SORVALL Legend MICRO 17R (Thermo Fisher Technology (China) Co., Ltd., Shanghai, China).

2.2. Materials and Reagents

Gold nanoparticles were purchased from Xi'an ruixi Biological Technology Co., Ltd. (Xi'an, China) (20 nm in diameter and spherical). These AuNPs are 20 nm in diameter, spherical, and coated with sodium citrate. The color of the AuNPs colloid is red, and the UV-vis shows a maximum absorbance at 520 nm and owns a uniform size distribution with an average size of 20 nm (Figure S1). Tris (2-carboxyethyl) phosphine hydro-chloride (TCEP) was purchased from Sigma (Shanghai, China). Tris-base was purchased from Sangon Biotech Co., Ltd. (Shanghai, China). Boronic acid and ethylenediaminetetraacetic acid tetrasodium (EDTA) were supplied by Sinopharm Chemical Reagent Co., Ltd. (Shanghai, China). DNA sequences (Table S1) were synthesized and purified by Sangon Biotech Co., Ltd. (Shanghai, China).

2.3. Preparation of DNA Walker

The DNA walker consists of three different types of ssDNA: a long walking strand, two probe strands, and a stem–loop strand. The 5′ ends of the long walking chain and the stem–loop chain are modified with sulfhydryl groups so that they can be attached to the AuNPs' surface. The 3′ end of the stem–loop chain is modified with fluorophores (FAM, ROX, and Cy5 were used for HAV, HBV, and HCV, respectively). The target sequence of the synthesized virus was the conserved region sequence, which was derived from NCBI (Table S1).

2.4. Preparation of AuNPs-DNA Walker

Long walking strands and blocking strands were mixed in one tube with a molar ratio of 1:3 in annealing buffer. The reaction solution was heated to 95 °C for 10 min and gradually cooled to room temperature to ensure that the blocking strands could completely bind with the walking strands. The stem–loop strands were also annealed from 95 °C to room temperature to form secondary structures. The above two groups of solutions were incubated with TCEP at a 1:50 molar ratio for 2 h at room temperature to reduce the formation of disulfide bonds. The AuNPs, locked walking strands, and stem–loop strands were mixed at a 1:20:200 molar ratio and incubated in darkness at 4 °C for 16 h. Then, the sodium chloride solution was gradually added to the above mixture at 40 min intervals to achieve a final concentration of 0.2 M of NaCl. The solution was incubated further in the dark at 4 °C for 24 h. After that, the solution was centrifuged at 4 °C at 13,000 rpm for 30 min to separate the AuNPs-DNA walker from the effluents. The AuNPs-DNA walker was washed three times with washing buffer and re-suspended in a reaction buffer as a working solution.

2.5. Colorimetric Detection and Fluorescence Detection of Virul Fragments

First, 3 µL of 10 µM of the target fragments was added to 160 µL of working solution containing 1 nM of the AuNPs-DNA walker and performed at room temperature for 3 h. Then, the above solution was centrifuged at 13,000 rpm (16,600 rcf) for 30 min at 4 °C, and 110 µL of the supernatant was measured for fluorescence study. Take HAV, for example: the AuNPs-DNA walker for detecting HAV target sequences is labeled with FAM, whose excitation light is 490 nm and emission light is 520 nm. Fluorescence spectra were detected at an excitation light of 490 nm and emission light of 520 nm. The remaining 50 µL of solution was re-suspended, and 600 mM of $MgCl_2$ was added, followed by visual observation and UV-vis absorption measurements.

3. Results

3.1. Operating Principle of the AuNPs-DNA Walker

We designed a DNA walker using gold nanoparticles (AuNPs) and DNA complementary strands to detect viral DNA fragments using a colorimetric assay and fluorescence detection (Figure 1). Attached to an AuNP, the DNA walker consists of a long walking strand with two probe strands bound to it and stem–loop structural strands with a modified fluorescent molecule at the 3' end, which contains the DNAzyme structural domain. These DNAs were adsorbed on the AuNP by the sulfhydryl group at the 5' end. Due to the repulsion between DNA, the electrostatic field force between the AuNPs became more extensive, so the aggregation phenomenon did not occur. The color of the AuNPs-DNA walker working solution was still pink. After the virus fragment was added, the target sequences would be wholly bound to the probe. The long walking chain would be released and continuously hybridized with the stem–loop structural chain. The DNAzyme would be hydrolytically cleaved by Mg^{2+}, cutting the stem–loop structural chain into two linear single strands. One of the linear single chains was still attached to the AuNP, while the other containing the fluorescent molecule was released into the solution. When stem–loop structure chains were bound to the surface of the AuNP, fluorescence molecules were quenching due to the close distance between them. Then, the long walking chain would be released again and automatically found the next stem–loop chain. The previous step would be repeated until the DNAzyme hydrolyzed all the stem–loop chains on the AuNP into two short linear chains. After the step of DNAzyme cleaving, the part of the single chain containing fluorophores was released into the solution, and the quenching effect then disappeared. Changes in the DNA of AuNPs led to weaker repulsion between them, resulting in the aggregation of AuNPs when added to the salt solution. The color of the solution would change from pink to blue or bluish-purple. The colloidal stability before and after assembling the DNA walker was determined by the transmission electron microscopy (TEM) image and dynamic light scattering characterization (Figure S2). It can be seen that the AuNP before and after the function has a good dispersion, no large area aggregation phenomenon, and the AuNPs' size increases.

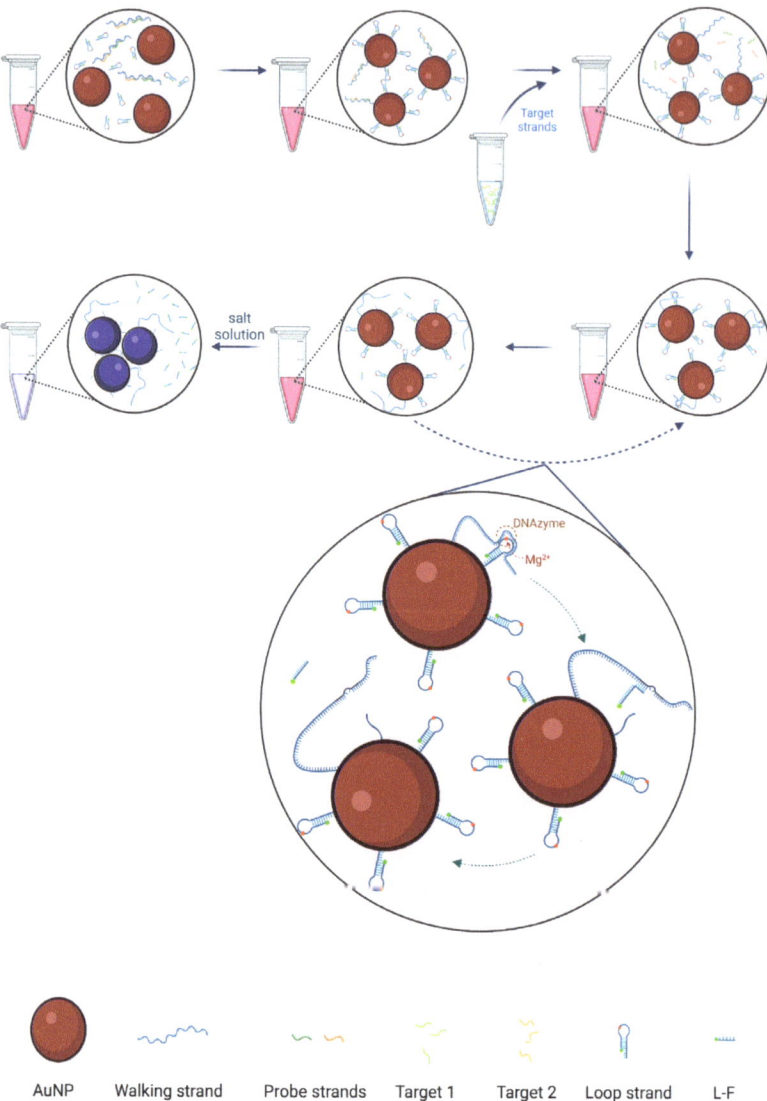

Figure 1. Schematic diagram of AuNPs-DNA walker for detecting viral DNA/RNA. AuNPs solution is pink. The color of the solution did not change after ssDNA coupling. After adding the target sequence, the color of the solution changes from pink to blue or bluish-purple visually after DNA strand replacement and salt addition.

3.2. Colorimetric Response of the AuNPs-DNA Walker to HAV Target Sequences

To test whether the AuNPs-DNA walker functioned as expected, we added the HAV target sequence to the AuNPs-DNA walker working fluid. Then, we observed the color change of the solution, UV-vis detection, and fluorescence spectrum detection of the supernatant. As a result, as shown in Figure 2A, after we added the HAV target sequence and $MgCl_2$ solution, the color changed from pink to bluish-purple compared to the solution without the virus target sequence. UV-vis showed a red shift in the maximum absorbance from 520 nm to 550 nm, and a new absorption band appeared at 610 nm (Figure 2B). As a blank control, the color of the ordinary AuNPs solution without ssDNA was red, and

UV-vis showed a peak at 520 nm (Figure S1). The results of the fluorescence spectrum detection showed that the AuNPs-DNA walker worked after adding the target sequence. The solution contained a linear single chain containing fluorophore FAM released after the DNAzyme hydrolyzed the stem–loop chain, and the emission light peak appeared at 519 nm (Figure 2C). The transmission electron microscopy (TEM) image showed the dispersion of the AuNPs-DNA walker with or without the addition of target sequences (Figure 2D,E). With this method, we also successfully detected the inclusion of HBV (Figure 3) and HCV target sequences (Figure S3): the visual observation and UV–vis absorption spectra of normal AuNPs.

To determine the optimal NaCl concentration to promote DNA binding to AuNPs, we developed an NaCl concentration gradient experiment and screened it using UV-vis detection and fluorescence spectroscopy. After mixing ssDNA with AuNPs, a certain amount of 2 M of NaCl solution was added to make the final concentrations of 0.05 M, 0.1 M, 0.2 M, 0.3 M, and 0.4 M, respectively. We explored the color changes of adding 600 mM of $MgCl_2$ solution into the AuNPs-DNA walker with different final concentrations of NaCl. The color turns purple at 0.05 M and 0.1 M, there is no significant change in color at 0.2 M, and the color becomes lighter at 0.3 M and 0.4 M (Figure S4A). In addition, we added 1.5 µL of 10 µM of the target sequence chains into the AuNPs-DNA walker with different final concentrations of NaCl, and performed at room temperature for 3 h, centrifuged at 13,000 rpm, and the supernatant was taken to detect the fluorescence spectrum. The results show that the fluorescence intensity was the highest at 0.2 M concentration, which meant that the binding amount of ssDNA and AuNPs was the highest at this concentration (Figure S4B).

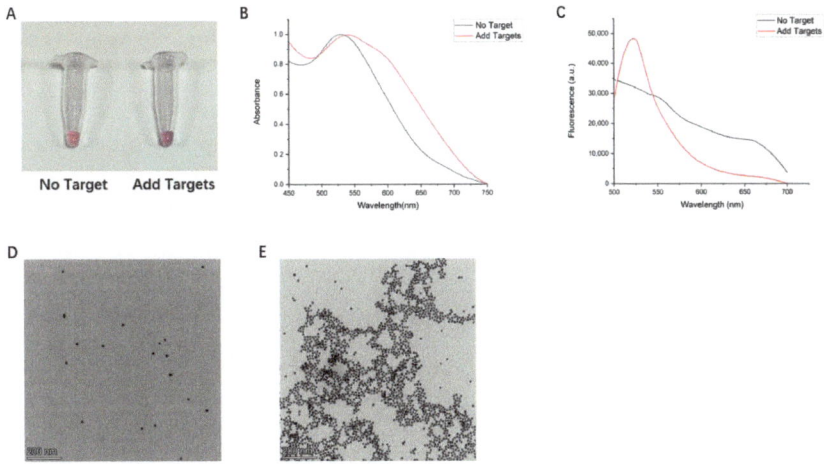

Figure 2. (**A**) The visual observation of AuNPs-DNA walker treated with/without target sequences. (**B**) UV–vis absorption spectra and (**C**) fluorescence spectra of the AuNPs-DNA walker in the absence or the presence of target sequences. Transmission electron microscopy (TEM) image of AuNPs-DNA walker treated with (**D**)/without (**E**) target sequences.

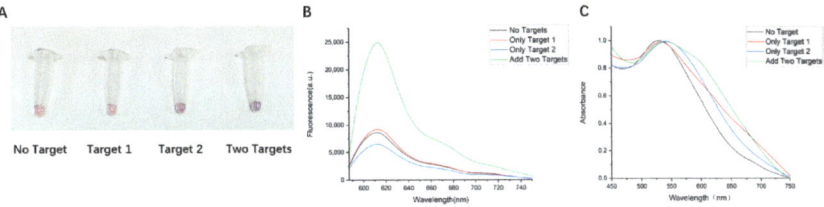

Figure 3. Cont.

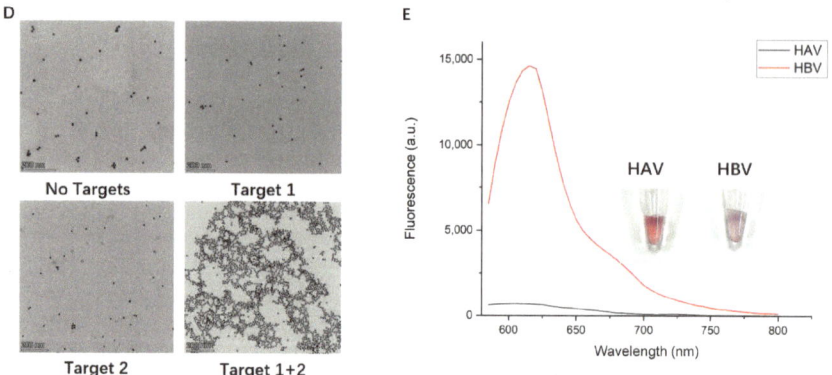

Figure 3. (**A**) Visual observation, (**B**) fluorescence spectra, (**C**) UV–vis absorption spectra, and TEM images (**D**) of AuNPs-DNA walker treated with no target, only one target, and both targets. (**E**) Fluorescence spectra and visual differentiation from HAV target sequences and HBV target sequences.

3.3. Specific Detection of the AuNPs-DNA Walker

To detect the specificity of the AuNPs-DNA walker, we added target sequence 1, target sequence 2, and both target sequences to the AuNPs-DNA walker working solution, respectively. After the salt solution was added, the color of the solution without the target sequence and the solution with target sequence 1 did not change significantly. In contrast, the color of the solution with target sequence 2 changed slightly, and the color of the solution with both target sequences turned bluish-purple (Figure 3A). The fluorescence spectra showed a noticeable intensity difference between the four tubes (Figure 3B), while UV-vis detection was consistent with the color change of the solution (Figure 3C). Although there were color changes and red shifts in the tube added to target sequence 2, there was no apparent fluorescent luminescence. This might be related to the location of the long walking chain exposed after the target sequence binds to the probe chain. TEM showed the dispersion of the AuNPs-DNA walker with or without the addition of target sequences (Figure 3D). We also used the HAV target sequences as a confusion chain and added them into the HBV-specific AuNPs-DNA walker solution. The results show that the AuNPs-DNA walker could distinguish the non-target sequence from the target sequence with reasonable specificity (Figure 3E).

3.4. Test of the AuNPs-DNA Walker to Virus Target Fragments

Firstly, we examined the minimum detection concentration of the target sequence by this method. We designed a concentration gradient from 0 nM to 50 nM for the HCV target sequence and tested it with the AuNPs-DNA walker. The results show that as the concentration of the double target sequences increased, the color of the detection solution changed from red to purple at 1 nM (Figure 4A). The low concentration mixture was further detected by transmission electron microscopy. It was found that AuNPs only partially coagulated at a low concentration, which was consistent with the change in color of the solution under visual observation, which was not as obvious as that under a high concentration (Figure S5). To verify the success of this method on the viral fragments extracted from the blood samples of infected patients, we collected blood samples from HCV patients. We obtained HCV single-stranded DNA samples by RT-PCR and asymmetric amplification. HCV target fragments were added into the solution containing the AuNPs-DNA walker, shaken for 30 min, and then performed at room temperature for 2 h, followed by salt addition and visual observation. The concentration of the HCV sample used was 200 copies/mL, and the results show that a significant color change in the solution could be seen visually (Figure 4B). The detection limit was estimated to be 200 copies/mL, which could meet the requirements of individuals with a high diagnostic transmission rate and short duration of symptom onset.

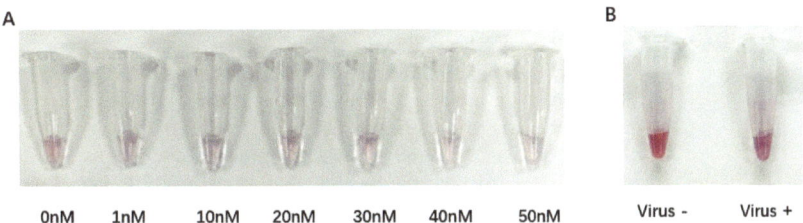

Figure 4. (**A**) Visual observation of the AuNPs-DNA walker treated with different concentrations of target chains within the range from 0 to 50 nM. (**B**) Visual observation of the AuNPs-DNA walker treated with/without HCV virus fragments.

4. Discussion

In this study, we combined gold nanoparticles and DNA strand replacement technology to design an AuNPs-DNA walker for the rapid detection of hepatitis virus. We took conserved region sequences of different types of hepatitis viruses as target sequences, connected corresponding different fluorophores at the 3' end of stem-loop chain (HAV for FAM, HBV for ROX, HCV for Cy5), and used fluorescence spectrum detection technology to distinguish hepatitis virus types. The method does not require the amplification/reverse transcription of viral DNA/RNA in the blood samples of hepatitis patients, and the minimum detection threshold is set to 200 copies/mL.

The electrochemical method was proposed for the detection of the hepatitis C virus (HCV) RNA level and identification of the HCV-1b genotype based on the site-specific cleavage of BamHI endonuclease combined with gold nanoparticles (AuNPs) signal amplification [39]. This method's procedures include reverse transcription, PCR amplification, and electrochemical detection. In 2014, it was reported that a new assay using magnetic nanoparticles and unmodified cationic gold nanoparticles was developed for detecting hepatitis C virus in serum samples and tested in clinical samples [40]. The specificity and sensitivity were 96% and 96.5%, respectively, and the detection limit was 15 IU/mL. Meanwhile, in 2017, Sherif Shawky et al. used RT-PCR and nano-assay to quantitatively detect HCV RNA samples with detection limits as low as 4.57 IU/mL [41]. In response to the low abundance of viral nucleic acid in the blood of early patients, Clarke et al. obtained a high-quality, reproducible surface-to-enhanced raman spectroscopy (SERS) with report-modified gold nanoparticles to detect anti-HCV antibodies in the blood samples of HCV-infected patients [42]. Feng Tao et al. then combined the DNA walker and catalytic hairpin assembly (CHA) to perform the targeted detection of HBV DNA. Their method achieves a wide detection range of 0.5 nM to 50 nM, with detection limits as low as 0.20 nM [37]. An isothermal amplification technique based on digital ring street was developed in 2020, and the HCV viral nucleic acid in the blood samples was detected by silica coating and AuNPs [43]. This system could detect HBV-DNA at a concentration of 10 to 1×10^4 copies/μL. During the COVID-19 pandemic, there have been many reports of viral RNA detection using colorimetric sensing methods of nanoparticles to detect whether patients are infected with COVID-19 quickly [36,44]. Jiafeng Pan et al. combined DNA logic gates with various fluorophores to identify COVID-19, SARS-CoV, and Bat-SL-CoVZC45 simultaneously [15]. However, this method needs to add exonuclease III to assist operation, which is inconvenient in practical application. To provide more accurate detection, Maha Alafeef et al. used a dual-targeted approach to detect early infected samples with low viral loads, reducing the detection limit to 10 copies/μL [45]. Kai Zhang et al. optimized DNA probes and nanomaterials, and lowered the detection limit to 59 aM based on the electrochemical luminescence detection method [46]. Recently, Laibao Zheng et al. utilized 3D-DNA walking nanomachines for the sensitive detection of hepatitis C virus. This method has shown excellent sensitivity in detecting HCV with a detection limit of 42.4 pM and a linear range of 100 pM to 2 nM [38].

Our method combines the DNA walker and AuNPs. After adding the target sequence, through DNA strand replacement and DNA walker operation, the mutual repulsion between AuNPs decreased, and the resistance to salt solution weakened, so that the color of the solution changed visually. This method is more practical for the preliminary screening of people in remote areas without instrument detection. In addition, our methods adopted AuNPs and single-strand DNA instead of the traditional ELISA test. Under the condition that the price is nearly the same, traditional ELISA kits can serve 48 people, whereas our AuNPs-DNA walker can accommodate over 200 people. In terms of time, traditional ELISA tests typically require 4–6 h, whereas ours only take 3 h and are much simpler to operate. Therefore, compared to traditional testing methods, our approach significantly enhances cost-effectiveness. Although our method can detect the type of hepatitis virus more easily, it cannot further detect its corresponding subtype, and for the double-DNA strand virus of HBV, it is still necessary to convert the viral DNA into ssDNA by asymmetric amplification technology before detection. The process can have conditions such as base mutations that can lead to false positives/false negatives. These are the limitations of the method, which can be further studied in the future.

5. Conclusions

In summary, we developed an AuNPs-DNA walker using gold nanoparticles (AuNPs) and DNA complementary strands for the detection of viral DNA fragments through colorimetric and fluorescence assays. Our AuNPs-DNA walker can detect viral target sequences with high specificity down to low concentrations (1 nM). Compared to commonly used methods, such as ELISA, our design offers several advantages. It eliminates the need for specialized equipment and does not require proteases. Additionally, it can simultaneously identify the presence of three viral genes. The AuNPs-DNA walker is straightforward to perform at room temperature, yielding test results quickly and producing clear, visualized outcomes. Furthermore, this method is cost-effective and easily scalable, making it accessible for widespread use. Overall, our approach provides a more convenient and reliable solution for the detection of hepatitis viruses.

Supplementary Materials: The following supporting information can be downloaded at: https://www.mdpi.com/article/10.3390/bios14080370/s1, Figure S1: The form and color of AuNPs; Figure S2: Size and stability before and after AuNP function. Figure S3: UV–vis absorption spectra and fluorescence spectra of the AuNPs-DNA Walker in the absence or the presence of HCV target DNA fragments. Figure S4: Color changes and fluorescence spectra of AuNPs-DNA walker aged with NaCl at different final concentrations after adding salt solution; Figure S5: TEM image of adding different concentration of target sequences. Table S1: Sequences of oligonucleotides used in this study.

Author Contributions: We would like to acknowledge all the participants in this study. Y.S., X.C. and Y.W. designed and supervised the whole research process. B.S., L.S., Y.W. and D.P. carried out all the experiments and managed the literature searches and analyses. L.S. and X.C. conducted the sample collection and verification. B.S., C.Z. and W.Z. undertook the statistical analysis. Y.W. and D.P. were responsible for the platform coordination and management. X.C. and Y.W. proofread the article. B.S. wrote the first draft of the manuscript. All authors have read and agreed to the published version of the manuscript.

Funding: This research was funded by the National Key R&D Program of China (2021YFC2702100, 2021YFC2100600, 2019YFA0905400, 2022FYC2503700), Natural Science Foundation of China (32070679, 32370724, 82070615), Shanghai Science and Technology Innovation Action Plan (23Y11905200), the fundamental research funds for the central universities (YG2021ZD2020, YG2023QNB12, YG2023QNB20).

Institutional Review Board Statement: Not applicable.

Informed Consent Statement: Informed consent was obtained from all subjects involved in the study.

Data Availability Statement: Data will be made available on request.

Conflicts of Interest: The authors declare no conflicts of interest.

References

1. Rasche, A.; Sander, A.-L.; Corman, V.M.; Drexler, J.F. Evolutionary biology of human hepatitis viruses. *J. Hepatol.* **2019**, *70*, 501–520. [CrossRef]
2. Gregorio, G.V.; Mieli-Vergani, G.; Mowat, A.P. Viral hepatitis. *Arch. Dis. Child.* **1994**, *70*, 343–348. [CrossRef]
3. Feinstone, S.M. History of the Discovery of Hepatitis A Virus. *Cold Spring Harb. Perspect. Med.* **2019**, *9*, a031740. [CrossRef]
4. Guvenir, M.; Arikan, A. Hepatitis B Virus: From Diagnosis to Treatment. *Pol. J. Microbiol.* **2020**, *69*, 391–399. [CrossRef]
5. Jeng, W.J.; Papatheodoridis, G.V.; Lok, A.S.F. Hepatitis B. *Lancet* **2023**, *401*, 1039–1052. [CrossRef]
6. Pol, S.; Lagaye, S. The remarkable history of the hepatitis C virus. *Genes Immun.* **2019**, *20*, 436–446. [CrossRef]
7. Nainan, O.V.; Xia, G.; Vaughan, G.; Margolis, H.S. Diagnosis of hepatitis a virus infection: A molecular approach. *Clin. Microbiol. Rev.* **2006**, *19*, 63–79. [CrossRef]
8. Trepo, C.; Chan, H.L.; Lok, A. Hepatitis B virus infection. *Lancet* **2014**, *384*, 2053–2063. [CrossRef]
9. Feinstone, S.M.; Kapikian, A.Z.; Purceli, R.H. Hepatitis A: Detection by immune electron microscopy of a viruslike antigen associated with acute illness. *Science* **1973**, *182*, 1026–1028. [CrossRef]
10. Al-Sadeq, D.W.; Taleb, S.A.; Zaied, R.E.; Fahad, S.M.; Smatti, M.K.; Rizeq, B.R.; Al Thani, A.A.; Yassine, H.M.; Nasrallah, G.K. Hepatitis B Virus Molecular Epidemiology, Host-Virus Interaction, Coinfection, and Laboratory Diagnosis in the MENA Region: An Update. *Pathogens* **2019**, *8*, 63. [CrossRef]
11. Delem, A.D. Comparison of modified HAVAB and ELISA for determination of vaccine-induced anti-HAV response. *Biol. J. Int. Assoc. Biol. Stand.* **1992**, *20*, 289–291. [CrossRef]
12. Purcell, R.H.; Wong, D.C.; Moritsugu, Y.; Dienstag, J.L.; Routenberg, J.A.; Boggs, J.D. A microtiter solid-phase radioimmunoassay for hepatitis A antigen and antibody. *J. Immunol.* **1976**, *116*, 349–356. [CrossRef]
13. Moritsugu, Y.; Dienstag, J.L.; Valdesuso, J.; Wong, D.C.; Wagner, J.; Routenberg, J.A.; Purcell, R.H. Purification of hepatitis A antigen from feces and detection of antigen and antibody by immune adherence hemagglutination. *Infect. Immun.* **1976**, *13*, 898–908. [CrossRef]
14. Farzin, M.A.; Abdoos, H.; Saber, R. AuNP-based biosensors for the diagnosis of pathogenic human coronaviruses: COVID-19 pandemic developments. *Anal. Bioanal. Chem.* **2022**, *414*, 7069–7084. [CrossRef]
15. Pan, J.; He, Y.; Liu, Z.; Chen, J. Tetrahedron-Based Constitutional Dynamic Network for COVID-19 or Other Coronaviruses Diagnostics and Its Logic Gate Applications. *Anal. Chem.* **2022**, *94*, 714–722. [CrossRef]
16. Gong, S.; Wang, X.; Zhou, P.; Pan, W.; Li, N.; Tang, B. AND Logic-Gate-Based CRISPR/Cas12a Biosensing Platform for the Sensitive Colorimetric Detection of Dual miRNAs. *Anal. Chem.* **2022**, *94*, 15839–15846. [CrossRef]
17. Mokni, M.; Tlili, A.; Attia, G.; Khaoulani, S.; Zerrouki, C.; Omezzine, A.; Othmane, A.; Bouslama, A.; Fourati, N. Novel sensitive immunosensor for the selective detection of Engrailed 2 urinary prostate cancer biomarker. *Biosens. Bioelectron.* **2022**, *217*, 114678. [CrossRef]
18. Goswami, B.B.; Burkhardt, W., 3rd; Cebula, T.A. Identification of genetic variants of hepatitis A virus. *J. Virol. Methods* **1997**, *65*, 95–103. [CrossRef]
19. Buti, M.; Jardí, R.; Bosch, A.; Rodríguez, F.; Sánchez, G.; Pinto, R.; Costa, X.; Sánchez-Avila, J.F.; Cotrina, M.; Esteban, R.; et al. Assessment of the PCR-Southern blot technique for the analysis of viremia in patients with acute hepatitis A. *Gastroenterol. Hepatol.* **2001**, *24*, 1–4. [CrossRef]
20. Jean, J.; D'Souza, D.H.; Jaykus, L.A. Multiplex nucleic acid sequence-based amplification for simultaneous detection of several enteric viruses in model ready-to-eat foods. *Appl. Environ. Microbiol.* **2004**, *70*, 6603–6610. [CrossRef]
21. Polish, L.B.; Robertson, B.H.; Khanna, B.; Krawczynski, K.; Spelbring, J.; Olson, F.; Shapiro, C.N. Excretion of hepatitis A virus (HAV) in adults: Comparison of immunologic and molecular detection methods and relationship between HAV positivity and infectivity in tamarins. *J. Clin. Microbiol.* **1999**, *37*, 3615–3617. [CrossRef]
22. Fujiwara, K.; Yokosuka, O.; Ehata, T.; Imazeki, F.; Saisho, H. PCR-SSCP analysis of 5′-nontranslated region of hepatitis A viral RNA: Comparison with clinicopathological features of hepatitis A. *Dig. Dis. Sci.* **2000**, *45*, 2422–2427. [CrossRef]
23. Liu, Y.P.; Yao, C.Y. Rapid and quantitative detection of hepatitis B virus. *World J. Gastroenterol.* **2015**, *21*, 11954–11963. [CrossRef]
24. Arikan, A.; Sayan, M.; Sanlidag, T.; Suer, K.; Akcali, S.; Guvenir, M. Evaluation of the pol/S Gene Overlapping Mutations in Chronic Hepatitis B Patients in Northern Cyprus. *Pol. J. Microbiol.* **2019**, *68*, 317–322. [CrossRef]
25. Sayan, M.; Arikan, A.; Sanlidag, T. Comparison of Performance Characteristics of DxN VERIS System versus Qiagen PCR for HBV Genotype D and HCV Genotype 1b Quantification. *Pol. J. Microbiol.* **2019**, *68*, 139–143. [CrossRef]
26. Ferrari, E. Gold Nanoparticle-Based Plasmonic Biosensors. *Biosensors* **2023**, *13*, 411. [CrossRef]
27. Medici, S.; Peana, M.; Coradduzza, D.; Zoroddu, M.A. Gold nanoparticles and cancer: Detection, diagnosis and therapy. *Semin. Cancer Biol.* **2021**, *76*, 27–37. [CrossRef]
28. Al-Radadi, N.S. Green Biosynthesis of Flaxseed Gold Nanoparticles (Au-NPs) as Potent Anti-cancer Agent Against Breast Cancer Cells. *J. Saudi Chem. Soc.* **2021**, *25*, 101243. [CrossRef]
29. Naz, F.; Kumar Dinda, A.; Kumar, A.; Koul, V. Investigation of ultrafine gold nanoparticles (AuNPs) based nanoformulation as single conjugates target delivery for improved methotrexate chemotherapy in breast cancer. *Int. J. Pharm.* **2019**, *569*, 118561. [CrossRef]
30. Mirkin, C.A.; Letsinger, R.L.; Mucic, R.C.; Storhoff, J.J. A DNA-based method for rationally assembling nanoparticles into macroscopic materials. *Nature* **1996**, *382*, 607–609. [CrossRef]

31. Saha, K.; Agasti, S.S.; Kim, C.; Li, X.; Rotello, V.M. Gold Nanoparticles in Chemical and Biological Sensing. *Chem. Rev.* **2012**, *112*, 2739–2779. [CrossRef]
32. Omabegho, T.; Sha, R.; Seeman, N.C. A bipedal DNA Brownian motor with coordinated legs. *Science* **2009**, *324*, 67–71. [CrossRef]
33. Jung, C.; Allen, P.B.; Ellington, A.D. A stochastic DNA walker that traverses a microparticle surface. *Nat. Nanotechnol.* **2016**, *11*, 157–163. [CrossRef]
34. Xu, M.; Tang, D. Recent advances in DNA walker machines and their applications coupled with signal amplification strategies: A critical review. *Anal. Chim. Acta* **2021**, *1171*, 338523. [CrossRef]
35. Chai, H.; Miao, P. Bipedal DNA Walker Based Electrochemical Genosensing Strategy. *Anal. Chem.* **2019**, *91*, 4953–4957. [CrossRef]
36. Ge, J.; Song, J.; Xu, X. Colorimetric detection of viral RNA fragments based on an integrated logic-operated three-dimensional DNA walker. *Biosens. Bioelectron.* **2022**, *217*, 114714. [CrossRef]
37. Tao, F.; Fang, J.; Guo, Y.; Tao, Y.; Han, X.; Hu, Y.; Wang, J.; Li, L.; Jian, Y.; Xie, G. A target-triggered biosensing platform for detection of HBV DNA based on DNA walker and CHA. *Anal. Biochem.* **2018**, *554*, 16–22. [CrossRef]
38. Zheng, L.; Jin, M.; Pan, Y.; Zheng, Y.; Lou, Y. 3D-DNA walking nanomachine based on catalytic hairpin assembly and copper nanoclusters for sensitive detection of hepatitis C virus. *Talanta* **2024**, *269*, 125478. [CrossRef]
39. Liu, S.; Wu, P.; Li, W.; Zhang, H.; Cai, C. Ultrasensitive and selective electrochemical identification of hepatitis C virus genotype 1b based on specific endonuclease combined with gold nanoparticles signal amplification. *Anal. Chem.* **2011**, *83*, 4752–4758. [CrossRef] [PubMed]
40. Shawky, S.M.; Guirgis, B.S.; Azzazy, H.M. Detection of unamplified HCV RNA in serum using a novel two metallic nanoparticle platform. *Clin. Chem. Lab. Med.* **2014**, *52*, 565–572. [CrossRef]
41. Shawky, S.M.; Awad, A.M.; Allam, W.; Alkordi, M.H.; El-Khamisy, S.F. Gold aggregating gold: A novel nanoparticle biosensor approach for the direct quantification of hepatitis C virus RNA in clinical samples. *Biosens. Bioelectron.* **2017**, *92*, 349–356. [CrossRef] [PubMed]
42. Clarke, O.J.; Goodall, B.L.; Hui, H.P.; Vats, N.; Brosseau, C.L. Development of a SERS-Based Rapid Vertical Flow Assay for Point-of-Care Diagnostics. *Anal. Chem.* **2017**, *89*, 1405–1410. [CrossRef] [PubMed]
43. Zhang, Z.; Zhao, S.; Hu, F.; Yang, G.; Li, J.; Tian, H.; Peng, N. An LED-Driven AuNPs-PDMS Microfluidic Chip and Integrated Device for the Detection of Digital Loop-Mediated Isothermal DNA Amplification. *Micromachines* **2020**, *11*, 177. [CrossRef] [PubMed]
44. Moitra, P.; Alafeef, M.; Dighe, K.; Frieman, M.B.; Pan, D. Selective Naked-Eye Detection of SARS-CoV-2 Mediated by N Gene Targeted Antisense Oligonucleotide Capped Plasmonic Nanoparticles. *ACS Nano* **2020**, *14*, 7617–7627. [CrossRef] [PubMed]
45. Alafeef, M.; Moitra, P.; Dighe, K.; Pan, D. RNA-extraction-free nano-amplified colorimetric test for point-of-care clinical diagnosis of COVID-19. *Nat. Protoc.* **2021**, *16*, 3141–3162. [CrossRef]
46. Zhang, K.; Fan, Z.; Huang, Y.; Ding, Y.; Xie, M.; Wang, M. Hybridization chain reaction circuit-based electrochemiluminescent biosensor for SARS-CoV-2 RdRp gene assay. *Talanta* **2022**, *240*, 123207. [CrossRef]

Disclaimer/Publisher's Note: The statements, opinions and data contained in all publications are solely those of the individual author(s) and contributor(s) and not of MDPI and/or the editor(s). MDPI and/or the editor(s) disclaim responsibility for any injury to people or property resulting from any ideas, methods, instructions or products referred to in the content.

MDPI AG
Grosspeteranlage 5
4052 Basel
Switzerland
Tel.: +41 61 683 77 34

Biosensors Editorial Office
E-mail: biosensors@mdpi.com
www.mdpi.com/journal/biosensors

Disclaimer/Publisher's Note: The title and front matter of this reprint are at the discretion of the Guest Editors. The publisher is not responsible for their content or any associated concerns. The statements, opinions and data contained in all individual articles are solely those of the individual Editors and contributors and not of MDPI. MDPI disclaims responsibility for any injury to people or property resulting from any ideas, methods, instructions or products referred to in the content.